DATA ACQUISITION
AND
CONTROL

MICROCOMPUTER APPLICATIONS
FOR
SCIENTISTS AND ENGINEERS

JOSEPH J. CARR

TAB Professional and Reference Books

Division of TAB BOOKS Inc.
P.O. Box 40, Blue Ridge Summit, PA

FIRST EDITION

FIRST PRINTING

Library of Congress Cataloging in Publication Data

Carr, Joseph J.
Data acquisition and control : microcomputer applications for
scientists and engineers / by Joseph J. Carr.
p. cm.
Includes index.
ISBN 0-8306-2956-4
1. Physical measurements—Data processing. 2. Mensuration—Data processing. 3. Microcomputers. I. Title.
QC39.C36 1988
530.8′5—dc19
88-2205
CIP

Questions regarding the content of this book
should be addressed to:

Reader Inquiry Branch
TAB BOOKS Inc.
Blue Ridge Summit, PA 17294-0214

Contents

Introduction

DATA-ACQUISITIONS SYSTEMS ARE USED TO electronically monitor or gather data from the external physical environment. These systems differ from regular data processing systems in that the data is typically derived from transducers, sensors, and other devices automatically, or at least semiautomatically. In this same category of equipment is are also included simple control systems, process controllers, and others.

Most of the systems discussed as examples in this book are oriented towards medicine, the life sciences, and biology. This emphasis is because of my own professional involvement with biomedical engineering for many years. The principles, however, are almost universally applicable and so may be extrapolated to your own area of professional involvement rather easily.

There is a collection of computer programs in the Appendix A of this book. For information on diskettes for these programs on IBM-PC or Apple IIe formatted disks, write to:

FTA
Joseph J. Carr
c/o TAB BOOKS, Inc.
Blue Ridge Summit, PA 17294-0214

Chapter 1

Data-Acquisition Systems: Scope and Approach

DATA-ACQUISITION SYSTEMS ARE DESIGNED TO interface a computer to the correct circuits or equipment in order to collect data from external sources. I limit this activity to those that collect data from external transducers, as opposed to data-processing systems. In data acquisition systems, the input devices will be sensors and transducers of several assorted types, and will measure parameters such as temperature, pressure, displacement, nuclear radiation, biopotentials, and so forth.

In this book I will cover a wide variety of topics. I assume that the reader is not an electronics engineer but is a person who is sophisticated in one or the other sciences or other technical disciplines. Examples of this model reader vary from the plant engineer keeping track of process control parameters, to the life scientist making measurements of bipotentials, to the physicist who savours odd exotica and wants to make a computerized study of it, to the chemist performing a laboratory experiment. You will need to know a certain amount about electronics to use this book, but true professional depth in that discipline is not critical.

Part of the book is devoted to a discussion of typical transducers and other sensors used in collecting analog data. We will take a look at temperature transducers, including the newer forms of integrated circuit pn-junction device. We will also look at pressure/force transducers, light transducers, flow transducers and other kindred types. Two chapters are devoted to biomedical-signals acquisition. Part of this emphasis evolved out of my own professional involvement in biomedical engineering, and also from the fact that a lot of scientists are involved in this type of research.

A section of this book is devoted to some elementary analog electronic circuits. The ubiquitous operational amplifier is covered in detail. It is impossible to be really knowledgeable about data acquisition problems without appealing to operational amplifiers to solve some of them. The staying power of the op amp over the years is due, in large part, to the fact that we can set gain on an operational amplifier with only two resistors. Just a little more knowledge allows you to contrive many devices using operational amplifiers.

There is a chapter on differentiators and integrators. Although modern data-acquisition systems tend to perform some of these functions in software, there is still quite a need for differentiator and integrator circuits in the analog subsystem of your computer. The design of practical circuits of this type is not as straightforward as some textbooks would have you believe, so some effort was made to let you know how to actually get the circuit working using regular, practical, operational-amplifier devices.

Signals-acquisition problems can make any effort a waste of time and money. Problems such as ground loops, noise, and extraneous high-voltage signals can foul up the best-laid plans. Therefore, some effort is needed to help you understand these problems and their usual solutions.

Because data-acquisition systems are built around programmable digital computers, I also cover the basics of interfacing these computers to other devices. Again, the emphasis is on practical circuits. Also covered in the same section of the book is material on analog-to-digital (A/D) con-

verters and digital-to-analog converters (DAC). These devices allow the computer to be interfaced to what is still essentially an analog world. The A/D converts analog voltages or currents to a binary digital word that the computer can digest. Alternatively, the DAC converts binary digital words from the computer to a proportional analog voltage or current.

Finally, in the back of the book is material on certain peripherals that you will encounter in building data-acquisition systems. Topics covered include paper chart recorders, oscilloscopes, printers, and various forms of test equipment.

In the Appendix you will find a collection of BASIC computer programs suitable for data acquisitions use. Although these programs are designed to run on the IBM-PC and its clones, the BASIC language is broadly used, so conversion to other dialects should be an easy task.

Chapter 2

Basics of Instrumentation for Data Acquisition

AN ELECTRONIC INSTRUMENT IS AN APPARATUS consisting of a collection of circuits and mechanical and electromechanical devices that do some particular job of measurement or control. In the present context of data acquisition, both applications are typically found. The instrument might acquire a signal from a transducer of some sort and then process the signal to produce some numerical display or analog waveform at the output. Alternatively, it might take several different inputs, make some computations or comparisons, and then display the result or use them to perform some decision in a control circuit.

TYPES OF INSTRUMENTATION

Instrumentation can be analog, digital, or a synthesis of the two separate types. When high precision is mandatory or when extreme complexity might be required, the design is best left to qualified electronics engineers with experience in that area. The amateur, novice, or student designer can, though, perform chores in instrumentation that were once regarded as too difficult. So if you are the kind of person who enjoys designing and building his own from the electronics projects, there are a lot of sophisticated things you could do.

Two main areas of opportunity present themselves as ripe for such people: the creation of simple to moderately complex instruments, and interfacing two or more existing, commercially produced instruments or systems that were previously not compatible.

You will be surprised at the level of complexity that can be achieved by those whose competency is in other areas of expertise, although probably at the expense of some level of frustration in their earliest attempts. The frustration quotient, though, can be reduced considerably if the person is willing and able to seek advice from the professionals and the companies (too often overlooked!) who make and sell the products you plan to use.

Even where sophistication is modest, there is room for clever application of electronic circuits to save money. A physiologist, for example, might want to use a ×10 preamplifier to acquire cell action potentials. This instrument costs over $300 when purchased from a scientific instrument supplier, but can be duplicated exactly for about $40, and (using other, more modern, parts) duplicated in "form-fit-function" for about $20. Its chief claim to being a physiological amplifier, as opposed to other forms, is the fact that it has a high CMRR (common-made rejection ratio) and an input impedance

Fig. 2-1. Basics of an instrumentation system.

on the order of 100 billion ohms. But modern BiFET and BiMOS operational amplifiers will do the same job, but offer input impedances on the order or 1,500 billion ohms (1.5 teraohms).

Interfacing two existing commercial instruments may be as simple as constructing an appropriate patch cord, which would have the connector for one instrument on one end, and the connector for the other instrument on the other end. Alternatively, interfacing might involve designing electronic circuitry that makes the instruments compatible.

In both industrial and university laboratory environments, as well as certain amateur science endeavors, one may be asked to make a specially-built instrument or a collection of commercial instruments. You might buy or reallocate (i.e., steal) a cabinet such as a 19-inch relay rack, and then go about mounting the instruments and subassemblies. But before much progress can be made, it will be advisable to study each instrument to ascertain its capabilities. Input/output requirements and specifications, the function and range of each control, and the signal-flow path within the instrument. Any attempt at producing an instrumentation system without doing this will result in a less than optimum device, or at worst, might destroy one or more of the subassemblies.

Figure 2-1 shows the generalized block diagram of a "typical" instrument. There are three sections, which in real life might be anything from one device to a whole rackfull of devices. The input device is some sort of signals-acquisition device. It might be a set of biopotentials electrodes, a transducer for measuring a given parameter, or some other device that originates the signal for the system. Once the signal is acquired, it must be processed. This stage almost always consists of amplification, but might also contain filtering, logarithmic compression, or other functions. In today's microcomputer world, the signals processing might be in the form of digital-signal-processing program routines. Alternatively, it might be a sophisticated analog module. Burr-Brown Corporation, for example, produces a three-input device that has a transfer function on the order of:

$$V_o = V_x \times \frac{V_y}{V_z}^m$$

where:

    V_o is the output voltage
    V_x is the potential at the x-input
    V_y is the potential at the y-input
    V_z is the potential at the z-input
    m is a factor determined by an external resistor network

The last section of our universal instrument is the output stages. This section might consist of a display device (oscilloscopes, paper recorders, digital meters, etc.), or a control system actuator (furnace turn-on controller, etc). In general, the device will be either an indicator/display or an actuator of some sort.

In this book you will learn a little bit about electronic instrumentation design, but much more about that form of electronics instrumentation unique or particularly applicable to data-acquisition situations.

CLASSES OF SYSTEMS

There are various forms of electronic systems that can be generically classed as data-acquisition systems. In this section I will classify these instruments into a few coherent categories.

Measurement. A measurement system is designed to examine some parameter and produce an output that indicates its value.

Control System. A feedback control system

is designed to examine a parameter, determine its value, and then actuate some device or process that forces the parameter to a preset value or level. Perhaps the most common example of a feedback control system (and also the least sophisticated) is the thermostat and furnace in your home.

Data Logging System. This type of system is designed to collect data and store it in a form that is retrievable later on. Classical data loggers were either AM, dc, or FM/FM tape recorders. Models existed that used every form of tape from a microcassette to 35 millimeter, 32-channel versions. Modern data loggers are digital versions that store data on either magnetic media (diskettes) or in a form of volatile memory that does not lose its data when the instrument is turned off (a keep-alive memory battery is provided).

Chapter 3

The Role of the Microcomputer

PERHAPS NO OTHER FACTOR HAS SPAWNED THE sharp increase in data-acquisition systems than the invention of the microcomputer. Previously, such systems were terribly costly because they had to be based on large mainframe computers or stand-alone minicomputers. While the minicomputers were an order of magnitude lower in cost while still being more than sufficient for most data-acquisition problems, they were still very expensive. The microcomputer revolution changed this situation. The desk-top microcomputer, especially those types that allow plug-in modules or printed circuit cards, are easily adaptable to data-acquisition problems. It is now possible to buy, for about the price of a ten-year old used car, a microcomputer that is capable of running a factory, performing a scientific experiment, or crunching the data acquired.

The key aspect of the microcomputer that makes it useful for our problem is the fact that it is programmable. This means we can make the computer do a lot of different jobs just by changing the software. In Chapter 22 I will discuss in general terms how to select such a computer.

The microcomputer can do different jobs for us. Let's suppose that we are life scientists that want to do an experiment that will win the Nobel prize next year. Let's look at the kinds of jobs that can be done by the same microcomputer in our Quest for the Grand Prize.

Word Processor. We have to write the grant that will fund our basic research. The microcomputer is a dandy word processor. This book was written on a word processor (FinalWord Ver. 1.15, run on an IBM-PC), and I am very familiar with its utility. Not only will you have to write your grant application, you will also have to write up the results of the experiment for the peer review journal.

The digital computer will not only make it easier to draft and edit your grant applications, professional papers, reports to management and so forth, it can make them look really spiffy. Modern software, such as PageMaker and PrintShop, allow you to lay out pages that look like they were professionally designed and typeset. Although such software is available for the IBM-PC, the hands-down master at this type of job is the Apple MacIntosh machine. When coupled with a laser printer (which should also be used on non-Apple machines used for this purpose), the Apple "Desktop Publisher" package is supurb for professional typesetting. Although the quality is not up to the best in electronically typeset pages, it is close enough that some publishers are now using Desktop Publishing.

Financial Records. Any project that requires funding will benefit from the computer's ability to keep records. Various programs are available, including spreadsheet and dedicated accounting programs. The same computer that helped you write your grant application will now be useful in keeping track of the grant money, and indeed, will even write the checks for you . . . if you want it to.

Program Management. Various program management software offerings allow you to keep track of a multiple critical path project, vary the parameters, and spot troubles on the way. An example of this form of program is the *Harvard Program Manager.*

Experiment Control. The programable computer can be married with various plug-in and external circuits to actually control the experiment. You can interface assorted instruments to the computer, and then write a program that will allow the computer to control the instruments and how they work in the given experiment.

This type of system has a distinct advantage over manually operated systems: consistency. I can recall a case where a physician complained that he didn't need a computerized laboratory system because his technicians, most of whom he had personally trained, were good at their jobs. Besides, the computer had a 15-percent error rate. In other words, 15-percent of the time the computer gives a result that is outside of a predetermined quality control band of acceptability. It was argued that the technicians could do the job almost error free. That, however, was wishful thinking! Inspection of the actual situation showed that technicians made less than 1% errors at 0700 when their shift started, but by the end of the day were trucking along at about a 25% error. The computer gives predictable and benign errors, while the human error is totally unpredictable.

Interestingly enough, the human/machine interface can affect the perception of the system. One of the first automated electrocardiogram (ECG) reading programs was deemed inadequate by several emergency-room physicians. The problem turned out to be one of perception. The program searched a database of 25,000 confirmed pathologies to find the one(s) that most nearly fit(s) the incoming ECG waveform for the present patient. During the search period (two to three minutes sometimes), the efficient programmer decided that the machine could be used to output an analog rendition of the digitized waveform. In other words, a DAC was used to output the waveform to a strip-chart recorder. The problem was that the ER doctors would examine the strip, and make a tentative diagnosis. If the computer disagreed with him or her, then the doctor arrogantly assumed that the machine has erred, and that his or her diagnosis was the correct one. That actually rarely turned out to be the case: the computer won hands down. The solution to the perception problem was to delay outputting the analog waveform until the printed diagnosis from the database search was completed. The analog waveform then merely confirmed what the computer, in its wisdom, had determined.

Controling an experiment or other data acquisition event requires collecting the correct plug-ins and external equipment. For this reason I prefer computer models that allow plug-in printed-circuit boards to expand the capability of the machine. For this reason, the Apple IIe and the IBM-PC are well-suited (note: not all PC-clones accept plug-in cards).

Data Logging. If the experiment is one that does not easily lend itself to control applications, then the computer can still be used for data collection. A suitable array of analog-to-digital converters to input analog data to the computer is necessary. The computer can then collect and store a large array of data on the experiment automatically. Note that the computer can collect a lot more data than manual methods.

Number Crunching. Raw data collected during an experiment is rarely useful in its initial form. We have to perform assorted forms of statistical analysis and other jobs on the data before it can be used. The computer does this job magnificently. In fact, you can use the same data logged (see above) during the experiment; in some cases the statistical massaging can be done as the data is collected.

Chapter 4

Designing a System

THE ONE UNMISTAKABLE SIGN OF THE NOVICE DE-signer of electronic instruments is the tendency to jump right into the construction and testing phase of the project without the benefit of the least little bit of planning. Don't make this expensive, time-consuming mistake.

This thing called "design" is a logical process. Indeed, the very word "design" suggests activities like planning, thought, intent, and procedure. There is little that is really arcane in elementary electronic design, and anyone with a little knowledge can design adequate electronic instruments.

This is not to diminish the first-class design engineer, but merely suggests that almost anyone can get something working that will perform the chore at hand. A good designer exhibits intelligence, insight, knowledge, and that subjective property known as cleverness. Most of these attributes are obtained through one process: experience, i.e., the art of surviving repeated attempts at nailing you to the wall.

You will have to learn good design and laboratory techniques if your efforts are going to be efficient. You will also have to learn some of the more objective things such as the upper limits of devices (mostly to prevent a costly method of con-verting silicon to carbon. If you follow true to form, you will notice your mistake approximately 1 milli-second before a puff of smoke indicates that your one-of-a-kind sample just evaporated). Don't worry if your early efforts seem futile. They are not totally without merit, if only because you are afforded the opportunity to learn from your mistakes. This book lets you learn from some of mine and those of others.

Do not be impatient to get started building hardware. This is actually one of the last steps in any proper design activity. Unfortunately, other people may not see it that way. If you are designing some electronic widget as part of your employment, or to fulfill a requirement in a school course, then there might be pressure from above to start producing something that can be seen, felt, smelled or heard almost immediately. There always seems to be an impatient supervisor, overlord, or nervous customer who is only too willing to believe that you are not producing anything if you are not constantly spritzing and fussing with wires, capacitors, ICs, and other electronic paraphernalia. So despite primordial urges to the contrary, resist the temptation to jump in prematurely.

The very first step in any design process involves <u>knowing and understanding the problem that you must solve</u>. This advice may seem at first glance like a case of runaway cynicism, but in reality it is common sense based on observation. A remarkable number of people will begin something they call "designing" without really knowing what the device is supposed to do. Studying the problem will involve any or all of several activities including literature search, thinking, interviewing the end user, and interviewing other people who solved the same sort of problem before.

Do not underestimate the value of a literature search. Too many alleged designers shun this step for some reason. I suspect that they suffer "N.I.H." (Not Invented Here) syndrome. This well-known malady afflicts those whose misplaced pride prevents them from using perfectly good solutions from prior art in favor of attempting the new and unknown. They are embarrassed to admit that somebody else once had a good idea. If prior art will solve your problem, then use it. Your job is to solve the problem, and not necessarily to prove how clever you are with state-of-the-art designs.

The next step is to formulate an approach to solving the problem. This will involve trying to figure out several methods or circuits that might do the trick. There is seldom a single "best" way to perform any electronic design job. So be sure to consider several possible alternatives. The word "contingencies" looms large in this area.

Some wise souls will tell you that your first design in any given project is usually the poorest one that you will invent. If this is true even some of the time, then it might be wise to collect several approaches before actually starting to build anything.

One other thing to do prior to building anything is to make a drawing of the entire circuit and every critical mechanical part. I know several people who allege themselves to be "electronic instrument designers," who are often seen in their workshops with a semiconductor manufacturer's data book propped up on a vise or pile of books, copying fragments of their total circuit, first from one page, then another. This is an extremely poor practice, and inevitably leads to burned-out ICs, ragged tempers, unhappy customers, and the well-

earned contempt of colleagues who are smarter and wiser. Even if you choose to duplicate a published electronic project, you should copy it *in toto* onto a working sheet of paper, especially if you plan to change the circuit, or any small part of it.

The drawing and other documentation will play a large role as you begin building and testing the first prototype. Keep detailed records of key voltages, signals, and other parameters that are important for the specific case. Change the master drawing to reflect any changes that you make in the circuit. Think with a pencil!

Also, please write an alignment and adjustment procedure that can be followed by someone less qualified than yourself. The procedure might be self-evident to you because you originated the concept, but to others it may be mysterious. Do not require your successors to use mental telepathy or the occult sciences to figure out how to adjust, align, or calibrate your creation.

Besides, six months or a year down the road, you may well be the one who is called on to make a repair or readjustment of the instrument. Guess who will then be neatly and properly nailed for not knowing how? Serves you right . . . good documentation is a fine save yourself tactic.

Use the laboratory notebook, or some other running design log. Most college bookstores, drafting supplies stores, or "engineering supplies" stores sell adequate, quadrilled laboratory notebooks. These, if kept properly, can be your file and may help you if a patentability question arises.

Another sign of the novice or inept designer is the tendency to commit even relatively complex or untried designs in final form without first breadboarding the circuit. Every new design cannot be considered "finished" until it has been tested properly and not found wanting. Every idea that you conceive must be considered merely hypothetical until it has been proven valid. It may appear that certain ideas will work, but when you connect them a big, nasty, smelly surprise is found waiting. This is the reason why laboratory breadboards are fast-selling items.

The final product will usually be built on a wireboard, DIP-board, PC board, or whatever works best and is most cost-effective. You will most likely want to package the circuit in as nice a cabinet as can be economically justified. Fancy cabinets add prestige value to your work, and are a valid source of pride to the good craftsman. But if you go ahead and build the circuit before testing it, then problems may appear that cannot be solved cheaply. You may, for example, drill an excess number of inappropriate holes. Or you may have selected a 4.5 × 6.5 inch DIP board, only to find that additional circuitry is required and that a 4.5 × 9 inch board would have been more appropriate. The little slivers of DIP board often found hanging onto such poorly laid-out projects are known contemptuously as "kluge boards."

It is a rather long road from concept to finished product, so don't add any more curves and detours through poor procedures or inept planning. At the very least you will use up costly resources and generate no small amount of aggravation.

Let's summarize some steps that will most often result in a viable electronic project design:

1. Study the problem that must be solved.

2. Formulate several approaches to solve the problem. Block diagram each of them.

3. Make records, drawings, and documents.

4. Prototype on a breadboard or some other medium that is either expendable or reusable.

5. Test the circuit under conditions that are as realistic as possible.

6. Build and test the final product using the best possible craftsmanship and materials.

Chapter 5

Transduction and Transducers

"TRANSDUCTION" IS THE PROCESS OF CHANGing energy from one form to another, and in our present context we could include the condition, "for purposes of measurement or control." Transducers are the eyes and ears of the data-acquisition system. These devices convert assorted forms of energy from physical systems (e.g., temperature, pressure, etc.) into electrical energy that makes sense to electronic systems. A typical transducer might convert, say, pressure to an analogous voltage or current. That voltage may be analog, or it may be a binary digital word.

In this section I discuss the basic types of transducers. My intent is not to provide a complete catalog of available types, but to introduce classes of generic types. Toward this goal we will start off with a discussion of piezoresistive strain gauges.

There are many different forms of transducer that use resistive strain gauge elements, and most of them are based on the Wheatstone bridge circuit. Various physical parameters can be measured with strain gauge transducers, including force, displacement, vibration, and both liquid and gas pressure. If you are so unfortunate as to require intensive-care hospitalization, then the doctor may order continuous blood-pressure monitoring through an in-dwelling catheter inserted in an artery. The transducer used to measure your blood pressure may well be a Wheatstone bridge strain gauge. In this section, I will discuss how strain gauges work and how to make them work in practical cases. Later, I will show you the electronic amplifiers used for processing transducer signals.

PIEZORESISTIVITY

All conductors possess electrical resistance, which is opposition to the flow of current; resistance is measured in ohms. The resistance of any specific conductor is directly proportional to the length (see Fig. 5-1), and inversely proportional to the cross-sectional area. Resistance is also directly proportional to a property of the conductor material called "resistivity." The relationship between length, area, and resistivity is shown in Fig. 5-1. The equation in Fig. 1 shows clearly that resistance is related to length and cross-sectional area.

Piezoresistivity is merely a ten-dollar word that means the resistance changes when length and/or area are changed. Figure 5-1A shows a cylindrical conductor with a length L_o and a cross-sectional area A_o. When a compression force is applied, as in Fig. 5-1B, the length reduces and the

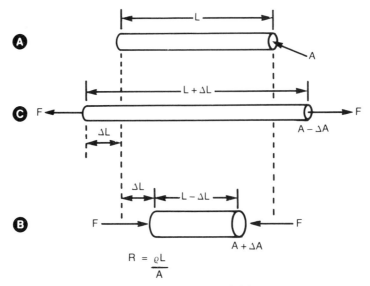

Fig. 5-1. Basis of piezoresistivity.

corss-sectional area increases. This situation results in an increase in the electrical resistance. Similarly, when a tension force is applied (Fig. 5-1C) the length increases and the cross-sectional area decreases, so the electrical resistance will increase. Provided that the change is small, the change of electrical resistance is a nearly linear function of the applied force, so it can easily be used to make measurements of that force.

STRAIN GAUGES

A strain gauge is merely a piezoresistive element — either wire, metal foil, or semiconductor, designed to create a resistance change when a force is applied. Strain gauges can be classified as either bonded or unbonded types. Figure 5-2 shows both methods of construction.

The unbonded strain gauge is shown in Fig. 5-2A, and consists of a wire resistance element stretched taut between two flexible supports. These supports are configured in such a way as to place a tension or compression force on the taut wire when external forces are applied. In the particular example shown, the supports are mounted on a thin metal diaphragm that flexes when a force is applied. Force F_1 will cause the flexible supports

to spread apart, placing a tension force on the wire and increasing its resistance. Alternatively, when force F_2 is applied, the ends of the supports tend to move closer together, effectively placing a compression force on the wire element and thereby reducing its resistance. In actuality, the wire at rest is already taut, which implies a tension force. So F_1 increases the tension force from normal, and F_2 decreases the normal tension.

The bonded form of strain gauge is shown in Fig. 5-2B. In this type of device a wire, foil, or semiconductor element is cemented to a thin metal diaphragm. When the diaphragm is flexed, the element deforms to produce a resistance change.

The linearity of both types of strain gauge can be quite good, provided that the elastic limits of the diaphragm and element are not exceeded. It is also necessary to ensure that the change of length is only a small percentage of the resting length.

In the past, the "standard wisdom" was that bonded strain gauges were more rugged but less linear than unbonded models. Although this may have been the situation at one time, recent experience has shown that modern manufacturing techniques can produce linear, reliable units of both types of construction.

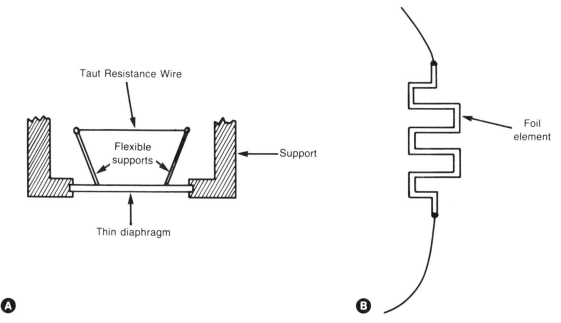

Fig. 5-2. A) Unbonded strain gauge, B) bonded strain gauge.

THE WHEATSTONE BRIDGE

The Wheatstone bridge is a 19th-century holdover that finds a home in many modern electronic circuits. The classic form of bridge is shown in Fig. 5-3. There are four resistive arms to the bridge, each arm being labeled R1, R2, R3 and R4. The excitation voltage (V) is applied across two of the nodes, while the signal is taken from the alternate two nodes (labeled "C" and "D"). We can consider this circuit as two series voltage dividers in parallel, one consisting of R1 and R4 and the other of R2 and R3 (see Fig. 5-4).

The output voltage from a Wheatstone bridge

Fig. 5-3. Wheatstone bridge circuit.

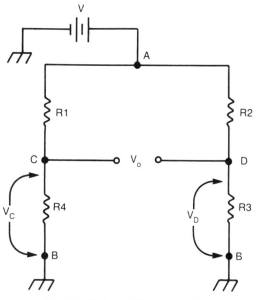

Fig. 5-4. Redrawn Wheatstone bridge.

is the difference between the voltages at points C and D. When all of the arithmetic is finished, we find that the output voltage will be zero when the ratio R4/R1 is equal to the ration R3/R2. If these ratios are not kept equal, as is the case when one or more elements in a strain gauge not at rest, then an output voltage is produced that is proportional to both the applied voltage and the change of resistance.

STRAIN GAUGE CIRCUITRY

Before the resistive strain gauge (or other form of resistive transducer) can be useful it must be connected into a circuit that will convert its resistance changes into a current or voltage output. Most applications are voltage-output circuits.

Figure 5-5 shows several popular forms of circuit. The circuit in Fig. 5-5A is both the simplest and least useful (although not unuseful!); it is sometimes called the "half-bridge" circuit, or "voltage divider" circuit. The strain gauge element of resistance R is placed in series with a fixed resistor, R1, across a stable dc voltage, V. The output voltage V_o is found from the simple voltage divider equation:

$$V_o = \frac{VR}{R + R1} \qquad \text{Eq. 5-1}$$

Equation 5-1 describes the output voltage V_o when the transducer is at rest (i.e., nothing is stimulating the resistive element). When the element is stimulated, however, its resistance changes a small amount, h. The output voltage in that case is:

$$V_o = \frac{V(R + h)}{(R -/+ h) + R1} \qquad \text{Eq. 5-2}$$

Another form of half-bridge circuit is shown in Fig. 5-5B, but in this case the strain gauge is connected in series with a constant current source (CCS) which will maintain current I at a constant level regardless of changes in the strain gauge resistance. In this case, $V_o = I(R -/+ h)$.

Both of the half-bridge circuits suffer from one major defect: output voltage V_o will always be present regardless of the value of the stimulus applied to the transducer. Ideally in any transducer system, we want the output voltage to be zero when the applied stimulus is zero. For example, when a gas pressure transducer is open to atmosphere, the gauge pressure is zero so the output voltage should also be zero. Second, the output voltage should be proportional to the value of the stimulus when the stimulus is not zero. A Wheatstone bridge circuit can have these properties. We can use strain gauge elements for one, two, three, or all four arms of the Wheatstone bridge.

Figure 5-5C shows a circuit in which two strain gauges (SG1 and SG2) are used in two arms of a Wheatstone bridge, with fixed resistors R1 and R2 forming the alternate arms of the bridge. It is usually the case that SG1 and SG2 are configured so that their actions oppose each other; that is, under stimulus, SG1 will have resistance R + h, and SG2 will have resistance R − h, or vice versa.

One of the most linear forms of transducer bridge is the circuit in Fig. 5-5D, in which all four bridge arms contain strain-gauge elements. In most such transducers all four strain-gauge elements have the same resistance (R), which will usually be a value between 50 and 1000 ohms.

Recall that the output from a Wheatstone bridge is the difference between the voltages across the two half-bridges. We can calculate the output voltage for any of the standard configurations from the equations. Let's work an example for a bridge with all four arms active (similar to Fig. 5-5D).

Example. A force transducer is used to measure the weight of small objects. It has a resting resistance of 200 ohms, and an excitation potential of +5 volts dc is applied. When a 1-gram weight is placed on the transducer diaphragm, the resistance of the arms changes by 4.1 ohms. What is the output voltage?

Solution
$V_o = Vh/R$
$V_o = [(5 \text{ volts}) (4.1 \text{ ohms})/(200 \text{ ohms})]$
$V_o = [20.5/200]$
$V_o = 0.103 \text{ volts, or } 103 \text{ millivolts}$

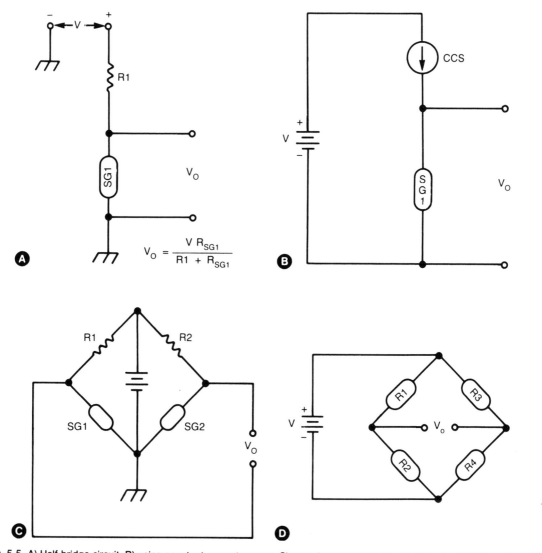

Fig. 5-5. A) Half-bridge circuit, B) using constant current source, C) two-element Wheatstone bridge circuit, D) four-element Wheatstone bridge circuit.

TRANSDUCER SENSITIVITY

Although there is some practical use for the example worked above, most readers will probably work with a known transducer for which the sensitivity is specified. The sensitivity factor (P) relates the output voltage (V) to the applied stimulus value (Q) and excitation voltage. In most cases, the transducer maker will specify a number of microvolts (or millivolts) output potential per volt of excitation potential per unit of applied stimulus. In other words:

$$P = V_o/V/Q_o$$

or, written another way:

$$P = \frac{V_o}{V \times Q}$$

where V_o is the output potential, V is the excitation potential, and Q is one unit of applied stimulus.

If we know the sensitivity factor, then we can calculate the output potential as follows:

$$V_o = PVQ$$

The equation above is the one that is most often used in circuit design. Let's work a practical example. A certain fluid pressure transducer is often used for measuring human and animal blood pressures through an in-dwelling catheter. It has a sensitivity (P) of $5\mu V/V/T$, which means "5 microvolts output potential is generated per folt of excitation potential per Torr of pressure (note: 1 T = 1 mmHg). Find the output potential when the excitation potential is +7.5 volts dc and the pressure is 400 Torr (the usual high-end limit for such transducers):

$$V_o = PVQ$$

$$V_o = \frac{5\mu V}{V \times T} \times (7.5 \text{ V}) \times (400 \text{ T})$$

$V_o = (5 \times 7.5 \times 400) \ \mu V$
$V_o = 15{,}000 \ \mu V$ (which is 15 millivolts, or 0.015 volts)

BALANCING AND CALIBRATING A BRIDGE TRANSDUCER

Few, if any, Wheatstone bridge transducers meet the ideal condition in which all four bridge arms have exactly equal resistances. In fact, the bridge resistance specified by the manufacturer is only a nominal value, and the actual value may vary quite a bit from the specified value. There will inevitably be an offset voltage (i.e., V_o is not zero when Q is zero). Figure 5-6 shows a circuit that will balance the bridge when the stimulus is zero.

Potentiometer R1 is usually a precision type with five to fifteen turns to cover the entire range. Alternatively, it is a one-turn potentiometer ganged to a multiturn vernier dial. The purpose of the potentiometer is to inject a balancing current (I) into the bridge circuit at one of its nodes. R1 is adjusted, with the stimulus at zero, for zero output voltage.

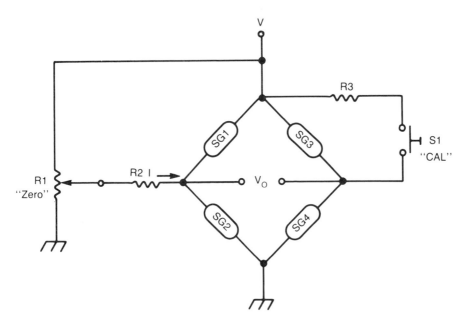

Fig. 5-6. Practical Wheatstone bridge with zero and calibrate.

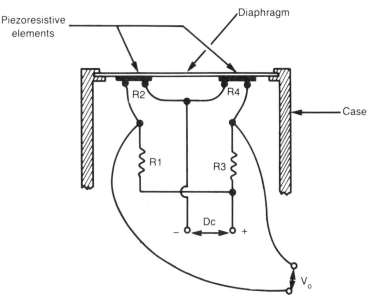

Fig. 5-7. Piezoresistive Wheatstone bridge transducer.

Another application for this type of circuit is injecting an intentional offset potential. For example, on the Heathkit digital scale such a circuit is used to adjust for the "tare weight" of the scale, which is the sum of the platform and all other weights acting on the transducer when nobody is standing on the scale. This is also sometimes called "empty weight compensation."

Calibration can be accomplished either the hard way or the easy (and less accurate way). The hard way is to set the transducer up in a system and apply the stimulus. The stimulus is measured and the result is compared with the transducer output. For example, if you are testing a pressure transducer, connect a manometer (pressure measuring device containing a column of mercury) and measure the pressure directly. The result is compared with the transducer output. All transducers should be tested in this manner initially when placed in service and then periodically thereafter.

In less critical applications, however, we can connect a calibration resistor to synthesize the offset and thereby allow the electronics to be calibrated. The resistor and CAL switch (S1) in Fig. 5-6 is used for this purpose. Resistor R3 should have a value of:

$$R3 = [(R/4QP) - (R/2)]$$

In the equation above we express the output voltage from the sensitivity factor (P) as volts instead of microvolts.

TRANSDUCER CONSTRUCTION

Although many forms of construction are used in transducer manufacture, we can examine a "generic" force/pressure transducer in order to get a general idea how it is done. Figure 5-7 shows a cut-away view of a typical bonded strain gauge force/pressure transducer. This particular model uses a pair of strain gauge elements (R2 and R4) and two fixed resistors in a Wheatstone bridge configuration. The case is a rigid structure that provides support for the thin metallic diaphragm, and protection for the internal circuitry. The piezoresistive elements, R2 and R4, are cemented to the thin metallic diaphragm. When a force or pressure is applied to the diaphragm, it distends and thereby applies strain to the strain gauge elements.

In addition to the components shown in Fig. 5-7, there may also be other components. In some models, for example, scaling resistors are used. These resistors normalize the output voltage to

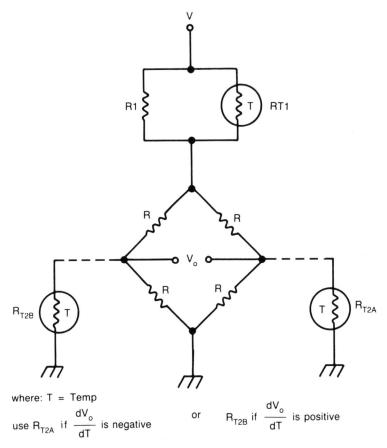

where: T = Temp

use R_{T2A} if $\dfrac{dV_o}{dT}$ is negative   or R_{T2B} if $\dfrac{dV_o}{dT}$ is positive

Fig. 5-8. Temperature-compensated bridge.

account for variations in strain gauge sensitivity. One commercial −100 to +450 Torr transducer, for example, advertises a sensitivity of 50 μV/V/ Torr. In actual practice, however, the manufacturer packs a calibration sheet with each model, and in one lot of ten transducers (all the same model) I found calibrations ranging from 30 to 60 μV/V/Torr.

Temperature-compensation components are also part of the internal circuitry. Uncompensated transducers tend to drift with changes in temperature. In one case, I recall a pressure-measuring servomechanism used in a medical research application in which the drift was so bad that the experimenter had to arrive three hours early to turn on the equipment to allow it to equilibrate. In that case, the temperature drift was due in large part to

the electronics, and to the transducer. The transducer portion could be compensated using a method similar to Fig. 5-8. The values and temperature coefficients of R_{T1} and R_{T2} depend upon the dV$_o$/dT experienced and the values of the transducer arm resistors (R).

AUTO-ZERO CIRCUITRY

Transducers are rarely perfectly balanced. Even when the arm elements have the same nominal at-rest resistance (R), we usually find an offset due to the fact that the bridge arm resistances each have a certain tolerance, and that causes a difference in the actual values of resistance. Of course, the tolerances are not all in the same direction and of the same value, but rather are malignantly addicted to Murphy's law. Because of this problem, the output voltage of the Wheatstone bridge transducer will be non-zero when there is no stimulus applied. Earlier I showed a manual means for zeroing or balancing the transducer. Now let's look at an autozero circuit.

Figure 5-9 shows the block diagram of an autozero circuit. The bridge and bridge amplifier (A1) are the same as in other circuits. The offset cancellation current (I) is generated by applying the voltage output of a digital-to-analog converter (DAC) to resistor R1. Monitoring the output of the bridge amplifier is a voltage comparator (COMP). When V_o is zero, the COMP input is zero, so the output is LOW. Alternatively, when the amplifier output voltage is non-zero, the COMP output is HIGH. The DAC binary inputs are connected to the digital outputs of a binary counter that is turned on when the EN line is HIGH. Circuit operation is as follows:

1. The operator sets the transducer to zero stimulus (e.g., for pressure transducers he or she will open the valve to atmosphere, or zero gauge pressure).

2. The ZERO button is pressed, thereby triggering the one-shot to produce an output pulse of time T, which makes one input of NAND gate G1 HIGH.

3. If voltage V_o is non-zero, then the COMP output is also HIGH, so both inputs of G1 are HIGH — making the G1 output LOW, thereby turning on the binary counter.

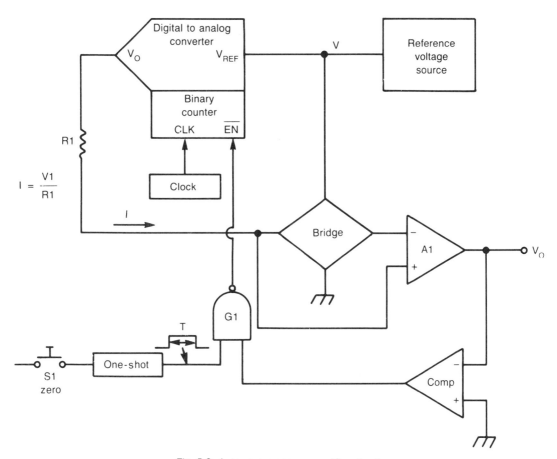

Fig. 5-9. Auto-zero pressure amplifier circuit.

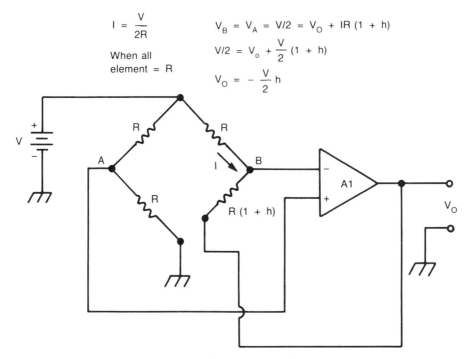

$$I = \frac{V}{2R}$$

When all
element = R

$$V_B = V_A = V/2 = V_O + IR\,(1 + h)$$

$$V/2 = V_o + \frac{V}{2}\,(1 + h)$$

$$V_O = -\frac{V}{2}\,h$$

Fig. 5-10. Linearized circuit.

4. The binary counter continues to increment in step with the clock pulses, thereby causing the DAC output to rise continuously with each increment. This action forces the bridge output towards null.

5. When V_o reaches zero, the COMP shuts off, stemming the flow of clock pulses to the binary counter, and stopping the action. The DAC output voltage will remain at this voltage level.

TRANSDUCER LINEARIZATION

Transducers are not perfect devices. Although the output function should be linear in a perfect world, real transducers are often highly nonlinear. For Wheatstone bridge strain gauges the constraints on linearity include making delta R (called "h" in some equations) very small (5 percent or less) compared with the at-rest resistance.

There are several forms of linearization techniques used in data-acquisition systems. An analog method is shown in Fig. 5-10. Here we modify the circuit of the usual single strain gauge Wheatstone bridge circuit. The ground end of one bridge resis-

tor is lifted, and applied to the output of a null-forcing amplifier, A1 (which is not the normal bridge amplifier as shown elsewhere). In this case, the resistor element R(1 + h) is in the feedback network of operational amplifier A1. Small amounts of nonlinearity are cancelled with this circuit.

For larger nonlinearities we must resort to other methods. Figure 5-11 shows a hypothetical transducer transfer function in which a voltage V_o is a function of applied pressure, P. The perfect transducer will obey the usual equation for a straight line, $V_o = MP + B$, in which M is the slope of the line and B is the offset. The actual curve may be a lot less straight.

Before digital computers were routinely used in instrumentation applications special function circuits or diode breakpoint generators were often used. In cases where a special function circuit was used, the assumption was that the equation of the actual curve was known. The special function circuit generated the inverse of that function and summed it with the input voltage. In the case of the diode breakpoint generator, an offset voltage was

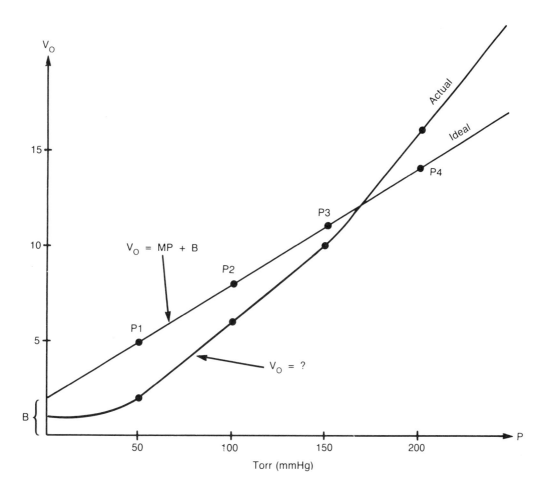

Fig. 5-11. Actual vs ideal transducer curves.

	Point	Output ideal	Voltage V_O actual
P1	50	5	2
P2	100	8	6
P3	150	11	10
P4	200	14	16

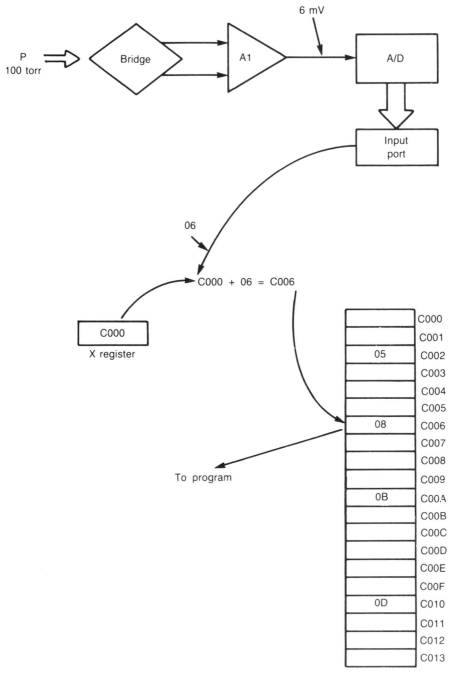

Fig. 5-12. Computerized linearization method.

added to or subtracted from the actual input signal to normalize it to the ideal. Neither of these methods was particularly good — and was often quite poor.

Now that microprocessors are used in instruments you can use a software method to correct the error. If the equation defining the actual curve is known, then you could write a software program that cancels the error. Alternatively, the look-up table method of Fig. 5-12 could be used. This example shows only a limited number of data points for simplicity sake, but the actual number would depend upon the bit length of the A/D converter.

The values for the ideal transfer function are stored in a look-up table that begins at location $C000 in memory. The value $C000 is stored in the X-register. When the A/D binary word is input to the computer it is added to the contents of the X-register. This value becomes the indexed address in the look-up table where the correct value is found. Although a pressure transducer example is shown here, it is useful for almost any form of transducer.

EXCITATION SOURCES

The dc Wheatstone bridge transducer requires a source of either ac or dc excitation volt-

age, with dc being the most common. Most transducers require an excitation voltage of 10 volts dc or less. This voltage is critical, and exceeding it will create a very short life expectancy for the transducer. A typical fluid pressure transducer requires +7.5 volts dc, and operates best (least thermal drift) at +5 volts dc. We must provide a source of dc that is stable, within specifications and (in some cases) precise.

Non-Dc Excitation

Although most Wheatstone bridge transducers are dc-excited, there are cases where non-dc sources are used. In Fig. 5-13 we see a transducer with pulsed excitation. A short duty-cycle pulse train (typically 1000 to 5000 pps) is applied to the transducer in lieu of the dc source. The amplifier output is also a pulse train, and is applied to an operational amplifier integrator. The voltage output of the integrator is a function of the repetition rate of the pulses (which is fixed) and the amplitude of the pulses.

Ac excitation is shown in Fig. 5-14. The principal advantage of this system is that ac amplifiers can be made a lot more stable at the signal levels delivered by transducers than dc amplifiers. It is often the case that amplifier drift is the same mag-

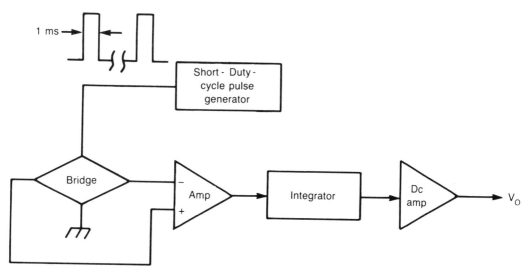

Fig. 5-13. Pulse-type transducer excitation.

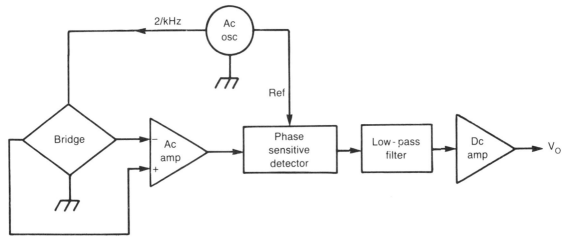

Fig. 5-14. Ac-excited carrier amplifier.

nitude as the stimulus signal, which obscures the reading by introducing considerable error. The ac amplifier eliminates some of that error using negative feedback to improve stability — a trick that's possible in dc amplifiers, but not as easily.

The ac excitation method of Fig. 5-14 typically requires a sensitive synchronous phase detector driven by the same ac signal as is used to excite the transducer bridge. The filtered output of the phase-sensitive detector is a dc voltage proportional to the applied stimulus.

Dc Excitation Sources

The simplest form of transducer excitation is the zener-diode circuit in Fig. 5-15A. A zener diode will regulate the voltage to a close tolerance that is sufficient for many applications. There are two problems with this circuit, however. The first is the fact that the zener potential is not a nice even value like 5.0 volts, but will have a value such as 4.7, 5.6, 6.2, or 6.8 volts. The second defect is thermal drift. The zener voltage will vary somewhat with temperature in all but certain reference-grade zener diodes. Unless the application is not critical, or the diode can be kept at a constant temperature, the method of Fig. 5-15A is not generally suitable.

Figure 5-15B shows a second method. In this case the regulator is a three-terminal IC voltage regulator (U1) of the LM-309, LM-340, 78xx, or similar families. In many cases, the "H" version of the regulator (100 mA) can be used, although in others the 750 mA "T" package devices must be specified. The selection depends upon the current normally drawn by the transducer, which is (V/R) where V is the regulator output voltage and R is the resistance of any one transducer element. In a typical case, the transducer will use a +5-volt excitation potential. If the resistance of the "R" elements is 50 ohms, then the current will be less than 100 mA. In that case, you can use a 100-mA LM-309H, LM-340H, and so forth.

The zener diode (D1) in Fig. 5-15B is not used for voltage regulation, but rather for protection of the transducer. If the regulator (U1) fails, then +8 to +16 volts from the V+. line will be applied to the transducer — which is fatal! The purpose of D1 is to clamp the voltage to a value that is greater than the excitation voltage, but less than the "groan voltage" rating of the transducer.

In some cases, a small fuse is inserted in series with the input (pin no. 1) of U1. The value of this fuse is set to roughly twice the current requirements (V/R) of the transducer, and will blow if the zener-diode voltage is exceeded. The fuse will add a certain amount of protection.

Some applications require a dual-polarity power supply. Figure 5-15C shows a version in which two zener diodes are used, one each for positive and negative polarities.

Neither the zener circuit nor the three-terminal regulator circuit will deliver precise output voltages. The voltage will be stable (that is, constant) but not precise. A typical three-terminal IC voltage-regulator output voltage, for example, may vary several percent from sample to sample. If we need a precise voltage, then a circuit such as Fig. 5-15D might be used. This circuit is basically a standard operational-amplifier voltage-reference circuit in which the op-amp is a high-current model, the National Semiconductor LM-13080.

The output voltage from Fig. 5-15D is deter-

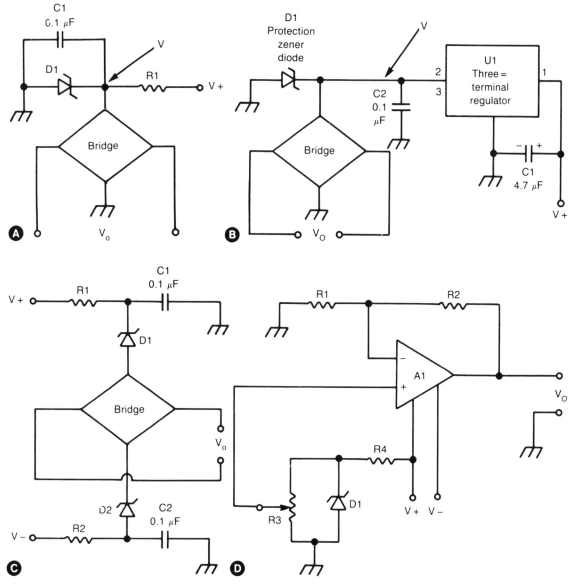

Fig. 5-15. A) Zener diode dc excitation, B) IC voltage regulator excitation, C) dual-polarity dc excitation, D) precision dc excitation.

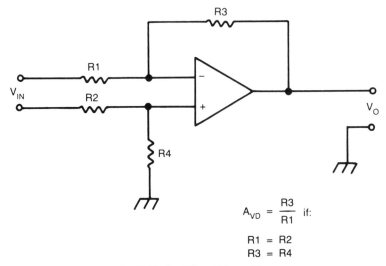

$$A_{VD} = \frac{R3}{R1} \text{ if:}$$

$$R1 = R2$$
$$R3 = R4$$

Fig. 5-16. Dc differential amplifier.

mined by R1, R2, the setting of R3, and the value of zener diode D1. The voltage V will be:

$$V = (V1) \times \frac{R2}{R1} + 1$$

or, since R1 = R2 = 10 kilohms,

$$V = 2 \times (V1)$$

The voltage V1, at the noninverting input of IC1, is a fraction of the zener voltage that depends upon the setting of potentiometer R3. You can adjust V1 from 1.13 vdc to 6.8 vdc, so the transducer voltage can be set at any value from 2.26 to 13.6 volts dc. In most cases you would probably set V at 5.00 volts, 7.50 volts, or 10.00 volts depending upon the nature of the transducer.

TRANSDUCER AMPLIFIERS

The basic dc differential amplifier is the most commonly used circuit for amplifying transducer signals. Fortunately, such amplifiers are easily constructed from simple operational amplifiers; Fig. 5-16 shows such a circuit. Assuming that R1 = R2 and R3 = R4, the gain of the amplifier will be R4/R2, or R3/R1. The amplifier output voltage will be found from:

$$V_o = (V_{IN}) \times (R3/R1)$$

where V_o is the amplifier output voltage, V_{IN} is the transducer output voltage, R1 and R3 are the resistors in the amplifier circuit.

The amount of gain required from the amplifier is determined from a scale factor, SF, which is the ratio between the voltage representing full-scale at the output of the amplifier, and the voltage representing full-scale at the output of the transducer:

$$SF = \frac{\begin{array}{c}\text{Voltage } V_o \\ \text{representing full scale}\end{array}}{\begin{array}{c}\text{Transducer output voltage} \\ V_{IN} \text{ representing full scale}\end{array}}$$

From the earlier discussion you may remember that the transducer output voltage is found from the excitation voltage, the applied stimulus, and the sensitivity factor (P). The sensitivity factor is given in terms of millivolts (or microvolts) output per volt of excitation potential per unit of applied stimulus:

$$P = \frac{V_o}{V \times Q}$$

The output voltage from the transducer is therefore found from

$$V_{IN} = P \times V \times Q$$

where V_{IN} is the transducer output voltage (or amplifier input voltage!), V is the excitation voltage, and Q is the applied stimulus (force, pressure, etc.). Let's again work a practical example.

Example. I once worked in a hospital/medical school where one project required a 0 to 1000 Torr fluid pressure transducer and amplifier. The transducer was rated with a sensitivity factor (P) of 50 μV/V/Torr, and a range of +200 to +1200 Torr. The available excitation source was 6.95 volts. At the full scale required (1000 Torr) the output voltage would be:

$$V_o = P \times V \times Q$$

$$V_o = \frac{50\ \mu V}{V \times T} \times (6.95\ V) \times (1000T)$$

$$V_o = (50\ \mu V) \times (6.95) \times (1000)$$

$$V_o = 347,500\ \mu V$$

which is 347.5 millivolts.

The amplifier output voltage required will depend upon the desired display method. For example, a strip-chart recorder might have a voltage range of 0.5 volts, 1.0 volt, or some such value. In my case I needed a digital panel meter for the output display. Most low-cost DPMs have a 0 to 1999 millivolt range, so I gained a great deal of utility by making the output voltage at full scale numerically the same at the DPM reading—for example, 1000 Torr being represented by 1000 millivolts. In that case, the DPM scale factor would be 1 mV/Torr, which is easy for humans to read. In that case "Voltage V_o Representing Full Scale" in the equation above is 1000 mV. The gain of the amplifier is the scale factor SF described earlier:

$$SF = \frac{\text{Voltage } V_o \text{ representing full scale}}{\text{Transducer output voltage}}$$
$$V_{IN} \text{ representing full scale}$$

$$SF = \frac{1000\ mV}{347.5\ mV}$$

$$SF = 2.878$$

Thus, a gain (SF) of 2.878 will provide the needed gain, so the ratio R3/R1 in Fig. 5-16 must be 2.878.

A PRACTICAL PROJECT

Now that we have gotten past the theoretical considerations, let's consider a practical project (see Fig. 5-17). This circuit is designed for a dc-excited Wheatstone bridge transducer, and is an outgrowth of the circuit of Fig. 5-16.

The transducer has a connector, P1, and will probably connect to the amplifier through a cable to jack J1. In most cases, the shielded housing of the transducer will be connected to the amplifier through a separate conductor and so will have to be grounded at the amplifier end. Because we are using a single polarity power supply, the −EXC voltage terminal (pin D) is also grounded (see J1, pins D & E).

The excitation voltage +EXC is applied through pin A. In this circuit the current for the transducer was less than 100 mA (the "R" value was 200 ohms), so an LM-309H three-terminal IC regulator was selected for a potential of +5.00 volts.

There are two stages of amplification in this circuit, A1 and A2. Amplifier A2 is simply a post-amplifier, so it will have a gain of one. This tactic makes it easier to control gain selection (R6) and zero (R8). The overall gain (when R6 is maximum) is set by R3/R1, assuming that R1 = R2 and R3 = R4 (the same as in Fig. 5-16). The output voltage V_o is given by the product of the sensitivity factor (P), the excitation voltage (5 volts), and the applied stimulus (Q):

$$V_o = 5 \times P \times Q$$

In the example given earlier the output would be

$$V_o = 5 \times 50\ \mu V \times 1000\ T$$

or 250,000 μV (250 mV) at full-scale (1000 torr).

Potentiometer R6 is used to set the gain and consists of a 10-kilohm potentiometer and a 2-kilohm fixed resistor. This combination allows

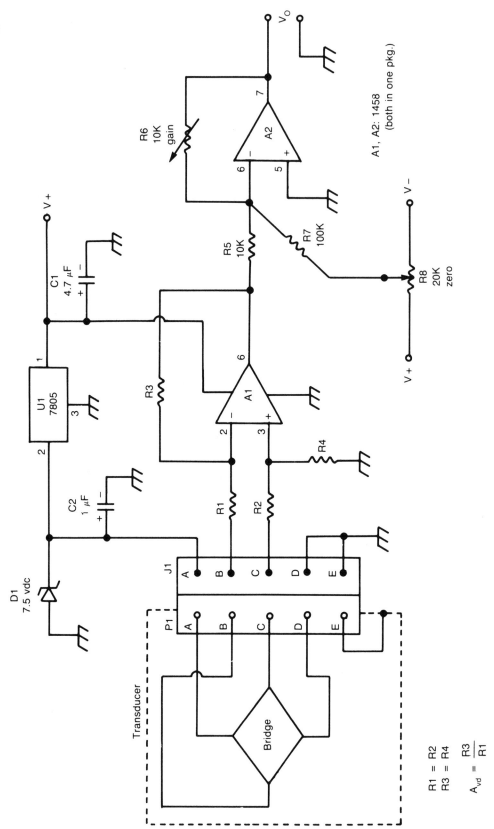

Fig. 5-17. Practical transducer amplifier.

A1, A2: 1458
(both in one pkg.)

$R1 = R2$
$R3 = R4$

$A_{vd} = \dfrac{R3}{R1}$

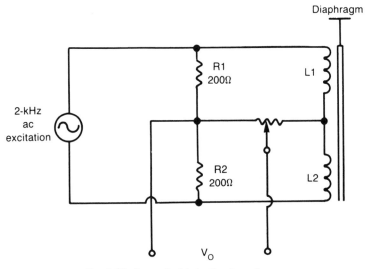

Fig. 5-18. Ac-excited inductive transducer.

you to adjust the output voltage from 0.2 to 1.2 of full-scale. Using the potentiometer you can compensate for errors in gain and transducer output voltage. Similarly, potentiometer R8 is used to adjust the output voltage to zero when the applied stimulus is zero. In the previous example, 0 to 1000 Torr pressure amplifier was needed. You would open the transducer to atmosphere (a gauge pressure of 0 Torr), and adjust R8 for 0.00 volts output. You would then apply a 1000 Torr pres-

sure (and measure it with a manometer), and then adjust potentiometer R6 for an output of 1.000 volts (i.e., 1000 millivolts).

INDUCTIVE TRANSDUCERS

The other transducers covered in this chapter are resistive devices based on piezoresistive strain gauges. In this section I will discuss two types of inductive transducer.

In both examples below the transducer con-

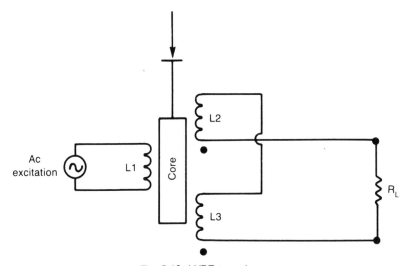

Fig. 5-19. LVDT transducer.

sists of a pair of inductors wound around a movable, permeable core. The core helps determine the inductance of each coil. When the core is evenly spaced between the two coils, the respective inductances are equal.

Figure 5-18 shows an inductive Wheatstone bridge transducer. The arms of the bridge comprise fixed resistors R1 and R2, plus inductors L1 and L2. The inductive reactances of L1 and L2 are a function of the applied ac excitation frequency and the inductance of L1 and L2. When L1 = L2, e.g., when the transducer is at rest, the bridge is at null and V_o is zero. When the coil core is moved, as when a pressure or force is applied, the relative inductances of L1 and L2 change, so the reactances are no longer equal — and the bridge is unbalanced an amount proportional to the applied stimulus.

A linear variable differential transformer (LVDT) transducer is shown in Fig. 5-19. In this case there are three coils because AC excitation is applied to the system via L1. Coils L2 and L3 are equal, and when the core is equally placed between the two their inductances are also equal. But when the core moves, the inductances are non-equal. The operation of the LVDT depends upon the fact that the two output coils are connected in series-opposing fashion such that the total output voltage is the algebraic difference. When $V_{L1} = V_{L2}$, the sum total voltage output is zero. Only when a stimulus is applied, when these voltages are not equal, will there be an output voltage.

Chapter 6

Temperature Measurements

TEMPERATURE IS PROBABLY ONE OF THE EARLIEST and most common forms of electronic measurement. Electronic thermometers are common in data-acquisition systems precisely because temperature is such an important parameter in so many cases. There are a variety of transducers, used in a still wider variety of instruments, now on the market. Figure 6-1 shows a simple consumer digital thermometer. This device, made by Heathkit, measures temperature from either of two transducers (one each for indoor and outdoor temperatures).

A somewhat costlier instrument is shown in Fig. 6-2. This particular instrument is used in medical applications and displays patient temperature in degrees centigrade. The type of probe shown in Fig. 6-2 is a rectal probe, so this instrument is configured to measure core temperature. Because it is a medical temperature monitor, alarms are provided for high and low limits.

Both types of instrument are capable of outputing both the digital display temperature and an analog voltage that is proportional to temperature. Thus, the temperature can either be charted on a strip chart recorder (see Chapter 34) or input to an A/D converter in a computerized data-acquisition system.

DIFFERENT MEASUREMENT SYSTEMS

There are several different temperature scales used in the measurement of heat. The familiar Fahrenheit and Celsius (or Centigrade), along with the less familiar Kelvin scales are frequently used in medical, scientific, and industrial applications. The Celsius and Fahrenheit scales are related through the equation:

$$F - 32 = 1.8C$$

where

F is degrees Fahrenheit
C is degrees Celsius

In the expression above you plug in the known temperature, and then use a little elementary algebra to find the unknown temperature. Trivia question: which is colder −40 degrees C or −40 degrees F? The point at which the indicated slopes of the F and C systems cross is −40 degrees. Thus, −40° C = −40° F.

TEMPERATURE TRANSDUCERS

A transducer is a device that converts a paramater such as temperature to either a voltage or a

Fig. 6-1. Homeowner's digital thermometer.

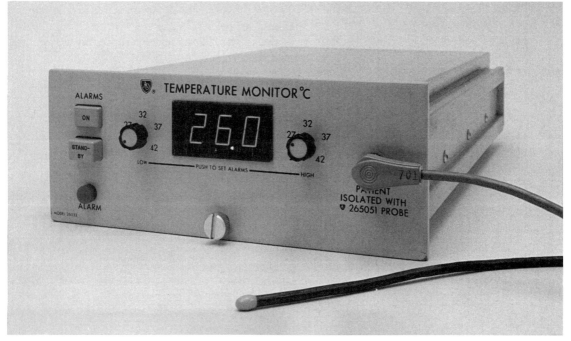

Fig. 6-2. Medical electronic thermometer.

current for purposes of measurement or control. Several different temperature transducers are commonly used: thermocouples, thermistors, and solid-state pn junctions.

An example of a thermocouple is shown in Fig. 6-3. This type of transducer consists of two dissimilar metals or other materials (some ceramics are used) that are joined at one end. Because the work functions of the two materials differ, there will be a potential generated across the open ends whenever the junction is heated. The potential is approximately linear with changes of temperature over wide ranges, although at the extreme limits of temperature (for any given pair of materials) nonlinearity increases markedly.

Thermocouples are typically used in pairs (Fig. 6-4). One junction will be used as the measurement thermocouple, while the other is the cold junction. Its name derives from the fact that some early systems required this junction to be bathed in an ice-water bath. In modern systems a synthetic cold-junction, or just room temperature, is used for the cold junction. The differential voltage between the two thermocouple junctions is proportional to the temperature difference.

Thermistors (i.e., thermal resistors) are resistors that are designed to change resistance value in a predictable manner with changes in applied temperature. A positive temperature coefficient (PTC) device (see Fig. 6-5A) increases in resistance with increase in temperature. Alternatively, a negative temperature coefficient (NTC) device decreases resistance with increase in temperature. The usual circuit symbol for thermistors is shown in Fig. 6-5B.

Most thermistors have a nonlinear curve when it is plotted over a wide temperature range, but when limited to narrow temperature ranges (such as human body temperatures) the linearity is not too bad. When thermistors are used, however, it is necessary to ensure that the temperature will not produce excursions outside of the permissible linear range. The range of linearity for thermistor sensors can be improved by using a resistance network consisting of PTC, NTC, and non-thermistor resistor elements, as shown in Fig. 6-5C.

The last class of temperature transducer, and perhaps the largest class of transducer used in the −55 to +125 degrees Celsius range, is the solid-state pn junction. If you take an ordinary solid-state rectifier diode (Fig. 6-6) and connect it across an ohmmeter, then you can see the temperature effect on diodes. Note the forward-biased diode resistance at room temperature. Next heat the diode temporarily with a lamp or soldering iron. The diode resistance drops dramatically as temperature increases.

Most temperature transducers, however, use the diode-connected bipolar transistor such as shown in Fig. 6-7. The base-emitter voltage (V_{BE}) of a bipolar transistor is proportional to temperature. For a differential pair, as in Fig. 6-7, the transducer output voltage is given by:

$$V_{BE} = \frac{KT \cdot \ln(IC1/IC2)}{q}$$

where

K is Boltzman's constant (1.38×10^{-23} joules/K)

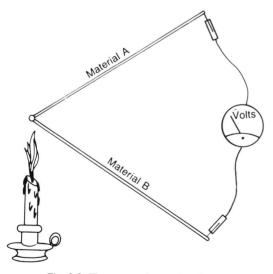

Fig. 6-3. Thermocouple construction.

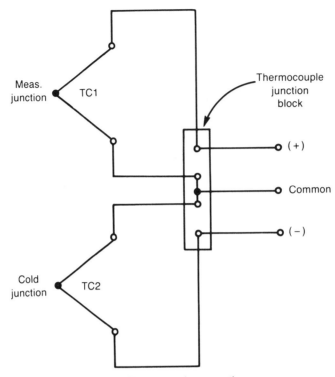

Fig. 6-4. Thermocouple connections.

T is the temperature in degrees Kelvin

ln is the natural logrithm

q is the electronic charge (1.6 × 10⁻¹⁹ coulombs per electron

The K/q ratio is constant under all circumstances. The ratio IC1/IC2 can be held constant artificially by making I3 a constant-current source. The only variable in the equation, therefore, is temperature. In the sections below we are going to take a look at some commercial integrated-circuit temperature devices.

COMMERCIAL IC TEMPERATURE MEASUREMENT DEVICES

Several semiconductor device manufacturers offer temperature measurement/control integrated circuits (TMIC). These devices are almost all based on the pn junction properties discussed earlier in this chapter, although at least one by Analog Devices, Inc., uses an external thermocouple. In this

section we will look at the semiconductor TMIC devices offered by National Semiconductor and Analog Devices, Incorporated. In addition, we will look at a method for converting temperature measurements to frequencies that can be transmitted along a communications link or recorded on a tape recorder.

National Semiconductor LM-335

The National Semiconductor LM-335 device shown in Fig. 6-8 is a three-terminal temperature sensor. The two main terminals are for power (and output), while the third terminal, shown coming out the body of the "diode" symbol is for adjustment and calibration. The LM-335 device is basically a special zener diode in which the breakdown voltage is directly proportional to the temperature, with a transfer function of close to 10 millivolts per degree Kelvin (10 mV/° K).

The LM-335 device, and its wider range cousins the LM-135 and LM-235 devices, operates

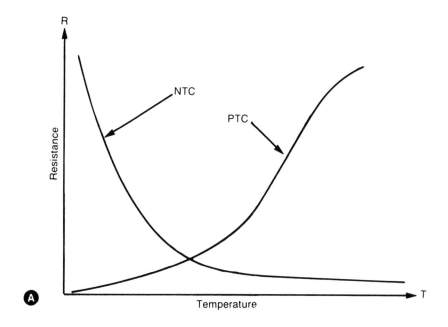

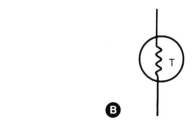

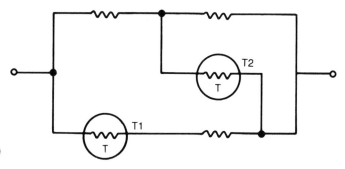

Fig. 6-5. A) Negative and positive temperature coefficient thermistors, B) thermistor symbol, C) linearized thermistor.

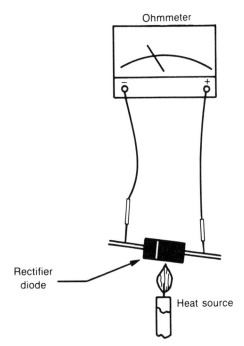

Fig. 6-6. Impromptu pn junction transducer.

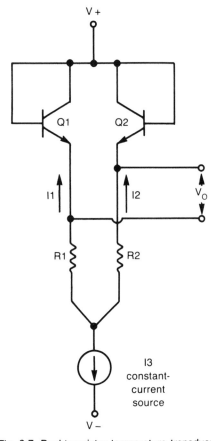

Fig. 6-7. Dual-transistor temperature transducer.

with a bias current set by the designer. This current is not super critical, but must be within the range 0.4 to 5 milliamperes. For most applications designers seem to prefer currents in the 1 mA range.

Accuracy of the device is relatively good, and is more than sufficient for most control applications. The LM-135 version offers uncalibrated errors of 0.5 to 1° C, while the less costly LM-335 device offers errors of 3° C. Of course, clever design can reduce these errors even further if they are out of tolerance for some particular application.

One difference between the three devices is the operating temperature range, which are as follows:

Device Type No.	Temperature Range (Centigrade)
LM-135	−55 to +150
LM-235	−40 to +125
LM-335	−10 to +100

There are two packages used for the LM-135 through LM-335 family of devices. The TO-92 is a small plastic transistor case "Z" suffix to part number, e.g., LM-335Z), while the TO-46 is the small metal can transistor package (smaller than the familiar TO-5 case).This case is identified with the suffix "H" or "AH" (for example, LM-335H or LM-335AH).

The simplest, although least accurate, method of using the LM-335 device is shown in Fig. 6-9A. The LM-335 is essentially a zener diode, and here

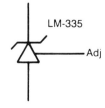

Fig. 6-8. LM-335 symbol.

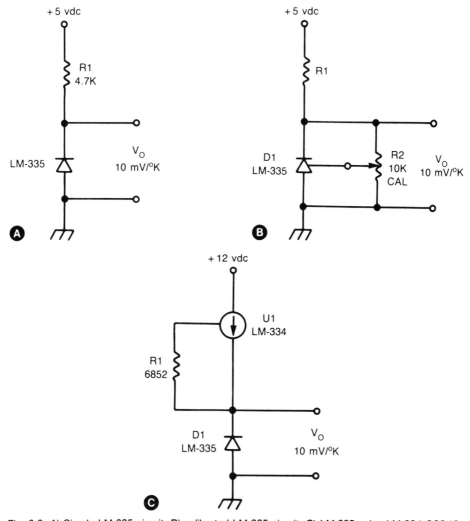

Fig. 6-9. A) Simple LM-335 circuit, B) calibrated LM-335 circuit, C) LM-335 using LM-334 CCS IC.

it is connected as a zener diode. The series current-limiting resistor limits the current to around 1 milliampere. This value of R1, i.e., 4700 ohms, is appropriate for +5 volt power supplies such as might be found in digital electronic instruments. The resistor value can be scaled upwards for higher values of dc potential according to Ohm's law (keeping $I = 0.001$ amperes):

$$R(ohms) = (V+) \times 1000$$

For example, when the power supply voltage is +12 volts dc, the value of the resistor in series with the LM-335 is:

$$R(ohms) = (V+) \times 1000$$
$$R(ohms) = (12\ Volts) \times 1000$$
$$R(ohms) = 12,000\ ohms$$

The output of the circuit in Fig. 6-9A is taken across the LM-335 device. This voltage has an approximate rate of 10 mV/°K. Recall from earlier that degrees Kelvin is the same as degrees Centigrade, except that the zero point is at absolute zero (close to $-273°$ C) rather than the freezing point of water. Using ordinary units conversion, arithmetic will show us how much voltage to expect at any given temperature. For example, suppose you want to know the output voltage at $78°$ C. The first thing you must do is convert the temperature degrees Kelvin. This neat little trick is done by adding 273 to the centigrade temperature:

$$°K = °C + 273$$
$$°K = 78° C + 273$$
$$°K = 351$$

Next, convert the temperature to the equivalent voltage:

$$V = \frac{10 \text{ mV}}{K} \times 351 \text{ K}$$

$$V = (10 \text{ mV})(351)$$
$$V = 3510 \text{ mV} = 3.51 \text{ volts}$$

One problem with the circuit of Fig. 6-9A is that it is not calibrated. While the circuit works well for many applications, especially those where precision is not needed, for other cases you might want to consider the circuit of Fig. 6-9B. This circuit allows single-point calibration of the temperature. The calibration control is obtained from the 10-kilohm potentiometer in parallel with the zener. The wiper of the potentiometer is applied to the adjustment input of the LM-335 device.

Calibration of the device is relatively simple. You only need to know the output voltage (a dc voltmeter will suffice), and the environmental temperature in which the LM-335 exists. In some less than critical cases, you might take a regular glass mercury thermometer and measure the air temperature. Wait long enough after turning on the equipment for both the mercury thermometer and the LM-335 device to come to equilibrium. After that, adjust the potentiometer (R2) for the correct output voltage. For example, if the room temperature is $25°$ C (i.e., $298°$ K), then the output voltage will be 2.98 volts. Adjust the potentiometer for 2.98 volts under these conditions.

Another tactic is to use an ice-water bath as the calibrating source. The temperature $0°$ C is defined as the point where water freezes, and is recognized by the fact that ice and water coexist in the same spot (the ice neither melts nor freezes, it is in equilibrium). A mercury thermometer will show the actual temperature of the bath. The potentiometer is adjusted until the output voltage is 2.73 volts (note: $0°$ C $= 273°$ K).

Still another tactic is to use a warmed oil bath for the calibration. The oil is heated to somewhat higher than room temperature (maybe $40°$ C), and stirred. Again, a mercury thermometer is used to read the actual temperature and the potentiometer is adjusted to read the correct value. The advantage of this method is that the oil bath can be at constant temperature. There are numerous laboratory pots on the market that will keep water or oil at a constant, preset temperature, a factor that avoids some problems inherent in the other methods.

Another connection scheme for the LM-335 is shown in Fig. 6-9C. In this variation on the theme, a National Semiconductor LM-334 three-terminal adjustable current source is used for the bias of the LM-335 device. Again, the output voltage will be 10 mV/° K.

All applications where the sensor is operated directly into its load have a potential problem or two, especially if the load impedance changes or if it is lower than some limit. As a result, you can sometimes justify using the buffered circuit of Fig. 6-10.

A "buffer" amplifier is one that is used for one or both of two purposes: 1) impedance transformation, and 2) isolation. The impedance transformation factor is used when the source impedance is high (not true of the LM-335). The isolation factor concerns us somewhat more here. The operational amplifier in Fig. 6-10 places an amplifier between the sensor and its load. The gain of the amplifier in this case is unity (i.e., 1), but a higher gain could be used if desired. In that case, simply substitute one of the gain amplifier circuits shown later in this book.

The operational amplifier shown here is an

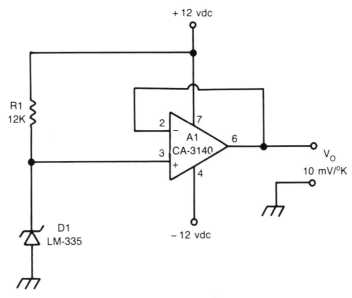

Fig. 6-10. Buffered LM-335 circuit.

RCA CA-3140 device. The reason for this is simply the freedom from bias currents exhibited by the BiMOS RCA operational amplifiers. The bias currents found on cheaper operational amplifiers could conceivably introduce error. The CA-4140 is not the only operational amplifier that will work, however; any low-input-bias current model will work nicely.

The noninverting input of the operational amplifier is connected across the zener diode-like LM-335. In this respect this circuit looks somewhat like the typical voltage reference circuits seen elsewhere. The bias for the LM-335 is from a 12-kilohm resistor, which is in keeping with our rule given earlier (Ohm's law, remember?).

Because there is no voltage gain in this circuit, the output voltage factor is the same as in previous designs, 10 mV/° K.

A circuit like that in Fig. 6-10 might prove useful in monitoring remote temperatures. If the operational amplifier is powered, a four-wire line is needed (V−, V+, ground, and temperature). The advantage is that the line losses are overcome by the higher output power of the operational amplifier. The LM-335 is a rugged little low-impedance device, however, and in many cases such measures would not be needed.

TEMPERATURE SCALE CONVERSIONS

The Kelvin scale is used extensively in scientific calculations, but is not always the most popular in practical measurement situations. In fact, I suspect that most readers of this book will want to make their temperature measurements in either degrees Centigrade, or degrees Fahrenheit. In this section I discuss the circuit methods used for both.

If the sensor is being input into a microcomputer, then it might be prudent to use the simplest circuit available, which is to measure in degrees Kelvin, and then let the computer do the neat trick of converting the units. The formulae below are useful for this purpose:

$$C = K - 273$$

and,

$$F = (1.8C) + 32$$

Of course, the first job will be to make the computer think it is seeing the correct kind of data. The analog-to-digital (A/D) converter will likely input a binary number between 00000000 and 11111111. This number must be properly scaled so that it represents a temperature value. Let's assume that we have an eight-bit A/D converter, and a temperature range of 0 to 100 degrees centigrade. The input voltage to the A/D will be 2.73 volts to 3.73 volts. If the A/D converter is able to offer offset measurements, then you can set the maximum range for 1 volt, and then offset it to 2.73 volts. In that unlikely case, 00000000 would represent 0° C, and 11111111 would represent 100° C. More likely, you would use a 5-volt, unipolar-input A/D converter to measure the narrow range of 2.73 to 3.73 volts, and suffer a resolution loss. Of course, this loss is not what it may seem because in many cases it will still be less than the nonlinearity of the transducer/sensor. In such a scheme, the voltage represented by a 1-LSB change in the A/D output data word represents approximately 20 mV, and so would represent 2° K. If all you need to measure is within two degrees, then you can use this system. Otherwise, some form of offset measurement is needed.

Figure 6-11 shows a scheme for converting the "degrees Kelvin" output of the LM-335 sensor (D1) into "degrees Centigrade." Because Centigrade degrees are the same size as Kelvin degrees, no change of slope in the output factor is needed:

the output is 10 mV/° C, and the circuit gain is unity.

The basic circuit of Fig. 6-11 is a dc differential amplifier based on a common operational amplifier (741-family devices work fine). The gain is set by R4/R2, assuming R2 = R3 and R4 = R5. The noninverting input of the dc differential amplifier receives the temperature signal, while the inverting input receives a dc offset bias. This circuit is adjusted by using potentiometer R6 to set the voltage at point "A" to +2.73 volts (use a 3-1/2 digit or more digital voltmeter). The result is that the output will be 2.73 volts less than it would have were the offset not placed in the circuit; thus the output potential is scaled in degrees centigrade.

Figure 6-12 shows a circuit for converting degrees centigrade to degrees Fahrenheit. The problem here is that the two types of degrees are 1) offset from each other (like Kelvin and Centigrade, they have different zero references), and 2) different sizes. Thus, the conversion amplifier must offer both an offset and a change of slope. The circuit in Fig. 6-12 does both. The offset is provided by potentiometer R5, which is used to set the ice-point (zero degrees centigrade) output level.

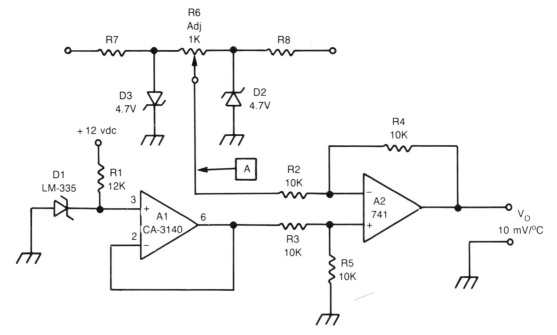

Fig. 6-11. Analog electronic thermometer.

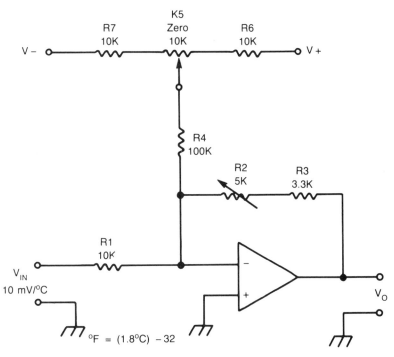

Fig. 6-12. C-to-F converter.

The feedback potentiometer is used to set a calibration point at some higher temperature (for example 25° C, or room temperature, i.e., 77° F).

Calibration of the two points is performed in a manner similar to above. The zero point is set using an ice bath and adjusting R5; the higher point is probably best set at room temperature. In both cases, the actual temperature could be measured with an ordinary mercury thermometer. Of course, the best accuracy is obtained with a laboratory grade mercury thermometer.

Analog Devices AD-590

The Analog Devices, Inc. AD-590 is another type of solid-state temperature sensor. This particular device is a two-electrode sensor that operates as a current source with a one microampere per degree Kelvin (1 μA/° K) characteristic. The AD-590 will operate over a temperature range of -55 to $+150$ degrees centigrade. It is capable of accepting a wide range of power-supply voltages, being happy with anything in the range $+4$ to $+30$

volts dc (this range is more than sufficient for most solid-state applications). Selected versions are available with linearity of $-/+0.3$ degrees centigrade and a calibration accuracy of $-/+0.5$ degrees centigrade.

The AD-590 comes in two different packages. There is a metal can (TO-52) that is recognized as the small-sized transistor package (smaller than TO-5). There is also a plastic flat-pack available.

Being a two-terminal current source, the AD-590 is simplicity itself in operation. Figure 6-13 shows the most elementary calibratable circuit for the AD-590. Because it is a current source that produces a current proportional to temperature, we can convert the output to a voltage by passing it through a resistor. In Fig. 6-13, the resistance is approximately 1000 ohms and consists of the resistance of R2 (950 ohms) and R1 (a 100-ohm potentiometer). From Ohm's law we know that 1 μA/K converts to 1 mV/K when passed through a 1000-ohm resistor. You can calculate the voltage output at any given temperature from the simple relationship below:

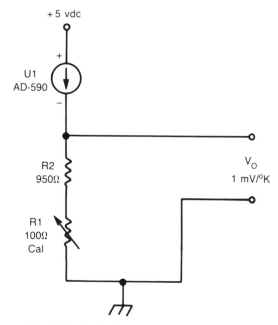

Fig. 6-13. Analog Devices, Inc., AD-590 circuit.

$$V_o = \frac{1\ mV}{K} \times Temp$$

Thus, if you have a temperature of 37° C, which is (37 + 273) or 310K, then the output voltage will be:

$$V_o = \frac{1\ mV}{K} \times Temp$$

$$V_o = \frac{1\ mV}{K} \times 310K$$

$$V_o = 310\ mV$$

Potentiometer R1 is used to calibrate this system. You can make a quick and dirty calibration with an accurate mercury thermometer (laboratory grade recommended) at room temperature. Connect a digital voltmeter across the output, and allow the system to come to equilibrium (should take about ten minutes). Once the system is stable, adjust the potentiometer for the correct output voltage. For example, assume that the room tem-

perature is 25° C, which is 77° F. This temperature converts to (272 + 25), or 298° K. The output voltage will be (1 mV × 298), or 298 millivolts (0.298 volts). Using a 3-1/2 digit voltmeter is sufficient to make this measurement.

In some cases it might be wise to delete the potentiometer and use a single 1000-ohm resistor in place of the network shown. There could be several reasons for this. First, the calibration accuracy is not critical for the application at hand. Second, potentiometers are points of weakness in any circuit. Being mechanical devices, they are subject to stress under vibration conditions and may fail prematurely. If the temperature accuracy is not crucial and reliability is, then consider the use of a single fixed 1-percent tolerance resistor in place of the network shown in Fig. 6-13.

The circuit of Fig. 6-13 is used sometimes to make a temperature alarm. By using a voltage comparator to follow the network and biasing the comparator to the voltage that corresponds to the alarm temperature, is over the limit. A "window comparator" will allow you to have an alarm of either under or over temperature conditions. Some electronic equipment designers use this tactic to provide an overtemperature alarm. In one application a commercial minicomputer generated a large amount of heat (it used a 35-ampere, +5-volt dc power supply!).The specification called for an air-conditioned room for housing the computer. An AD-590 device was placed inside at a critical point. If the temperature reached a certain level (45° C), then the comparator output snapped LOW and created an interrupt request to the computer. The computer would then sound an alarm and display an "overtemperature warning" message on the CRT screen.

The circuit of Fig. 6-13 suffers from one little problem: it allows calibration at only one temperature. Unfortunately, this situation does not allow for optimization of the circuit. You can, however, improve the situation using the two-point calibration circuit of Fig. 6-14. In this case we see an operational amplifier in the inverting follower configuration. The summing junction (inverting input) receives two different currents. One current is the output of the AD-590 (i.e., 1 μA/K), while the other current is derived from the reference voltage, V_{REF}, or 10.000-volts. Adjustment of this current provides the zero reference adjustment, while the overall gain of the amplifier provides the full-scale adjustment.

The operational amplifier selected is the LM-301 device, although almost any premium operational amplifier will suffice. The RCA CA-3140 BiMOS device, or some of those by either Analog Devices or National Semiconductor will also work nicely. If the LM-301 or similar device is used, then be sure to use the 30-pF frequency compensation capacitor.

The V− and V+ power supply lines are bypassed with 0.1-μF and 4.7-μF capacitor. The 0.1-μF capacitors are used for high-frequency decoupling and must be mounted as close as possible to the body of the operational amplifier. The values of these capacitors are approximate, and they may be anything from 0.1 μF to 1 μF.

Calibration of the device is simple, although two different temperature environments are required. The zero-degrees Centigrade adjustment (R1) can be made with the sensor in an ice-water bath (as described above). The upper temperature can be room temperature, provided that some means is available to measure the actual room temperature for comparison.

A differential thermometer circuit is shown in Fig. 6-15. This circuit uses a pair of AD-590 sensors to measure the difference between two temperatures. Rather than using a dc differential amplifier in this application, an inverting follower is used instead. The signal to the inverting input is derived from potentiometer R1, the current from sensor U1 (1 μA/K), and the current from sensor U2 (1 μA/K also). The output voltage of this circuit is proportional to the summation (or difference) of the input currents. Thus, the output voltage is given by:

$$V_o = (T1 - T2) \times (10 \text{ mV}/^\circ \text{ C})$$

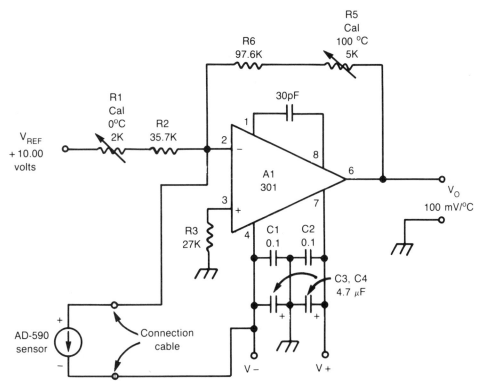

Fig. 6-14. AD-590 two-point calibration circuit.

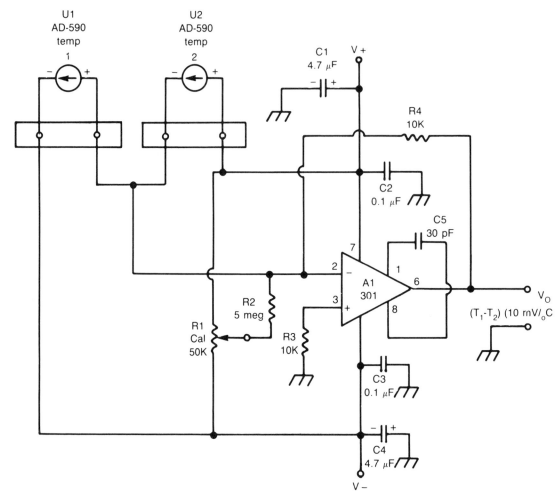

Fig. 6-15. Differential thermometer.

Potentiometer R1 is a balancing circuit that provides the offset that converts from Kelvin to Centigrade scales.

So what are differential thermometers used for? In some cases home environmental-control systems perform heating or cooling chores according to the difference between inside and outside temperatures. Solar water heating and space-heating devices sometimes use differential temperature sensors to decide whether to use the solar heat collectors or "fossil fuel" to heat the water.

Another application is in biofeedback studies. Some scientists in the biofeedback world measure the temperature between two points on the body (often the forehead and hand), and then train the subject with biofeedback methods to balance the temperatures. One reported application is in tension headaches, where the differential temperature indicates relative blood flow — and supposedly altering blood flow relieves the headache.

TEMPERATURE-CONTROLLED FREQUENCY GENERATOR

There are times when you might want to convert a temperature reading into a proportional audio tone. Examples of this need are transmission of temperature telemetry data over radio or telephone lines or in tape recording the data. You

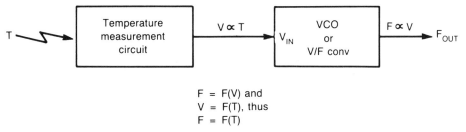

$$F = F(V) \text{ and}$$
$$V = F(T), \text{ thus}$$
$$F = F(T)$$

Fig. 6-16. Temperature to frequency converter.

might also want to use audio tones to represent temperature in cases where the sensor is connected by a long wire to the recording instrument, such that the voltage drop in the wire forms a significant error term. Such an application occurs in oceanography where a temperature sensor is lowered into the sea on a long tether (often hundreds or thousands of feet long). If the sensor output is converted into an audio tone, it can be sent to the surface over a wire, amplified and displayed; line losses play no part in the accuracy of the system.

Figure 6-16 shows a block diagram of a typical system. In this case one of the temperature sensor and/or amplifier systems shown earlier in this section are used. The voltage output of the sensor will be applied to the input of a voltage-controlled oscillator (VCO) or a voltage-to-frequency converter (VFC). The result is a frequency output (F_{OUT}) that is proportional to the applied input voltage, which in turn is proportional to temperature.

Figure 6-17 shows a slightly different tactic. The Analog Devices, Inc., AD-537 is an integrated circuit voltage-to-frequency converter that has a secondary output that is linearly proportional to temperature. If this output is fed back into the "voltage" input, then you have a signal generator that produces a frequency proportional to the applied temperature. The output frequency is found from:

$$F_o = 10 \text{ Hz}^\circ \text{ K}$$

with a maximum frequency of about 10 kHz. Calibration of this circuit is single point, using potentiometer R1. Perhaps the easiest method would be to use air temperature and a mercury thermometer. Set the frequency according to the formula

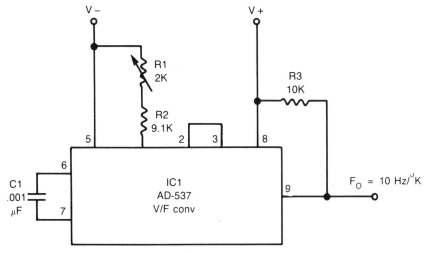

Fig. 6-17. IC T-to-F converter.

above and the reading on the mercury thermometer. For example, if the room temperature of 298 degrees Kelvin is used, then the output frequency is:

$$F = \frac{10 \text{ Hz}}{K} \times \text{Temp}$$

$$F = \frac{10 \text{ Hz}}{K} \times 298 \text{ K}$$

$$F = 2980 \text{ Hz}$$

The frequency is set in part by the 0.001-μF capacitor (C1) connected across pins 6 and 7. This must be either a silver mica or NPO disk ceramic capacitor in order to keep temperature drift from occurring.

THERMOCOUPLE SIGNAL PROCESSOR

Earlier in this chapter I discussed the thermocouple device. These temperature sensors are a junction of two dissimilar metals. The different "work functions" of the two metals are responsible for generating an output voltage that is proportional to the applied temperature. Figure 6-18 shows a simple integrated-circuit device that produces a linear output voltage of 10 mV/° C from a thermocouple input voltage. There are two devices available: AD-594 and AD-595. The AD-594 is used with Type J thermocouples, while the AD-595 works with Type K thermocouples.

The AD-594 and AD-595 devices operate with supply voltages ranging from a unipolar +5 volts dc to a bipolar −/+ 15 volts dc. The power dissipation is typically 1 milliwatt.

THERMISTOR CIRCUITS

The thermistor is little more than a resistor that changes value with changes in temperature. In this section I discuss how to produce a voltage that is proportional to the temperature. A half-bridge circuit is shown in Fig. 6-19A. In this circuit there is a voltage divider consisting of a fixed resistor (R1) and a thermistor (RT1). The output voltage is given by:

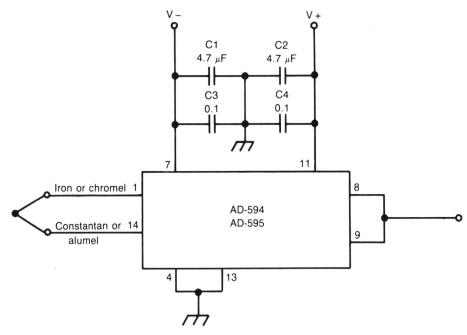

Fig. 6-18. AD-594 thermocouple IC processor.

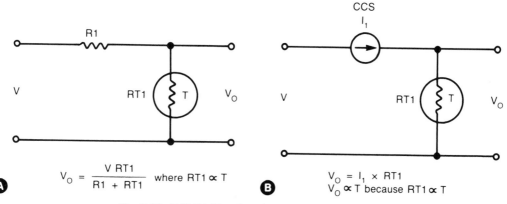

$$V_O = \frac{V \, RT1}{R1 + RT1} \quad \text{where } RT1 \propto T$$

A

$$V_O = I_1 \times RT1$$
$$V_O \propto T \text{ because } RT1 \propto T$$

B

Fig. 6-19. A) Half-bridge thermistor circuit, B) with CCS.

$$V_o = \frac{V \, RT1}{R1 + RT1}$$

where

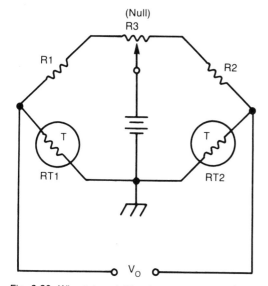

Fig. 6-20. Wheatstone-bridge temperature transducer.

V is the reference excitation voltage
V_o is the output voltage
R1 is the fixed resistance
RT1 is the resistance of the thermistor, which
   is a function of temperature

Some instruments replace the fixed resistor with a constant-current source, such as the LM-334 device. In these circuits (Fig. 6-19B) the output voltage is the product of the current and the thermistor resistance.

A disadvantage of the half-bridge circuit is that there is always a dc output voltage. A null circuit, on the other hand, can produce a zero output voltage when the temperature is zero degrees. An example is the Wheatstone bridge shown in Fig. 6-20. The output voltage is zero when the ratio of the resistances in the arms is equal: R1/RT1 = R2/RT2. When either thermistor changes resistance, then the output voltage goes up or goes down from zero. Although thermistor bridges are older (in fact, ancient) in design they are nonetheless used quite often.

Chapter 7

Pressure and Force Measurements

T HE MEASUREMENT OF PRESSURES IS WIDE-spread in medicine, science, and industry. Although many of the specifics vary from application to application, many of the underlying principles are the same. Indeed, when measuring fluid pressures the difference between a chemical pipeline in a plant, a water main under the street, or the blood in the artery of a surgical patient is less a matter of approach and basic theory than of specific hardware selections.

Gas pressures are treated similarly to fluid pressures, so are also included in this chapter. In fact, one often uses air pressure over atmosphere to calibrate fluid-pressure measurement instruments. Given a rigorous definition of "fluid," we must consider both liquids and gases. The difference between liquids and gases is that gases are compressible while the liquids are not. Compressibility affects the measurement technique. Gas pressure measurement is considered more closely in a later section, while fluid measurements are treated first.

MEDICAL PRESSURE MEASUREMENTS

Medical researchers and clinicians measure several different types of fluid and gas pressure.

The most common pressure measurement is arterial blood pressure, which is almost routinely monitored by electronic instruments in areas of the hospital such as ICU, CCU, Emergency Room, Trauma Unit, or Operating Room. Also of interest in special cases are venous pressures, central venous pressure (CVP), intracardiac blood pressure, pulmonary artery pressure, spinal fluid pressures, and intraventricular brain pressures. The principal differences between these pressures is primarily one of range; often the same instruments are used for all of these different pressures. The principal difference between medical pressures and nonmedical pressures is not merely one of range (and the fact that mmHg are used instead of lbs/in^2), but also of technique. The medical pressure measurement system needs to be as least invasive as possible and sterile (to prevent infection). In addition, electrical isolation from the ac power mains must be maintained at a higher level than usual for purposes of patient safety.

Figure 7-1 shows a typical medical monitor that includes a pressure monitor in the lower half-case (the upper unit is an electrocardiograph amplifier — see Chapters 11 and 12). Many of the features of this instrument are also incorporated in nonmedical pressure-measurement instruments. In fact, medical pressure monitors are often used for nonmedical applications. The readout is a digital meter that displays the pressure in millimeters of mercury (mmHg). The display can be used to select systolic, diastolic, mean, and venous. The systolic and diastolic pressures are taken from the arterial waveform by peak and valley (inverted peak) holder circuits, respectively. The venous pressure is merely a higher gain position that allows a lower pressure to be displayed full-scale. The mean pressure is merely the time average of the input pressure waveform, i.e., it is the time integral of the pressure waveform (more on this later).

The alarm section of the pressure monitor is used to set upper and lower limits through the use of slide controls underneath the meter. In analog circuits the alarm system ordinarily consists of a voltage window comparator in which the lefthand control sets the lower limit and the righthand control sets the upper limit. In microprocessor-controlled instruments, the controls are either digital in their own right, or are potentiometers connected to a reference voltage source input and an A/D converter output.

The oversize transducer connector at the input terminal is used to house certain calibrating components. Very few transducers have repeatable zero, range, and output potential from one unit to another. Thus, you must either provide calibrating controls on the electronic unit, or normalize all transducers with calibrating components (e.g., a potentiometer output-level control and Wheatstone-bridge linearization resistors). In the instru-

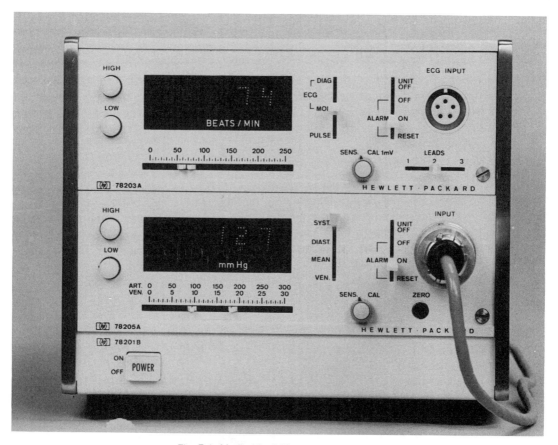

Fig. 7-1. Medical bedside pressure monitor.

ment of Fig. 7-1, a little of both is provided. The calibrating components in the transducer plug bring the output within range of the sensitivity and zero controls.

WHAT IS "PRESSURE?"

Most of us have some idea of how pressure is defined, but all too often we find even practicing engineers have a poor idea of what they are measuring in their pressure monitors. For example, a group of engineering students were asked to define "pressure" only one year after taking Physics I and II. Several of them gave the correct answer, but most wrote hazy, ambiguous statements that indicated to the professor that they did not really understand what is meant by the concept. Some students came close, indicating that pressure is a force. But this definition is still not correct: the proper definition is that a pressure is a *force per unit area*, or

$$P = F/A$$

where

P is the pressure in Newtons per square meter (N/m^2) or pascals (Pa); 1 N/m^2 = 1 Pa.

F is the force in Newtons

A is the area in square meters

Example. A small coin has a diameter of 1 cm and a mass of 1.5 grams. Find: A) the gravitational force (weight) of the coin in dynes and millinewtons. B) The pressure this coin exerts while lying flat on a table top in dynes per square centimeter and newtons per square meter (or pascals).

Solution:

A) Force in dynes

F = ma (Newton's Second Law)

F = (1.5 g) (980 cm/s^2)

F = 1470 g-cm/s^2

F = 1470 dynes

Force in millinewtons

F = 1470 dyne $\times$ (1N/10^5 dyne)

F = 1.47 $\times$ 10^2 N

F = 14.7 mN

B) Pressure in dynes per square centimeter

P = F/A

P = 1470 dynes/(3.14)(1 cm/2)^2

P = 1871 dyne/cm-cm

Pressure in pascals

P = (1872 dyne/cm-cm) $\times$ (10^{-5}N/dyne) $\times$(10^2 cm/m^2)

P = 187.2 N/m-m = 187.2 Pa

The pressure can be increased by either increasing the applied force, or by reducing the cross-sectional area over which the force operates. Alternate units for the pressure are 1) CGS System: dynes per square centimeter (dyne/cm^2), and 2) British Engineering System: pounds per square inch (lb/in^2 or psi.

When the force in any system is constant or static (that is, nonvarying), then that pressure is said to be *hydrostatic*. On the other hand, if the force is varying, the force is said to be *dynamic* or *hydrodynamic*. Pressure in a fluid pipeline, or physiological pressures (e.g., human arterial blood pressure) are examples of hydrodynamic pressures; the pressure head in a stoppered keg of beer is a hydrostatic pressure—at least until the bung is popped.

Pascal's Principle, named after French scientist and theologian Blaise Pascal (1623–1662) governs pressures in a closed system. This physical law states that "Pressure applied to an enclosed fluid is transmitted undiminished to every portion of the fluid and the walls of the containing vessel."

If a pressure is applied to the stoppered system (e.g., the syringe in Fig. 7-2), then the same pressure is felt throughout the interior of the syringe. Changing the applied pressure at the rear of the plunger causes the same change to be reflected at every point inside of the syringe.

A physics professor I had in engineering school had an experiment that amply illustrated Pascal's principle—even though over several years it caused the first floor lecture hall stage in the science building to rot away. He had a wooden water barrel that was sealed all around, except for a hole in the top where a water hose fitting was installed. A long garden hose was snaked out the door, up the outside of the building to the fourth floor, where a graduate student was posted. The

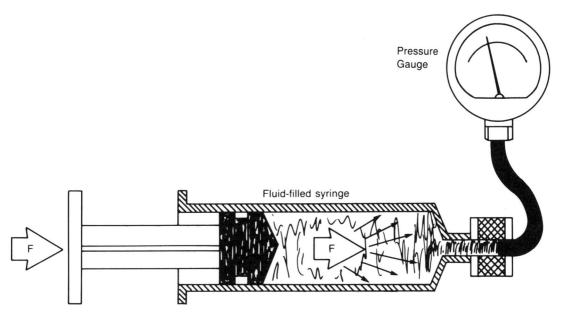

Fig. 7-2. Illustration of Pascal's principle.

system was filled with water, and a wooden dowel snugly fitted to the hose at the open end on the fourth floor served as the plunger (similar to Fig. 7-2). On command from the professor (issued over a pair of $10 CB walkie-talkies), the graduate student would drive the dowel into the hose with great force. The force of the student's hand produced a large force that was felt inside the barrel — causing the barrel to burst apart violently. Although the experiment forcibly illustrated Pascal's Principle, the twice annual inundation of the lecture stage with water from the barrel caused the wood beneath to rot out!

Pascal's Principle always holds true in hydrostatic systems. In hydrodynamic systems it holds true only for quasistatic changes. That is, when a very small change is made, and the turbulence allowed to die down before subsequent measurements are made. Pascal's Principle holds approximately true for those hydrodynamic systems where the flow is reasonably nonturbulent (no true nonturbulent flow exists) and the pipe lumen is small compared with its length. The simple model holds true in those cases, however, only in the center of the flow mass but not at the pipe wall boundaries.

In physiological systems, the situation is somewhat more complicated. Unfortunately, many students in life sciences lack the mathematics background to perform proper analysis of blood flow systems. Many medical students, for example, are taught a naive model that fails to take into account the fact that the situation is complicated by four factors a) the particulate nature of blood (it is not a strict fluid, but contain red cells and other material), b) that blood vessel walls are distensible and not rigid, c) blood viscosity changes under certain influences and d) the walls are not smooth, especially in older subjects. The naive model is analogous to Ohm's law, and states that:

$$R = RF$$

where

P is the pressure difference in Torr (mmHg)
F is the flow rate in milliliters/second (ml/s)
R is the blood vessel resistance in peripheral resistance units (PRU), such that 1 PRU allows a flow of 1 ml/s under a 1 Torr pressure

Example. Find the pressure if a blood flow of 1.7 ml/s flows through a 4 PRU resistance.

Solution:

$$P = FR$$
$$P = (1.7 \text{ ml/s}) \ (4 \text{ PRU})$$
$$P = 6.8 \text{ Torr}$$

The actual situation is physiological blood flow systems is a lot more complex because of the factors listed above. The actual vessel diameter changes (which is one way your body regulates blood pressure) both from systemic readjustments, and from the fact that the beating heart forms a pulsatile pressure wave. The flow rate (and hence the other parameters) is not a simple factor like the current flow in a simple resistance, but rather is better (but imperfectly) given by Poiseulle's law, which is:

$$F = \frac{P \times R^4}{8nL}$$

where

    F is the flow in cubic centimeters/second (cc/s)
    P is the pressure in dynes per square centimeter
    n is the coefficient of viscosity in dyne-seconds per square centimeter
    R is the vessel radius in centimeters
    L is the vessel length in centimeters

The study of pressure in turbulent or large lumen systems, or in the boundary area close to the pipe/vessel wall, is the subject of engineering mechanics and physics courses. We will assume that Pascal's Principle either holds true absolutely, or that the system can be made quasistatic for measurement or analysis purposes.

Pulsatile systems result from a pumping action that is not constant (which includes most mechanical pumps). A piston pump or bellows pump, for example, places a pulsatile pressure waveform on the system. In physiological systems, the heart of the subject animal (or man) beats in a manner

that produces a pulse flow (which can be felt with the fingertips where arteries run close to the surface — in the wrist for example). Figure 7-3 shows the human arterial blood pressure waveform, here used as an example of a pulsatile system. There are several values which can be measured in this system:

1. Peak pressure (called "systolic" in medical jargon)
2. Minimum pressure (called "diastolic" in medical jargon)
3. Dynamic average (one-half peak minus minimum)
4. Average pressure (i.e., time integral of P)

When you discuss "pressure" in a pulsatile system, you must also specify <u>what</u> pressure is intended! In a later section I will discuss the methods used for measuring these pressures electronically.

As an engineer designing a pressure measurement system for nonengineering personnel, you will want to consider the point of view of the client. During the time when I was in biomedical engi-

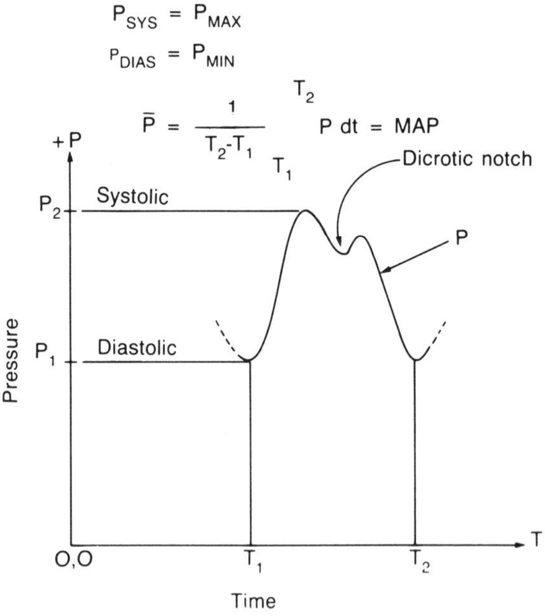

Fig. 7-3. Typical medical pressure waveform.

neering we had a constant problem with both clinicians and researchers with the average pressure readings on the electronic blood-pressure monitors. The problem was in their definition of "mean arterial pressure" (i.e., the time average). The correct definition, which is used in the design of the typical instrument, is written as:

$$\bar{P} = \frac{1}{T2\text{-}T1} \; P \; dt$$

In medical and nursing schools, and in a typical Intensive Care Unit nursing courses, however, a synthetic (and often incorrect) definition is used (referring to Fig. 7-3):

$$\frac{(P2 - P1)}{3} + P1$$

In medical terminology, this definition states that the mean arterial pressure (or MAP) is equal to the diastolic (P1) plus one-third the difference between systolic and diastolic (P2 − P1).

The problem faced by the biomedical engineer (and more often the Biomedical Equipment Technician who repairs the equipment for complaining nurses) is that the synthetic definition is merely an approximation of the functional definition (the integral) for healthy people! In many sick people, however, the portion of the waveform between the dicrotic notch and time T2 (see Fig. 7-3) is very heavily damped, so the actual MAP is less than the functional MAP actually measured by the electronic instrument (try telling that to an angry nurse at the patient's bedside!). The simple test, which is revealed by plugging values into both equations, is to place a constant pressure on the system and see what happens to the readings. In that case, P1 = P2 = MAP, so all three digital readouts should be the same.

In a later section I will deal with the electronic measurement of pressures, so I will want to return to Fig. 7-3 to illustrate the relationships of the various pressures.

BASIC PRESSURE MEASUREMENTS

The air forming our atmosphere exerts a pressure on the surface of the earth, and all objects on the surface (or above it). This pressure is usually expressed in atmospheres (atm), pounds per square inch (lb/in² or PSI), or other pressure units. The magnitude of 1 atm is approximately 14.7 lb/in² at mean sea level.

If a pressure is measured with respect to a perfect vacuum (defined as 0 atm), then it is called *absolute pressure;* and if against 1 atm ("open air") it is called a *gauge pressure.* Two gauge pressures, or a gauge pressure and an absolute pressure, can be measured relative to each other to form a single measurement called variously relative pressure, or differential pressure. Pressures in fluid pipelines, storage tanks and the human circulatory system are usually gauge pressures (if measured at a point) or differential pressures (if measured between two points along a length).

Figure 7-4 shows the Torricelli manometer, named after Evangelista Torricelli (Italian scientist, 1608–1647), which is used to measure atmo-

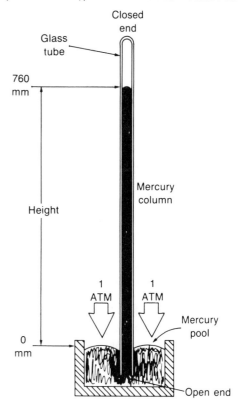

Fig. 7-4. Mercury manometer.

spheric pressure. An evacuated, small-lumen glass tube stands vertically in a pool of mercury. The end that is inside the mercury (Hg) pool is open, while the other end is closed. The pressure exerted by the atmosphere on the surface of the mercury pool forces mercury into the tube, forming a column. The mercury column rises in the tube until its weight (i.e., gravitational force) exactly balances the force of the atmospheric pressure. Torricelli found that a 760-mm column of mercury could be supported by atmospheric pressure at sea level. Thus, 1 atm is 760 mmHg (also sometimes given in weather reports and aviation in inches, i.e., 1 atm = 760 mmHg = 29.92 inHg).

The proper units of pressure, as established by scientists in an international agreement and adopted in the United States by the National Bureau of Standards, is the Torr (named after Torricelli), where 1 Torr is equal to 1 millimeter of mercury (i.e., 1 T = 1 mmHg). In medicine and medical science (e.g., physiology) the mmHg is still used instead of the more correct Torr.

Gauge pressures are usually given in mmHg (or inches) above or below atmospheric pressure. A manometer is any device that measures gauge pressure, positive or negative. By convention, pressures above atmospheric pressure are signed positive, and those below atmospheric pressure are signed negative. Also by convention, negative gauge pressures are called *vacuums* and negative-reading manometers are called *vacuum gauges* (positive reading manometers are also called *pressure gauges*). Both instruments are nonetheless properly called *manometers,* and in this chapter we will discuss both physical and electronic manometers.

All measurements require some form of reference point, and for gauge pressures the zero reference is a pressure of one atmosphere (1 atm). Even though the absolute value of the atmospheric pressure varies from one place to another, and in the same location over the space of a few hours, the zero point can be established by setting the zero scale on the indicator by opening the manometer to atmosphere.

Figure 7-5 shows a mercury manometer that

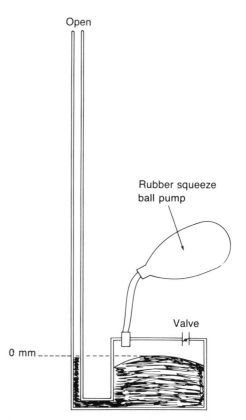

Fig. 7-5. Mercury manometer with squeeze ball pump.

is similar to those used to measure pressures and calibrate electronic pressure manometers. The open tube is connected to a mercury reservoir that is fitted with a rubber squeeze-ball pump that can be used to increase pressure in the system. A valve is used to either open the chamber to atmosphere or close it off.

If the valve is open to the atmosphere, then the pressure on the chamber is equal to the pressure on the column, i.e., one atmosphere. Under this condition the mercury column in the tube is at the same height as the mercury in the chamber. This point is defined as 0 mmHg. If the valve is closed, and the pressure inside the chamber is increased by operating the pump, then the mercury in the column will rise to a level proportional to the new pressure above atmospheric pressure.

If the rubber ball in Fig. 7-5 is replaced with a connection to a closed pressure system other than the squeeze ball, then the mercury will rise to a level proportional to the pressure in that system. You can use this manometer as a calibrating device by adding a tee-connector in the line between the rubber squeeze ball and the chamber. One port of the tee goes to the rubber ball, one port goes to the chamber, and the third port goes to the transducer or other instrument being calibrated.

Gauge pressure is used for measurement purposes because it is a lot easier to reference at the zero point (open the manometer to the atmosphere), and can be easily recalibrated for each use (no matter where in the world the measurement is made). In addition, for most practical applications the absolute pressure conveys no additional information content over gauge pressure. Should absolute pressure be needed, then it becomes a relatively simple matter to measure the atmospheric pressure (with a device like Fig. 7-4), and then add that value to the pressure measured on the gauge pressure manometer of Fig. 7-5.

BLOOD PRESSURE MEASUREMENT

Until 1905, the only method available for the measurement of blood pressure was to insert a manometer tube into the vein or artery being measured, and note how far up the tube the blood flowed. This method is still used for some spinal fluid and central venous pressure measurements, but was pioneered in 1773 by English physician Stephen Hales, who used an open-ended glass tube in the neck artery of an unanesthetized horse (which was presumably tied down securely!).

Sphygmomanometry is an indirect method of measuring the pressure. An inflatable cuff is placed over the arm, and inflated to a point where it closes (occludes) the underlying artery to a point where no blood can flow. A stethoscope placed downstream from the occlusion is used to monitor the onset of blood flow. The operator slowly releases the pressure in the cuff (3 mmHg/second is optimum) until a series of sharp, snapping *sounds of Korotkoff* are heard. These occur due to the turbulence of blood under pressure breaking through the occlusion into the downstream artery.

This event occurs when the cuff pressure equals the systolic pressure. The operator continues to monitor the pressure until the Korotkoff turbulence dies out, an event that occurs when the cuff pressure equals the diastolic pressure. Sphygmomanometry is an indirect measure of blood pressure.

The modern indirect measurement using the familiar blood pressure cuff is called *sphygmomanometry*. This method was pioneered in 1905 by Nicolas Korotkoff, but was not widely used until after 1935 when the Korotkoff readings were finally correlated to the Hale's method readings. It wasn't until well after World War II that nurses were permitted to take blood pressure readings, being considered "doctor's work" up until that time.

The modern electronic measurements consist of two types. One is an electronic version of the Hale's 1773 method, in which a thin, hollow catheter is introduced into the artery, and then connected to a transducer (see Fig. 7-6). The transducer outputs an electrical analog of the pressure waveform that can be directly calibrated in mmHg (Torr). The second type of electronic measurement is used by many all electronic home-type blood pressure kits. In these instruments a microphone (optimized for low-frequency sounds) is placed under the cuff and the pressure released. When the Korotkoff sounds are heard, the internal circuitry records the cuff pressure as the systolic, and when they disappear it records the diastolic pressure.

FLUID PRESSURE TRANSDUCERS

Figure 7-7 shows a cutaway view of a typical fluid-pressure transducer (including blood pressure). The body of the transducer contains the circuitry, which is separated from the fluid dome by the transducer's diaphragm. This thin metallic membrane feels the pressure in the dome and is distended an amount proportional to the applied pressure. The other side of the diaphragm will either be connected to the core of an LVDT transformer or a piezoresistive Wheatstone-bridge transduction element. The dome is used to contain

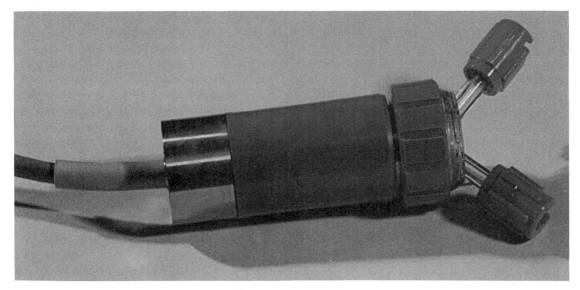

Fig. 7-6. Pressure transducer photo.

the fluid, and in medical applications will be disposable (in order to prevent cross infection between patients).

There are at least two ports on the transducer dome. These ports are controlled by stopcocks, either built-in or add-on. When the stopcock is opened, then fluid flows into or out of the transducer, but when it is closed the transducer is basically a closed system. In normal operations one port will be connected to the system being measured, while the other port opens to atmosphere. With the atmospheric port open, the transducer is at zero gauge pressure, so the electronic instrument it drives can be "zeroed."

Figure 7-8 shows the calibration set-up for an electronic fluid pressure-measurement instrument. The transducer is the sort discussed above, which is connected to an electronic manometer. In this

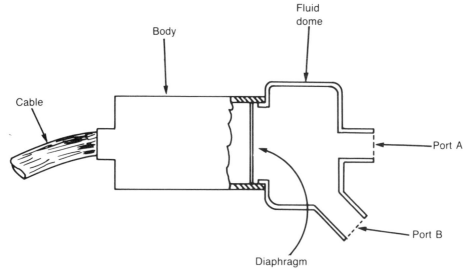

Fig. 7-7. Pressure transducer schematic.

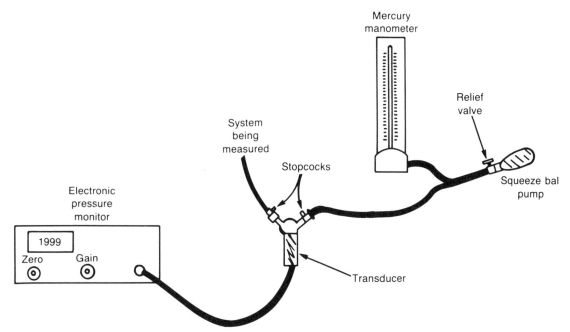

Fig. 7-8. Pressure calibration system.

test set-up, the atmospheric port is connected to a regular mercury (or in low pressure cases water) manometer. When the relief valve is open to atmosphere, the pressure monitor is set to zero. The valve is then closed, and a pressure pumped onto the system by the squeeze ball pump. In most cases it is a good idea to set the pressure (as read on the mercury manometer) to some standard value such as 100 Torr, 200 Torr, etc. You can then adjust the SPAN or GAIN control on the electronic monitor to the same reading as obtained on the mercury.

In biomedical applications, it is critical that the operator be instructed in proper technique. One reason is to prevent contamination of the patient's lines or other problems. Another is that incorrect settings of sometimes complicated stopcock arrangements can cause the system to be pumped with air, which could enter the patient as a potentially fatal air embolism — get instruction!

FLUID MEASUREMENT SYSTEM FAULTS

The measurement of fluid pressures with electronic apparatus may not always be the simple matter that it seems if certain precautions are not observed. There are several problems that can affect the data acquired. For example, although it seems like a trivial observation, in multibranch or varying diameter systems you have to understand which pressure is being measured. One problem engineers have with medical personnel in pressure measurement situations is that the pressure measured by the instrument is not the same pressure measured by the blood pressure cuff. Typically, the blood-pressure cuff measures a pressure in the upper arm, while the electronic instrument catheter is placed distal (i.e., downstream) from that point. The blood pressure catheter is typically placed in the radial artery in the wrist. The problem is that these two pressures are normally different, so a nurse who accurately takes the blood pressure manually (accuracy is a problem with the manual method) will normally obtain a slightly higher reading than the downstream reading — taken after the artery has branched. The same observation also holds true for nonmedical multibranch systems.

Another problem frequently encountered is either resonance or damping of the pressure in nonstatic systems. Of course, if the pressure is a dead pressure, i.e., one which does not vary or varies quasistatically, then the problem does not exist. But if the pressure has a waveform, as in a blood pressure system, then there are certain problems with the plumbing that can yield error (also called "artifact" in the life sciences and medicine).

There are several causes of damping in fluid-pressure systems. One is clogging of the tubing to and from the transducer. This problem is especially common in medical systems where the blood enters the tubing and clots. Other forms of fluid will show the problem either due to phenomena like clotting or particulate matter blocking the sampling catheter lumen. The result of damping is to add a certain degree of inertia to the system, with a resultant loss of frequency response and peak data.

Another cause of damping is purely a procedural problem. Sometimes the wrong form of tubing is used for the transducer plumbing system. The correct tubing is stiff-walled. If the tubing is rubber, neoprene, or some other distensible material, then you have a problem. The reason is that increases and decreases of the pressure waveform causes the tubing diameter to increase or decrease in response to the changes. Unfortunately, changing the diameter of the measurement system also changes the pressure, so the measurement interferes with the system being investigated. This problem is especially common in medical or life sciences areas because users will put various types of medical tubing into service. Both surgical tubing and intravenous (IV) set tubing are sometimes used — erroneously — and will result in a damped waveform. The end result is an incorrect waveshape and artifactual readings.

Resonance or ringing in the system is caused by two main phenomena. First, there is the possibility of air or other gas bubbles in the system. It is almost impossible to rig a pressure system without having air enter the plumbing. It is therefore necessary to purge the system of air bubbles. Placing the transducer and plumbing in an attitude where the air can rise to the top near a port or stopcock and be removed will clear the system.

The second cause of resonance or ringing is improper length of tubing. Like any dynamic mechanical or electrical system, the transducer plumbing has a given resonant frequency. If the length and diameter of the tubing is such that the system is resonant for the applied waveform, then ringing will result. Typically, a ringing system produces a jagged waveform, or one in which the peak pressure indication is substantially larger and sharper in shape than can be justified by the application.

Another problem is transducer placement when the transducer is not physically part of the system being measured. When measuring the pressure at the bottom of the tank, for example, the transducer should be placed at the same height as the bottom of the tank. This line is called a *datum reference line*. In the human blood pressure application, the nurse should place the transducer at the level of the patient's heart — or not complain to the Biomedical Equipment Technician that the readings are "wrong." The problem is hydrostatic pressure head differences — which are nullified by proper transducer placement.

Another example of hydrostatic pressure head problems is shown in Fig. 7-9. The fluid in the tubing has weight, so will add a pressure of its own due to gravity. Figure 7-9A shows a positive pressure head, in which the tubing approaches the transducer from above. Similarly, a negative pressure head (Fig. 7-9B) is produced when the plumbing approaches from below. Figure 7-9C shows the proper scheme (where it can be accomplished) in which the tubing is routed equally above and below the datum reference line, so the positive and negative head cancel each other out.

In the event that you cannot place the transducer in a manner that permits a system such as Fig. 7-9C, it is sometimes possible to either live with the positive or negative offset or make electronic corrections in the amplifier or processor circuits.

PRESSURE MEASUREMENT CIRCUITS

There are several basic forms of pressure-amplifier circuit. Some of them are so simple that it is easy to build your own from operational amplifiers or other devices. Common types include dc, iso-

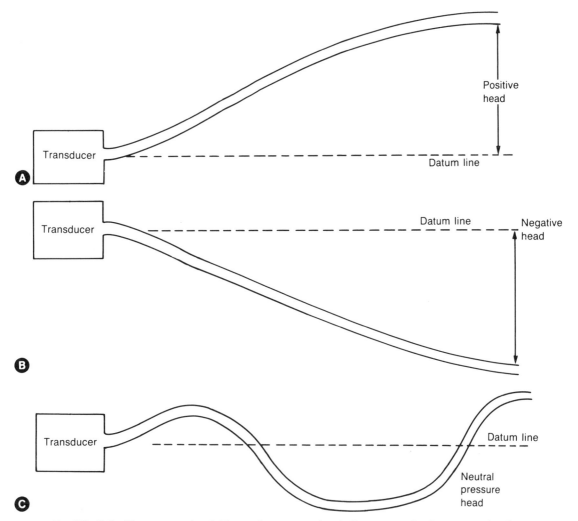

Fig. 7-9. A) Positive pressure head, B) negative pressure head, C) compensation for pressure head.

lated dc, pulsed excitation, and ac carrier amplifiers. The dc amplifiers work only with resistance strain-gauge transducers or the newer types of inductive transducer that output a rectified dc signal. Ac carrier amplifiers work with both resistive strain gauges and inductive transformers (inductive Wheatstone bridges and LVDT—see Chapter 5). The pulsed excitation works with resistive strain gauges, but only with some inductive transducers (which depends upon the inductance, duration of the pulse and other factors).

Regardless of the design, there are certain features common to all forms of pressure amplifier.

Some devices are narrowly limited in range for special purposes, while others are capable of a wider range for more general applications. In cases where moderate accuracy is needed, we can rely on internal calibration methods, but where superior accuracy is a requirement the user should calibrate the system against a mercury manometer.

When designing pressure amplifier/display systems be sure to provide both zero control and gain control. There are no perfect transducers on the market, and all will exhibit an offset and a gain problem. The offset is caused by errors in the strain gauges, errors in the other circuitry, and distension of the diaphragm. The gain error is caused by variations in the sensitivity of the transducer. Although some transducers contain internal circuitry (sometimes in the connector), all will exhibit to some extent both offset and gain problems. For example, I once received twelve Statham transducers (a good brand), nominally rated at a sensitivity of 5 μV/V/Torr, that came with factory calibration certificates showing the real sensitivity figures were 3.7 to 6.5 μV/V/Torr.

There are three approaches to standardizing the transducer for your own applications. One is to tightly specify the offset and sensitivity to the supplier, and force them to hand-select or high-grade their product. Another is to provide an internal balance and sensitivity adjustment. Figure 7-10 shows a transducer with both zero offset and sensitivity trimmer potentiometers installed. The final method is to provide a "Calibration Factor" for the transducer (which is printed on a label on the body of the transducer) that is used to adjust the external amplifier circuitry.

Figure 7-11 shows the simplified circuit of a dc pressure amplifier that uses the calibration fac-

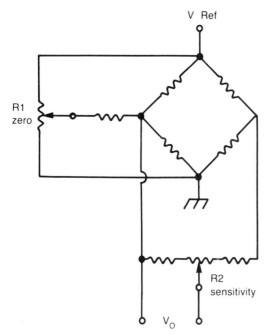

Fig. 7-10. Wheatstone-bridge pressure circuit.

tor method. The pressure amplifier (A1) is a dc amplifier, so the pressure transducer is a resistive Wheatstone-bridge strain gauge. Diode D1 provides the 7.5-volt dc excitation to the transducer and the potentials for the BALANCE and CAL FACTOR controls. The calibration factor for the transducer will sometimes change, so it is necessary to provide a procedure for measuring the new factor:

1. The dc pressure amplifier is first calibrated with an accurate mercury manometer.

2. Set switch S1 to the OPERATE position, open the transducer stopcock to atmosphere, and adjust R3 for zero volts output (i.e., 0 Torr reading on the display).

3. Close the transducer stopcock and then pump a standard pressure (e.g., 100 Torr, or at least half-scale). Adjust gain control R6 until the meter reads the correct (standard) pressure. Check for agreement between the meter and the manometer at several standard pressures throughout the range (e.g., 50, 100, 150, 200, 250, and 300 Torr). This last step is needed to ensure the transducer is reasonably linear, i.e., to check that the diaphragm was not strained by out-of-range pressures or vacuums.

4. Turn switch S1 to the position corresponding to the applied standard pressure, and then adjust CAL FACTOR control (R4) until the same standard pressure is obtained on the meter. The CAL FACTOR control is ganged to a turns counting dial. The number appearing on the dial at the position of R4 that creates the same standard pressure signal is the Calibration Factor for that transducer. Record the turns counter reading for future reference.

For a period of time (usually 6 months) the calibration factor will need not be redetermined. The calibration factor is entered into the amplifier by turning R4 (or digitally in modern instruments). The following procedure is normally used:

1. Open the transducer stopcock to atmosphere, place switch S1 in the 0 Torr (0 mmHg)

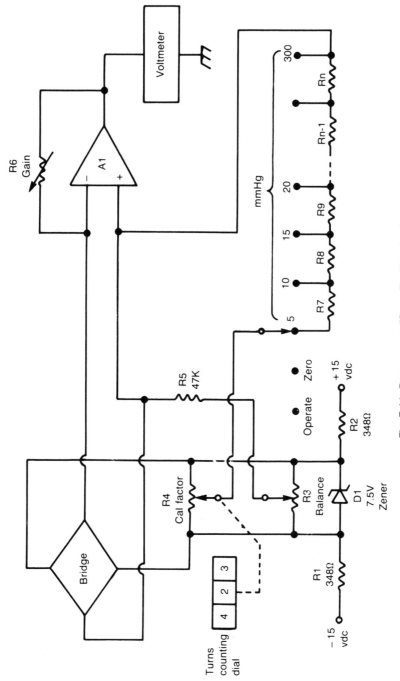

Fig. 7-11. Pressure amplifier calibration circuit.

position and adjust R3 for a 0-volts output indication (0 Torr on the display).

2. Set switch S1 to a position that is most convenient for the range of pressures to be measured. In general, select a scale that places the reading mid-scale or higher.

3. Set the CAL FACTOR knob to the figure recorded previously.

4. Adjust the gain control (R6) until the meter reads the standard pressure used in the original calibration.

The circuit in Fig. 7-11 is an example of a dc pressure amplifier, while Fig. 7-12 is a more detailed version of the actual amplifier block. Note that there are only two operational amplifiers in this circuit. There is little that is complex about the simple dc amplifier. Amplifier A1 is the input amplifier, and it should be a low-drift, premium model (in many off-the-shelf commercial instruments A1 in Fig. 7-12 was a uA725, but better and cheaper devices are now available). Both gain and zero controls are provided, so the amplifier will work with a wide variety of transducers.

The excitation voltage of the transducer is determined (as a maximum) by the transducer manufacturer, with a value of 10 volts being common. In general, it is best to operate the transducer at a voltage lower than the maximum in order to prevent drift due to self-heating. Pressure amplifier manufacturers typically specify either 5 volts or 7.5 volts for a 10-volt (max) transducer.

The required gain can be calculated from the required output voltage that is used to represent any given pressure. Because digital voltmeters are used extensively for readout displays, the common practice is to use an output-voltage scale factor that is numerically the same as the full-scale pressure. For example, 1 millivolt or 10 millivolts per Torr is common. Let's assume a maximum pressure range of 1000 Torr if we use a scale factor of 1 mV/Torr, when 1000-Torr is represented by 1000 mV, or 1 volt. No further scaling of the meter output is needed.

The sensitivity and the excitation potential give us the transducer output voltage at full-scale.

For example, let's assume that a 1000-Torr pressure amplifier, scaled at 1 mV/Torr is desired. What is the output voltage of the transducer if a +5 vdc excitation is applied, and the sensitivity is 5 μV/V/Torr?

$$V_t = \frac{5\ \mu V}{V\ Torr} \times (5\ volts) \times (1000\ Torr)$$

$$V_t = (5\ \mu V)\ (5)\ (1000) = 25{,}000\ \mu V = 25\ mV$$

The output of the transducer at full-scale will be 25 millivolts. This potential is the amplifier input voltage, so the gain can be calculated as:

$$Av = V_o/V_{IN}$$
$$Av = 1000\ mV/25\ mV$$
$$Av = 40$$

A variation on the dc-amplifier scheme is the *isolated dc amplifier*. In these circuits the input amplifier is a special isolation amplifier. These devices provide a very high impedance (10^{12} ohms or more) between the input and the dc power-supply terminals. This electrical isolation is required in medical applications for patient safety reasons. In certain industrial environments the isolation allows measurement of pressures in fluids or gases that are at dangerous potentials, for example, a liquid in a charged vessel, as in electrophoresis operations.

A example of a pulsed-excitation amplifier is shown in Fig. 7-13A. Instead of dc, the Wheatstone-bridge strain gauge is excited with a short-duration biphasic pulse. This method allows the transducer to be excited to a voltage high enough to make a measurement possible, without a constant flow of current to aggravate the self-heating problem inherent in dc transducers. The pulse typically has a 1-ms duration and a 25-percent duty cycle (which translates to a 4-ms period, or a frequency of 250 Hz). An advantage of the short duty cycle is that operations like amplifier drift cancellation can be incorporated.

Amplifier A1 is a dc pressure amplifier, while A2 is a unity-gain summation stage. The output signal indicator is a digital voltmeter that will update the display only when the STROBE line is

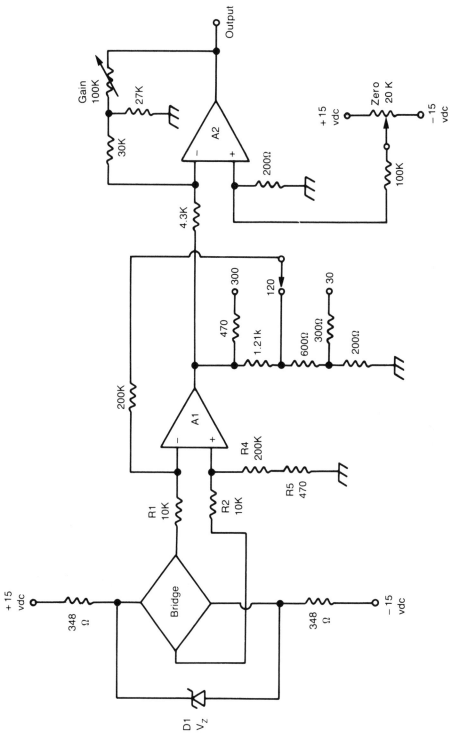

Fig. 7-12. Multirange pressure amplifier.

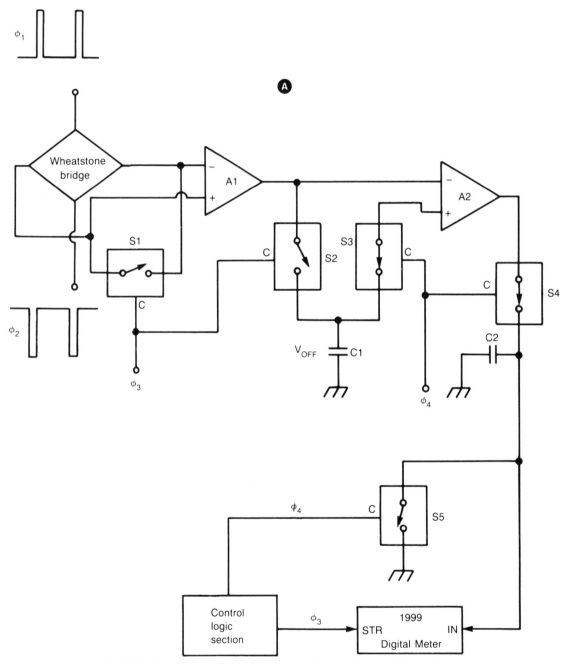

Fig. 7-13. Pressure amplifier with offset cancellation. A) circuit, B) waveforms.

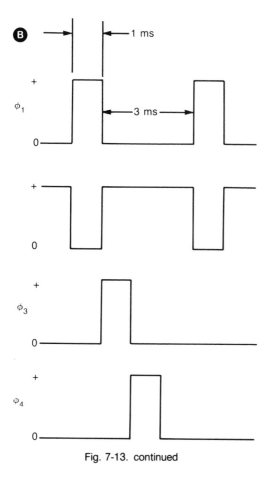

Fig. 7-13. continued

HIGH. Switches S1 through S3 are CMOS electronic switches which close when the control line (C) is HIGH. All circuit action is controlled by a four-phase clock. Phases P1 and P2 excite the transducer and operate the amplifier drift-cancellation circuit. Phase P3 updates the display meter, and phase P4 resets the circuit following the update.

All dc amplifiers tend to drift (although modern premium IC op amps have sharply reduced drift), that is, they create output-voltage offset voltages due to thermal changes. Capacitor C1 and amplifier A2 serves as a drift cancellation circuit (see Fig. 7-13A).

The transducer is excited only when P1 is HIGH positive and P2 is HIGH negative. At all other times the transducer is not excited, which keeps transducer self-heating to a minimum. Amplifier A1 will drift, however, because of its high gain and inherent offset voltages. The idea in this circuit is to charge capacitor C1 during the nonexcited period (during which A1 input is shorted by S1), and then add the capacitor voltage algebraically to the signal and thereby removing the amplifier drift component.

Ac Carrier amplifiers (Fig. 7-14) use an ac signal for transducer excitation and so will operate equally well with resistive strain gauges and inductive transducers. The carrier frequencies are typically 200 to 5000 Hz, with 400 and 2400 Hz being the most common. The Hewlett-Packard 8800-series pressure carrier amplifiers, for example, produce a 2400-Hz, 5-volt, rms signal. Carrier amplifiers are probably the most stable on the market, a result of using narrow bandwidth, heavy-feedback design.

Some cheap carrier amplifiers use simple envelope detectors to extract the pressure waveform, but all proper instruments use a variant of the circuit shown in Fig. 7-14. This circuit uses a quadrature detector, or phase-sensitive detector, to extract the signal information.

PRESSURE PROCESSING

Only rarely will you need a simple pressure amplifier for dynamic measurements (the same is not true where static pressures are involved). You will want a system such as in Fig. 7-15. The pressure waveform, P, is the analog output of a pressure amplifier, and it is fed to four different circuits: a peak detector (maximum pressure), inverted-peak detector (minimum pressure), a time integrator, and a differentiator (dP/dt).

The peak detectors and integrator can be calibrated by applying a constant value of pressure, P. But since the differentiator measures dP/dt you need a varying signal. Typically, a squarewave of the same amplitude as a standard value of the P signal is applied, as in Fig. 7-16. This signal produces an output from the differentiator (as shown) which is a linearly rising slope that can be measured.

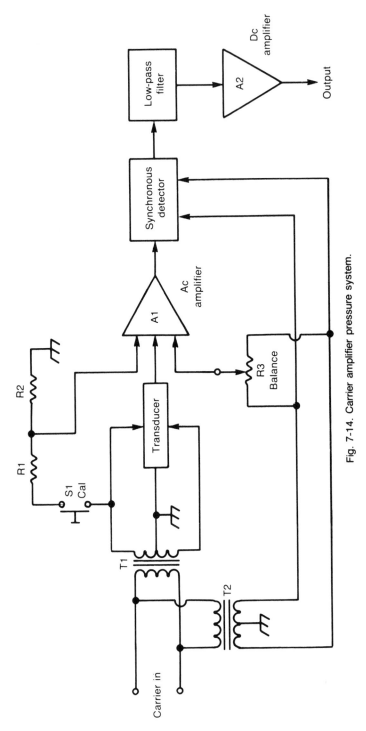

Fig. 7-14. Carrier amplifier pressure system.

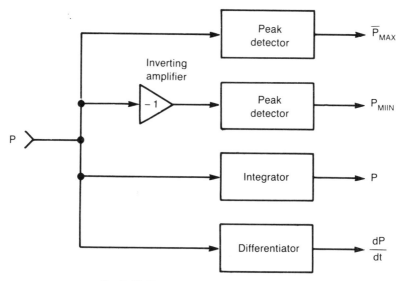

Fig. 7-15. Pressure amplifier block diagram.

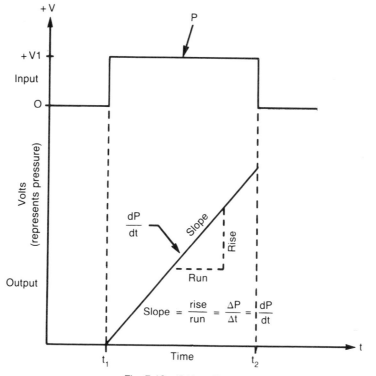

Fig. 7-16. dP/dt calibration.

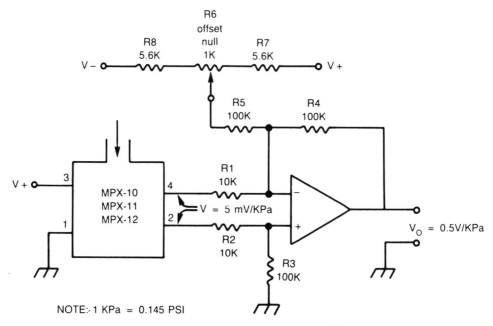

Fig. 7-17. Semiconductor pressure amplifier.

SOLID-STATE PRESSURE TRANSDUCERS

Figure 7-17 shows a modern solid-state pressure transducer. The piezoresistive element is internal, as are some of the electronics. The Motorola MPX-1x series shown in Fig. 7-17 produces an output voltage of 5 mV/kPa (note: 1 kilopascal equals 0.145 psi).

Chapter 8

Measurement of Position, Displacement, Velocity and Acceleration

T HE PARAMETERS OF POSITION, DISPLACEMENT, velocity, and acceleration are related according to some very simple equations. Thus, when you measure one of these parameters, you can measure all of them by indirect implication. Let's consider each of these in turn. Displacement is merely a change in position from one point to another, without regard for time or other variables. Figure 8-1 shows such one-dimensional displacement. The system defined in Fig. 8-1 is a Cartesian coordinate system, even though it could have easily been a single dimensional line. The black ball in Fig. 8-1 was initially at position x_1, but a hidden, invisible, adiabatic faerie came along and pushed it to position x_2. The difference, $x_2 - x_1$, is its displacement.

Velocity is nothing more than a measure of displacement per unit of time. In other words,

$$V = \frac{\Delta X}{\Delta T} = \frac{x_2 - x_1}{t_2 - t_1}$$

where

V is the velocity
Δ is the Greek upper-case letter "delta," which denotes "change"

In the notation of calculus, when Δx and Δt are very tiny, the above expression is written:

$$V = dx/dt$$

In other words, velocity is the first derivative of displacement with respect to time. You can then differentiate the velocity expression and find the acceleration:

$$a = dV/dT$$

If acceleration is the first derivative of velocity, and velocity is the first derivative of displacement, then we can conclude that acceleration is also the second derivative of displacement:

$$a = d^2x/dt^2$$

Thus, you can compute acceleration and velocity from positional displacement by using either a differentiator circuit or a computer to differentiate the collected positional data.

POSITION TRANSDUCERS

A position transducer will create an output that is proportional to the position of some object

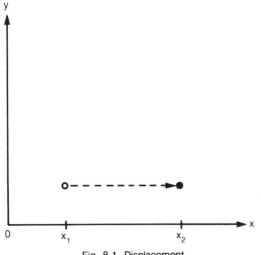

Fig. 8-1. Displacement.

along a given axis. For very small position ranges, one could use a simple strain gauge, although one must realize that the range is extremely small. Some amplification of the available range is available by placing the strain gauge at one end of a lever arm and measuring the displacement of the opposite end by actually measuring the forces on the strain-gauge end. Keep in mind that strain gauge position transducers are sometimes permanently damaged by overrange displacements; care is in order.

The linear variable differential transformer (LVDT) is also usable as a small-position trans-

ducer. Figure 8-2 shows a typical LVDT. It consists of three coils, L1, L2A, and L2B. The transformer thus formed is excited by applying an ac carrier signal, usually in the 200 to 5000-hertz range, to the primary (L1). The two L2 coils (L2A and L2B) are identical and share the same permeable core. Thus, when the core shades both coils equally, they will possess exactly the same inductive reactance, x; x(L2A) = x(L2B). But when the core is moved, i.e., displaced, one coil is shaded more than the other, so the inductive balance is upset and (x(L2A) does not equal x(L2B)).

Therefore, the voltage drops across these coils are equal for zero displacement, and nonequal for displacement conditions. Note in Fig. 8-2 that the two secondary coils (L2) are connected in a series-opposing configuration. Thus, the two voltages are out of phase with each other. If the voltages are equal (as they are at zero displacement), then they cancel each other and the net output V_o is zero. But under displacement conditions, the cancellation is not complete because of the inequality of the voltages. Under this condition V_o is nonzero and proportional to the displacement.

The output signal from an LVDT can be processed in ordinary ac voltage amplifiers, or rectified and processed in dc amplifiers. But the type of inductive position transducer shown in Fig. 8-3 requires a different tactic. In this case there is no primary coil even though there are two identical

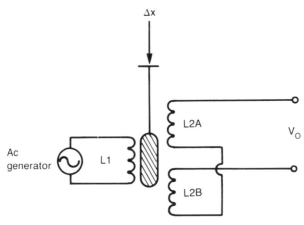

Fig. 8-2. LVDT displacement transducer.

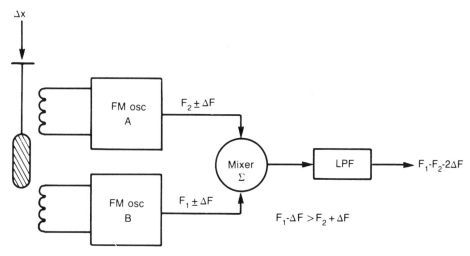

Fig. 8-3. Inductive displacement transducer.

coils analogous in construction to L2 of Fig. 8-2. The inductance of these coils is changed by the core position. The heart of this circuit is a pair of FM oscillators, A and B (which produce frequencies F_2 and F_1, respectively). The inductances of the transducer windings are part of the frequency-determining circuits of these oscillators. When the displacement is zero, the inductances are the same —so the output frequencies are the same.

The mixer circuit produces an output spectrum of $F_1 + F_2$, and $F_1 - F_2$ (i.e., sum and difference frequencies). The low-pass filter (LPF) following the mixer selects only the difference frequency, $F_1 - F_2$, and rejects the sum frequency. The difference frequency produced by this circuit is proportional to the displacement of the transducer core. A hospital in-bed patient weighing system once used this method. By measuring the downward displacement of the transducer when the bed was rolled onto the spring-loaded platform, the sum of the patient's weight and the tare weight of the bed could be measured.

Perhaps the most common form of displacement transducer is the potentiometer. For applications tolerant of error, almost any ordinary linear-taper potentiometer can be pressed into service. For other applications, either a precision potentiometer or special potentiometers designed as position transducers are used instead. For rotary ("curvilinear") displacement, a rotary potentiometer is used, while for rectilinear displacement a slide potentiometer is used.

Figures 8-4 and 8-5 show two possible circuits

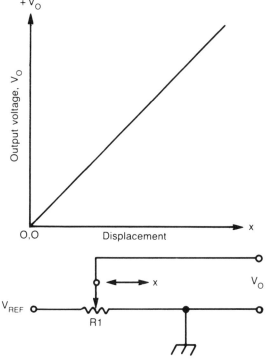

Fig. 8-4. Single-quadrant displacement transducer.

for potentiometer displacement transducers. Figure 8-4 shows a single-quadrant system. In this case there is a single potentiometer with one end grounded, and the other end connected to a reference voltage, V_{REF}. It is terribly important to make this voltage both stable and precisely known, or error will creep into the system. Chapter 33 gives details of suitable reference-voltage circuits. The output voltage V_o is a function of displacement x, as measured by the position of the potentiometer wiper along its element. The graph in Fig. 8-4 shows the voltage relationship to displacement.

A two-quadrant system is shown in Fig. 8-5. It is functionally the same as the single-quadrant version, except that it uses two reference voltage supplies, one positive to ground ($+V_{REF}$) and the other negative to ground ($-V_{REF}$). The output voltage will be negative for negative displacements and positive for positive displacements about a zero reference point.

It is possible to make a four-quadrant system by using either four single-quadrant devices (Fig. 8-4) or two two-quadrant devices (Fig. 8-5) mechanically arranged at 90° angles to each other. Perhaps the most common form of these types of transducers is the so-called "joystick," simple examples of which are found in computer game applications.

A circuit for mating a position/displacement transducer (potentiometer type) to an operational amplifier is shown in Fig. 8-6. In this circuit the potentiometer transducer forms the feedback network of an operational amplifier. The input resistor (R1) is a precision type. Alternatively, it could be

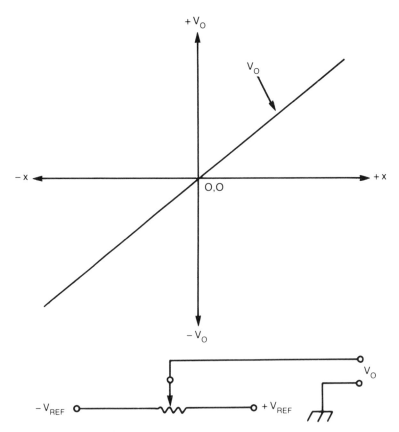

Fig. 8-5. Two-quadrant displacement transducer.

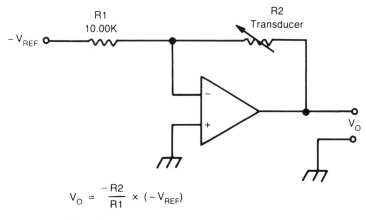

$$V_O = \frac{-R2}{R1} \times (-V_{REF})$$

R2 ∝ dx

Fig. 8-6. Operational-amplifier circuit for transducer.

another (trimmer) potentiometer and be used to trim out errors in the main potentiometer. A $-V_{REF}$ voltage reference source is applied in lieu of an input signal. The gain of an inverting-follower operational-amplifier circuit is the ratio of the feedback and input resistors:

$$A = -R2/R1$$

The transfer equation of the system is:

$$V_o = (-V_{REF}) \times \frac{-R2}{R1}$$

When the potentiometer is at zero displacement, R2 = 0, so V_o = 0 also. The output voltage is proportional to displacement.

Chapter 9

Flow Measurements

THE MEASUREMENT OF FLOW IS THE ART OF DE-termining how much fluid or gaseous material is passing though a given pathway. You might be measuring blood in an artery, respiratory gases into and out of a ventilator tube, the amount of water passing through a pipe, or any of a number of other similar situations. From an engineering conceptual point of view, the blood in the artery and the petroleum in the pipe-line are merely variants of the same problem.

Flow can be either turbulent or smooth. The measurement of turbulent flow is notoriously difficult, so we will concern ourselves mostly with the measurement of laminar-style (or nearly so) flow situations. Fortunately, most of our transducers have inertia and so tend to integrate out small variations due to turbulence. Not so, however, with ultrasonic systems, which are covered later in this chapter. In these systems the transducer (at least) should be designed to minimize introduced turbulence.

FLOW VOLUME VERSUS FLOW RATE

Most transducers (but not all) measure flow rate, that is, the amount of material per unit of time. For example, a popular medical respiratory flow meter measures the patient's inspiration or expiration in liters per minute (l/min). Similarly, a certain fluid transducer measures fluid flow in cubic centimeters per second (cc/s) (note: some such transducers measure in milliliters per second —ml/s—which is the same thing. For water at standardized temperature and pressure, 1 cc = 1 ml by definition).

You can obtain the flow volume from flow rate data by the simple expedient of integrating the flow signal, as in Fig. 9-1. The output of the flow transducer is amplified (and possibly further processed) in amplifier A1. The output of A1 is proportional to the flow rate. This same signal is applied to an integrator. Although an operational-amplifier Miller integrator is shown here, it is likely that modern computerized systems would contain the integration as a software algorithm.

FLOW DETECTORS

A flow detector is a circuit that indicates that flow is present, but does not necessarily quantify either the flow rate or the flow volume. Two such systems are shown in Fig. 9-2. Figure 9-2A shows a thermistor bridge version. Two thermistors, RT1 and RT2, are placed inside of the flow con-

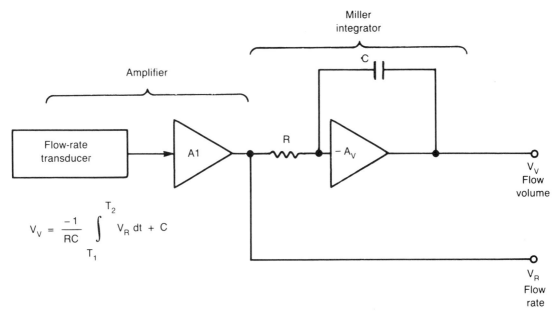

Fig. 9-1. Flow rate/volume system.

tainer. In the particular case shown, the container is a tee-connector used in a respirator pneumatic hookup (a medical device). The idea here is to detect whether or not air is getting to the patient; "how much" is less important because this transducer has a simplified alarm scheme. Thermistor RT1 is placed in the air stream, while RT2 is in a dead space off the main flow; it performs the ambient measurement.

The two thermistors in Fig. 9-2A are connected in an ordinary Wheatstone bridge circuit. When R1/RT2 = R2/RT1, the output voltage V_o is zero (R3 balances inequalities, and is adjusted for $V_o = 0$ when no air is flowing in the tube). The (V+) voltage biases the thermistors to the point of self-heating (but not greater). At this point the thermistor resistance is the most sensitive to changes in air flow, which cools the surface of the device. When air flow changes the device resistance, the output voltage will be:

$$V_o = (V+) \times \left[\frac{RT2}{R1 + RT2} - \frac{RT1}{R2 + RT1} \right]$$

The output voltage is proportional to the temperature change caused by the flowing air. While it

is not impossible to calibrate this temperature change over a narrow range of flow rates, it is too difficult to consider in most applications. There are more suitable transducers on the market.

Figure 9-2B shows a photo-optical device that is used to detect flow. Here there is a light source (LED in this case) shining across the flow path to a detector (phototransistor in this case). Either the fluid or gas must be opaque to the light, or the light frequency chosen for maximum attenuation. For example, ordinary visible red LEDs (or other light sources) can be used for an opaque liquid. For a gas like carbon dioxide, on the other hand, you can use the fact that CO_2 absorbs infrared energy. By making the light source and the detector IR-sensitive, you can detect the presence of CO_2. Other gases and liquids may absorb other wavelengths, so each system must be developed specifically with that liquid or gas in mind.

POTENTIOMETER SYSTEMS

Figure 9-3A shows a system that uses a potentiometer to measure flow volume (not flow rate). This system is used in certain syringe pumps in medical devices. A glass or plastic syringe is placed in a saddle with a worm-gear pump and

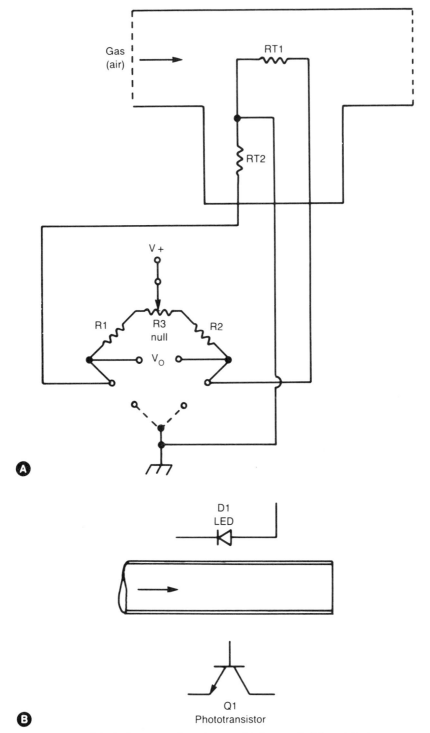

Fig. 9-2. A) Thermistor flow detector B) photo-optical flow detector.

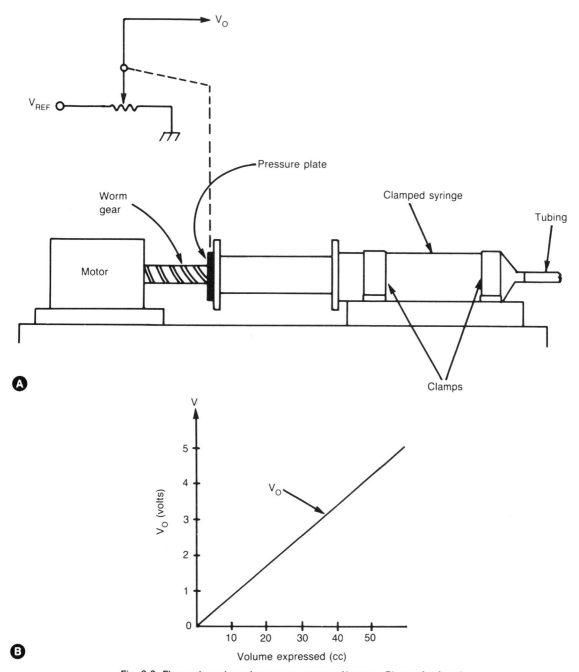

Fig. 9-3. Flow volume in syringe pump system A) setup, B) transfer function.

motor system. As the worm gear advances, it pushes the syringe plunger, thereby pushing fluid out of the tubing.

A rectilinear potentiometer is ganged to the pressure plate on the end of the worm gear. Thus, you can use the displacement of this plate as a measure of volume expressed. Consider an example in which a 50-cc syringe is connected to a

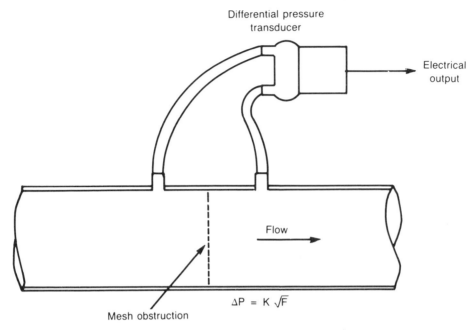

ΔP = K √F

Mesh obstruction

Fig. 9-4. Flow-rate monitor (constriction type).

potentiometer that is in turn connected to a $+V_{REF}$ voltage of +5.00 volts. Figure 9-3B shows the output function of this system. The output voltage is 5 volts at 50 cc, so the scaling factor is 5 V/50 cc, or 100-millivolts per cc (100 mV/cc).

OTHER SYSTEMS

Figure 9-4 shows a common form of flow-rate transducer. An obstruction placed in the gas path will create a pressure drop that is proportional to the square root of the flow rate:

$$Pd = K^2(F^3)^{1/2}$$

where

Pd is the pressure drop
K is a sensitivity constant
F is the flow rate
½ denotes that we take the square root of F

In some transducers the obstruction is a narrowing of the path at a point. In others, including most medical transducers, the obstruction is a wire or plastic cloth mesh stretched across a constant-diameter airway. The usual mesh for medical respiratory measurements is 400 grid/inch.

The pressure drop is measured by a differential-pressure transducer. These transducers have two ports, one on either side of the diaphragm. Most pressure transducers measure gauge pressure, and so are open on one side to the atmosphere. Those devices are not suitable for this type of measurement. We need, instead, a differential-pressure transducer as shown in Fig. 9-4.

Another gas flow system is shown in Fig. 9-5. Two versions are presented; Fig. 9-5A is a magnetic system, while Fig. 9-5B is optical.

In the magnetic transducer in Fig. 9-5A, a small magnet is introduced into the flow stream. This form of transducer is used for both liquids and gases, and is also usable in closed systems where it is difficult to introduce other forms of transducer. A pair of coils, L1 and L2, are placed at right angles to the flow path, at the point where the magnetic rotor is placed. When a moving magnetic field cuts across the turns of a coil, a current is introduced into the coil. Thus, a voltage, V_o, is found at the output of the series-aiding connected coils. The amplitude of the voltage is proportional

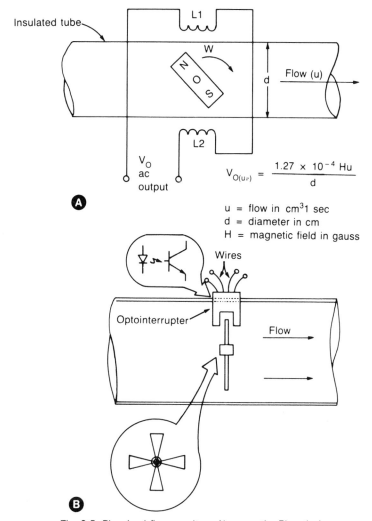

u = flow in cm^{3}1 sec
d = diameter in cm
H = magnetic field in gauss

$$V_{O(u r)} = \frac{1.27 \times 10^{-4} \, Hu}{d}$$

Fig. 9-5. Pin-wheel flow monitors A) magnetic, B) optical.

to the magnetic field, while its frequency is proportional to the rotational frequency of the magnet. The rotational frequency of the magnet is related to the flow rate. Transducers of this type are used in a wide variety of instrumentation applications.

Figure 9-5B shows an optical version. An *opto-interrupter* is a device that places an LED and a phototransistor across an open path. When the path is blinded, then the phototransistor is darkened; when the path is not blinded, then the phototransistor is illuminated. Similar interrupters are used in applications such as the "PAPER OUT"

sensor in computer printers, tape sensors in recorders, and others. The light path in the transducer of Fig. 9-5B is interrupted by a multibladed fan placed in the flow path. Again, the frequency of the output signal is proportional to the flow rate.

With the right circuitry, a rotating flow-rate transducer can also produce a flow volume signal. By using the ac signal to trigger a one-shot multivibrator, you obtain a pulse train of pulses that are of constant duration and constant amplitude; the only variation is the pulse repetition rate (which is equal to the ac signal frequency from the trans-

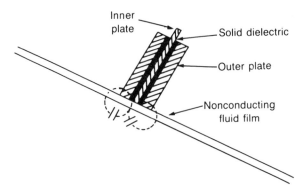

Fig. 9-6. Dielectric flow detector.

ducer). Integrating these pulses produces a dc voltage that is proportional to the total area under all of the pulses per unit of time — in other words the flow volume (i.e., the integral of the rate).

Figure 9-6 shows a method for measuring the flow of a thin liquid film. In this case the transducer is a coaxial, cylindrical capacitor. The nonconducting film flowing across the surface changes the dielectric constant of the capacitor, hence its capacitance. You can measure the actual capacitance in any of several ways. For example, you can use an electrometer, an FM oscillator, etc.

ULTRASONIC SYSTEMS

Ultrasonic waves are acoustical waves (i.e., vibrations in a medium) that have a frequency above the range of human hearing. In measurement applications, "ultrasonic" can mean anything from 20 kHz to 20 MHz. The important parameter is that they are acoustical waves, not electromagnetic (i.e., radio) waves. There are several types of ultrasonic transducers available, but most of them are either dynamic or piezoelectric. The dynamic form are analogous to dynamic microphones: a thin diaphragm stretched over an electromagnet. These transducers are used for relatively low frequencies. The piezoelectric types are analogous to the crystal microphone. A piezoresistive crystal vibrates when an ac signal at its resonant frequency is applied. This vibrational energy is applied either to a diaphragm or directly to the media being measured.

Figure 9-7 shows a medical blood-flow detector based on piezoelectric crystals. Two crystals are used, one for transmitting and one for receiving. The frequency of the incident signal (F) is changed by the doppler effect as blood flows underneath the transducer. The reflected energy will contain frequency components of F $-/+$ ΔF, where ΔF is the doppler shift.

In blood flow monitors it is nearly impossible to calibrate this transducer for blood flow rate. The system is used to check for vessel patency, i.e., whether or not blood is flowing, but not to measure

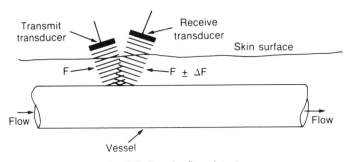

Fig. 9-7. Doppler flow detector.

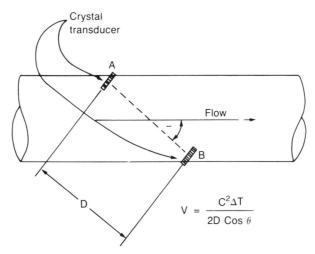

Crystal transducer

A

Flow

B

D

$$V = \frac{C^2 \Delta T}{2D \cos \theta}$$

Fig. 9-8. Ultrasonic flow meter (type 1).

blood flow rate. In other systems the calibration might be possible, however, because the doppler frequency shift is proportional to the fluid velocity. By filtering out these doppler cells it ought to be possible to calibrate the system. The problems in medical systems are that: 1) blood vessels are distensible, i.e., flex at blood flows, 2) lood flow is pulsatile, and 3) blood vessels are not located at a constant distance beneath the surface.

An example of a transit-time flow transducer is shown in Fig. 9-8. This system depends upon the fact that the upstream and downstream transit times of acoustical pulses is different. A pair of piezoelectric crystal transducers are aimed obliquely at each other across the flow path. The angle between the crystal path and the flow path is θ [theta]. Both crystals are used for both transmit and receive functions. You first fire a pulse from A

to B and measure its transit time; you next fire an upstream pulse from B to A and measure its transit time (in reality, both measurements are made continously). The average flow velocity is:

$$V = \frac{C^2 \times \Delta T}{2\,D \cos \theta}$$

where

V is the average flow velocity

C is the speed of the signal in the media

ΔT [delta T] is the difference between downstream and upstream transit times.

D and θ are as defined in Fig. 9-8.

A doppler system is shown in Fig. 9-9. This system uses the frequency change of a wave scat-

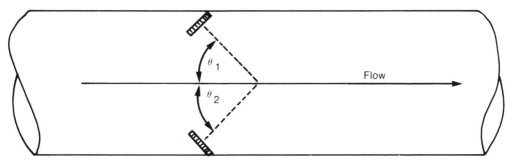

Flow

θ_1

θ_2

Fig. 9-9. Ultrasonic flow meter (type 2).

tered from particulate matter flowing in the fluid path and is particularly useful in blood flow meters. It has been shown that the change in frequency [delta F] ΔF, is given by:

$$\Delta F = \frac{V\, Fs\, (\cos \Theta_2 + \cos \Theta_2)}{c}$$

where

ΔF [delta F] is the doppler shift frequency
Fs is the source frequency
V is the fluid velocity

c is the velocity of sound in the media
Θ_1 is the angle of the receive crystal to the flow axis
Θ_2 is the angle of the transmit crystal to the flow axis

Ultrasonic flow meters are considered noninvasive in most situations. Medically, there is no evidence that sound energy harms the patient. In other applications the designer must consider whether or not the small sound pressures involved are influential.

Chapter 10

Light Transducers

V ARIOUS CLASSES OF INSTRUMENTS USE LIGHT AS a transducible element. In Chapter 13 you will find a typical colorimetry instrument based on the transducers used in this chapter. Various instrumentation techniques may depend on different properties of light and light transducers. In some cases only the existence or nonexistence of the light beam is important. Examples of this use are the PAPER OUT sensor on your computer printer, and the light beam that counts entrances and exits from a building or controlled space. In other cases the color of the light is important. In still others, the absorption of particular colors is the important factor. Nonetheless, the light beam must be sensed before it can be used — and that is the subject of this chapter.

LIGHT

Light is a form of electromagnetic radiation, and in the ultimate sense is the same as radio waves, infrared (heat) waves, ultraviolet, and X-rays. The principal difference between these types of electromagnetic radiation is the frequency and wavelength. The wavelength of visible light is 400 to 800 nanometers (1 nm = 10^{-9} meters); infrared has longer wavelengths than visible light and ultra-violet has wavelengths shorter than the 400 of visible light; X-radiation has wavelengths even shorter than ultraviolet. We know that frequency and wavelength are related by the equation:

$$\lambda = \frac{c}{f}$$

where

c is the velocity of light (300,000,000 meters/second)
λ is wavelength in meters
f is frequency in hertz

From the above equation you can see that light has a frequency on the order of 10^{14} hertz (compare with the frequencies of AM and FM broadcast bands in the radio portion of the spectrum).

Because IR, UV and X-radiation are similar in nature and wavelength to visible light, many of the sensors and techniques applied to visible light also work to one extent or another in those adjacent regions of the electromagnetic spectrum. Although performance varies somewhat, and some devices aren't even useful in certain areas, it is nonetheless true that workers dealing with those spectra may find these devices useful.

The photosensors described in this section depend upon quantum effects for their operation. Quantum mechanics arose as a new idea in physics in December, 1900, the very dawn of the 20th century, with a now-famous paper by the German physicist, Max Planck. He had been working on thermodynamics problems, and found the experimental results reported in 19th-century physics laboratories could not be explained by classical Newtonian mechanics; the then-prevailing world view of physics. The solution to the problem turned out to be a simple, but terribly revolutionary idea: energy existed in discreet bundles, not as a continuum. In other words, energy comes in packets of specific energy levels; other energy levels are excluded. The name eventually given to these energy levels was *quanta*, thus was born quantum mechanics. The name given to energy bundles that operated in the visible light range was *photons*.

The energy level of each photon is expressed by the equation:

$$E = \frac{ch}{\lambda}$$

or alternatively,

$$E = h\nu$$

where

E is the energy in electron volts (eV)
c is the velocity of light
λ (Greek lambda) is the wavelength
h is Planck's constant (6.62×10^{-34} J-s)
ν is the frequency of light

(Note: the constant *ch* is sometimes combined and expressed as 1240 eV/nm). The basis of light sensors is to construct a device that allows at least one electron to be freed from its associated atom by one photon of light. Materials in which the electrons are too tightly bound for light photons to do this work will not work well as light sensors.

You may detect in the alternate form of this equation the reason why certain forms of radiation cause cancer, while lower wavelengths (radio and microwaves) do not. The high-frequency X-rays and UV have tremendous energy that is not present in radio waves. Thus, they are able to cause cellular damage that would be impossible with lower frequency radiation.

THE PHOTOTUBE

The phototube is a vacuum-tube device that operates on the photoelectric effect. Interestingly, physicist Albert Einstein won the Nobel Prize in physics for his explanation of the photoelectric effect, not for either Special or General Theories of Relativity as is commonly assumed. Einstein wrote three seminal papers for the 1905 edition of Annalen der Physik, any of which could have qualified him for the Nobel: explanation of Brownian motion as a molecular effect, explanation of the photoelectric effect, and Special Relativity. Because of the immensely important nature of these discoveries, souvenir hunters have stolen most of the 1905 volumes of that journal from the university libraries of the world. Although the 1905 volume is available in reproduction form, it is extremely rare in the original.

The effect that Einstein explained in 1905 had long perplexed physicists. If you shine a light onto certain types of metallic plate in a vacuum, then electrons are emitted from the surface of the plate. Oddly, increasing the brightness of the light does not increase the current level. If this were purely a mechanical kinetic event, then one would assume from the classical point of view that increasing the intensity of the light would increase the current emission from the surface. It turned out, however, that changing the color of the light affected the current flow! Red light produced fewer electrons than violet light; the level of current flow was color-sensitive. Once Planck's principle was known, however, Einstein was able to explain this effect by quantum principles. From the above equations you can see that the higher frequency

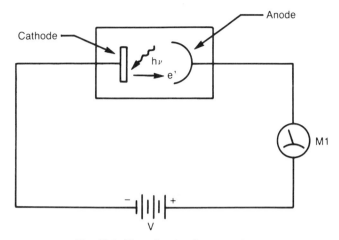

Fig. 10-1. Photoelectric effect experiment.

(shorter wavelength) violet-colored light has significantly more energy than the red light.

Figure 10-1 shows a phototube based on the photoelectric effect. The cathode is a metallic plate that is exposed to light; it is made of a material that will easily emit electrons when light is applied. The anode is positively changed (with respect to the cathode) by an external power supply. Electrons emitted from the photocathode are collected by this plate, and can be read on an external meter (or other circuit).

The photoemission process is less efficient than is needed in many cases. We can make the system more efficient by using a photomultiplier (PM) tube (Fig. 10-2). In this type of photosensor,

there are a number of positively charged anodes, called *dynodes,* that intercept the electrons. When light smacks into the cathode, electrons are emitted. They are accelerated through a high-voltage potential (V1) to the first dynode. They acquire substantial kinetic energy during this transition, so when each electron strikes the metal it gives up its kinetic energy. In the giving up of this energy, some is converted to heat, while some is converted by dislodging additional electrons from the dynode surface. Thus, a single electron caused two or more additional electrons to be dislodged. These electrons are accelerated by high-voltage potential V2, and reproduce the same effect at the second dynode. The process is repeated several times, and

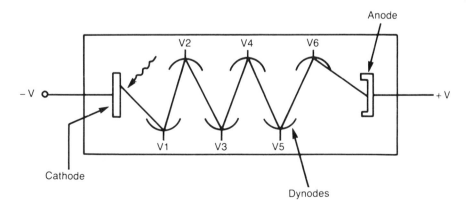

Fig. 10-2. Photomultiplier tube.

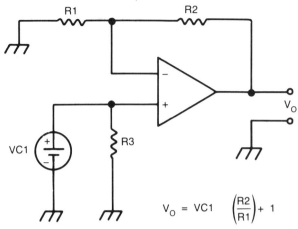

$$V_O = VC1 \quad \left(\frac{R2}{R1}\right) + 1$$

Fig. 10-3. Photovoltaic cell.

each time several more electrons join the process for each previously accelerated electron. Finally, the electron stream is collected by the anode, and can be used in an external circuit.

PHOTOVOLTAIC CELLS

A photovoltaic cell is one in which a potential difference is generated, and thus a current flow, by shining light onto its surface. The common solar cell is an example of the photovoltaic cell. There are only a few instrumentation applications for this type of cell, although some are used. Figure 10-3 shows a typical circuit for instrumentation applications of the photovoltaic cell. The cell is connected across the input of a high-impedance amplifier, such as the noninverting operational amplifier shown. The output voltage is found from:

$$V_o = VC1 \times \frac{R2}{R1} + 1$$

The photovoltaic cell is nonlinear, so it is of limited usefulness for data-acquisition systems.

PHOTORESISTORS

A photoresistor is a device that changes electrical ohmic resistance when light is applied. Figure 10-4 shows the usual circuit symbol for photoresistors; it is the normal resistor symbol enclosed

within a circle, and given the greek lambda (λ) symbol to denote that it is a resistor that responds to light. When photoresistors are specified, it is typical that a dark resistance is given, and a light/dark ratio. In most common varieties, the resistance is very high when dark, but drops very low under intense light. The intensity of the light affects the resistance, so they can be used in photographic lightmeters, densitometers, colorimeters and so forth.

Figure 10-5 shows three circuits in which photoresistors can be used. The half-bridge circuit is shown in Fig. 10-5A. In this circuit the photoresistor is connected across the output of a voltage divider made up of R1 and PC1. The output voltage is given by:

$$V_o = \frac{V \, PC1}{R1 + PC1}$$

Fig. 10-4. Photoresistor symbol.

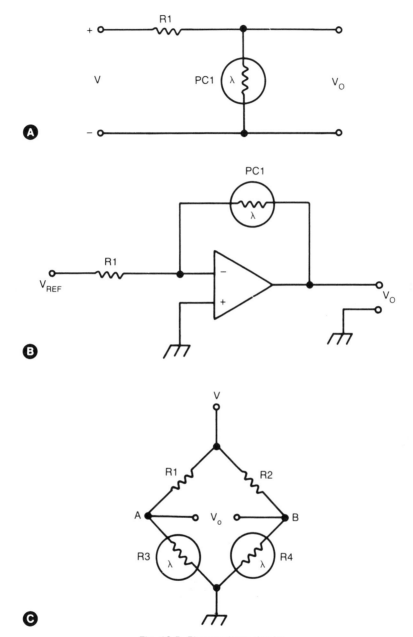

Fig. 10-5. Photoresistor circuits.

where

V_o is the output potential
V is the applied excitation potential
R1 and PC1 are in ohms

A problem with this circuit is that the output potential does not drop to zero, but always has an offset value.

A second way to use the photoresistor is shown in Fig. 10-5B. Here the photoresistor is the

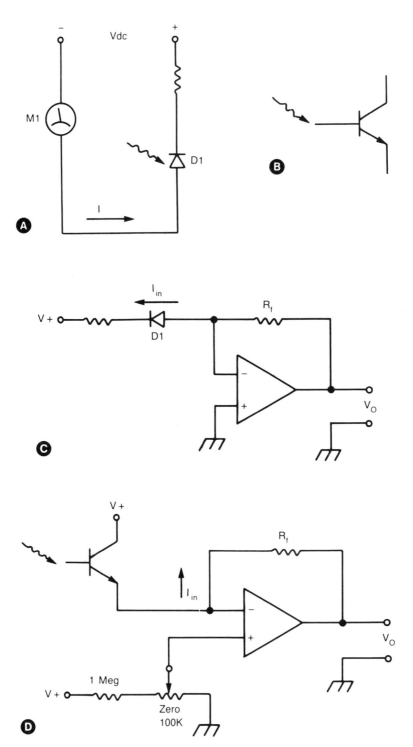

Fig. 10-6. Photodiode and phototransistor circuits.

feedback resistor in an operational-amplifier inverting-follower circuit. The output voltage, V_o, is found from:

$$V_o = -(-V_{REF}) \times (PC1/R1)$$

The circuit of Fig. 10-5B provides a low-impedance output, but like the other half-bridge circuit (Fig. 10-5A), the output voltages does not drop to zero. In addition, with some photoresistors the dynamic range of the operational amplifier may not match the dark/light ratio of the photoresistor at practical values of $-V_{REF}$ potential.

Our last photoresistor configuration is the Wheatstone bridge shown in Fig. 10-5C. This circuit allows the output voltage to be zero under the right circumstances and is the circuit favored by most designers. If a low-inpedance output is required or additional amplification is needed, then a differential dc amplifier can be connected across output potential, V_o.

The Wheatstone bridge can be considered as two half-bridges in parallel. The output voltage is equal to the difference between the respective half-bridge output voltages, i.e., the voltages at points A and B. The voltages at these points are found from the same equation as served for the half-bridge in Fig. 10-5A, so you may conclude (after minimal algebra) that the output voltage from the bridge is equal to:

$$V_o = (V) \left[\frac{R3}{R1 + R3} - \frac{R4}{R2 + R4} \right]$$

The Wheatstone-bridge circuit figures prominently in the circuits of Chapter 13 and will be revisited at that time.

PHOTODIODES AND TRANSISTORS

Perhaps the most modern light sensor is the pn junction in the form of special photodiodes. When the pn junction is illuminated in certain types of diode, the level of reverse leakage current available will increase. Figure 10-6A shows the basic circuit used for these sensors. The diode is normally reverse biased, with a current-limiting resistance in series. Microammeter M1 measures the reverse leakage current that crosses the pn junction during this type of operation. When light strikes the pn junction, the reading on M1 will increase.

The same principle applies to a class of npn or pnp transistors called *phototransistors* (Fig. 10-6B). In these devices, collector-to-emitter current flows when the base region is illuminated. These devices are the heart of optoisolator and optocoupler integrated circuits, as well as being used in various instrumentation sensor applications.

Figures 10-6C and D show one way in which photodiodes and phototransistors are used. The current flow through the device under light conditions is the transducible event, so you must connect them into a circuit that makes use of this property. The inverting-follower operational-amplifier circuit will do the trick neatly. In all such cases, the output voltage V_o is equal to the product of the input current and the feedback resistance ($I_{IN} \times Rf$). In the case of Fig. 10-6D, a zero control is added, and can also be added to the other circuit as well.

Chapter 11

Biophysical Signals and Their Acquisition

T HE ACQUISITION AND PROCESSING OF BIOLOGI-
cal signals constitutes a large part of instrumentation design efforts in equipment used in clinical medicine, and also by researchers in biology,
physiology, biochemistry, and certain aspect of
psychology. In this chapter I discuss the origins and
acquisition of these signals, why they are important, and some of the problems encountered.

BIOELECTRICAL PHENOMENA

A considerable number of biophysical signals
are derived directly from electrical activity of the
body. Examples are the electrocardiograph (ECG),
which measures heart activity, the electroencephalograph (EEG), which measures brain-wave
activity, and the electromygraph (EMG), which
measures skeletal muscle activity. These signals
have their origin in the electrical properties of the
elementary cell.

All mammalian organisms, including humans,
are constructed from basic building blocks called
cells. Each body is composed of a number of different types of cells. The human body, for example, is
believed to contain 75 trillion individual cells, of
which about 25 trillion are red blood cells. The

types of cells vary considerably as to size, shape,
construction, and properties according to their
function. Cell size varies from 200 nanometers (1
nm $= 10^{-9}$ m) to several centimeters (cm). The
ostrich egg is a single cell that reaches 20 cm (1
cm $= 10^{-2}$ m) in size. Most cells in the human
body fall into the range 0.5 to 20 micrometers (1
μm $= 10^{-6}$ m).

The cell can be viewed as a closed body in
which internal material (cytoplasm and, in some, a
nucleus) is separated from the surrounding body
fluid by a cell wall membrane (Fig. 11-1). The key
to cell bioelectricity is in the fact that the cell wall
is a *semipermeable membrane.* That is, its structure is such that it is permeable (i.e., will pass) only
certain ions, and no others. As a result of a sodium-
potassium phenomena associated with this membrane, the concentration of ions inside and outside
the cell are different.

The effect of the sodium-potassium effect is to
pump sodium outside the cell, and potassium inside
the cell. Because the rate of sodium pumping is two
to five times the rate of potassium pumping, the
net result is a cell interior that is largely potassium
(K) ions, and an external environment that is
largely sodium (Na) ions. The difference in ionic

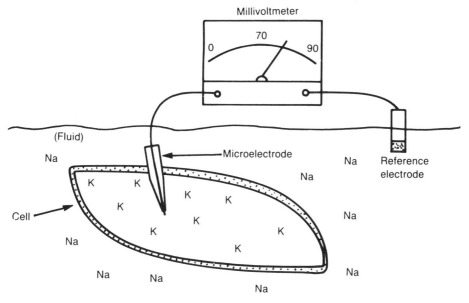

Fig. 11-1. Transmembrane potential across living cells.

concentration produces a membrane potential across the cell wall (again, Fig. 11-1).

The normal potential of the cell with respect to its environment is between −70 and −90 millivolts. This variation is somewhat dependent on the cell and its situation, but also indicates some disagreement amongst the various authorities on cell structure. An expression called the Nernst equation governs cell potential:

$$V = -/+ \ 60 \ nl \left[\frac{C_o}{C_i} \right] \qquad \textbf{Eq. 11-1}$$

where

V is the cell potential in millivolts

C_o is the extracellular concentration of ions in moles per cubic centimeter (moles/cm³)

C_i is the intracellular concentration of ions in moles per cubic centimeter (moles/cm³).

Find the membrane potential of a cell when the extracellular concentration of K⁺ ions is 4.6 moles/cm³, and the intracellular concentration of K⁺ ions is 2.1×10^{-5} moles/cm³.

$$V = -/+ \ 60 \ nl \ [C_o/C_i]$$

$$V = -/+ \ 60 \ nl \left[\frac{4.6 \times 10^{-6} \ \text{moles/cm}^3}{2.1 \times 10^{-5} \ \text{moles/cm}^3} \right]$$

$$V = -/+ \ 60 \ nl \ (0.219)$$
$$V = 60 \ nl \ (0.219)$$
$$V = [(60)(-1.52)] = -91.2 \ \text{millivolts}$$

A cell at rest is said to be "polarized." Certain types of cell can be stimulated in certain ways (depending upon function and type), and when stimulated becomes depolarized. During the depolarization period the character of the cell membrane changes radically; the change forces potassium out of the cell and allows sodium inside. The depolarization phenomenon typically lasts for several milliseconds. During this period, the change of ionic concentrations forces a change in membrane potential from −70/90 millivolts to +20/40 millivolts. This latter potential is called the *action potential.*

The duration of the action potential is typically 1.5 to 4 milliseconds. During most of this time, the cell cannot be retriggered by repeated episodes of the stimulus. This segment is called the *refractory period.* The membrane potential need not return all the way to its resting potential before retriggering is possible, so the refractory period is measured from initiation of the stimulus until the potential again reaches the retriggering threshold (a little above the resting potential, or about −60 mV in Fig. 11-2).

The action potential phenomenon is analogous to the electronic circuit called a *monostable multivibrator,* or *one-shot.* The one-shot circuit has but one stable state, corresponding to the cell resting potential. When the one-shot is triggered (i.e., stimulated), it reverts temporarily to its unstable state (action potential). During the unstable state period, the one-shot (of ordinary design) may not be retriggerable — the same term "refractory period" is used for both cells and one-shot circuits. Following return to its stable (resting) state the one-shot remains dormant until retriggered.

The "all or nothing" operation of the cell action potential explains why restimulation does not cause an increase in the intensity of a phenomenon. For example, if you stub your toe, pain sensation results from depolarization of the right sensor cells. But if you immediately re-stub your toe before the refractory period expires, the second injury is not felt to increase the pain. An increase in pain is felt only if some or all of the pain sensors have completed their refractory period.

Conduction of the action-potential signal occurs by two mechanisms. Ohmic conduction occurs through tissue because the action potential is an electrical potential difference — like a small battery — acting on conductive tissue in accordance with Ohm's law. This phenomenon is an example of simple ionic conduction and is mostly a localized event. Ionic or ohmic conduction results in very low signals at distance because the electri-

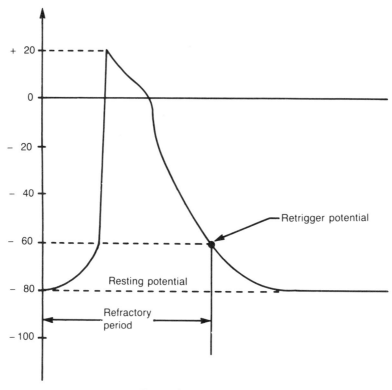

Fig. 11-2. Action potential.

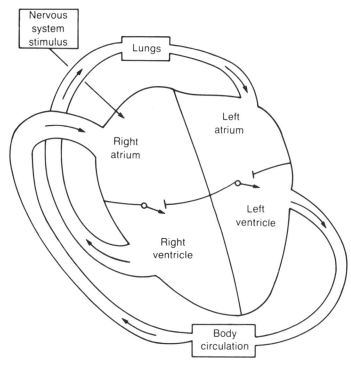

Fig. 11-3. Human circulatory system.

cal energy is integrated over the body area. Thus, a large action potential in the heart muscle produces a weak one millivolt (1 mV) signal at the surface. The alternate means of conduction is found in nerve cells and certain other cases. This means of conduction results from a successive depolarization of adjacent cells. In other words, the depolarization of one cell forces the spontaneous depolarization of the adjacent cell, much the same as a chain of cascaded one-shot circuits propagates the orginal pulse unattenuated. By this same means the action potential is propagated over a relatively long distance without being diminished. Each successive cell depolarization effectively reconstitutes the pulse amplitude.

By way of example, let's take a look at a well-known biopotential: the electrocardiogram ("ECG," or "EKG" after the German spelling). The ECG signal is the surface view of the collected cell action potentials from the heart (which is a muscle). Figure 11-3 shows a grossly simplified schematic of the heart. Note that the heart con-

sists of four chambers that together form a dual two-chamber pump. Each pump consists of an atrium (input chamber) and ventricle (output chamber).

Blood returning from the body circulatory system is at low pressure (2 to 4 millimeters of mercury, or mmHg), and is drawn into the right atrium. When the pressure of returned blood builds up to a certain point, a valve between upper and lower chambers opens and allows the blood to enter the lower chamber or ventricle. The pressure on the blood inside the atrium is increased to the point where the valve opens by a gentle contracting action that starts in the upper right atrium from a nervous system stimulus, and then spreads across the upper portion of the heart. This same electrical stimulation travels along an internal conduction system faster than through other tissue, and so synchronizes the depolarization of the ventricle muscle cells. This contraction occurs when the ventricles are about full and forces blood out of the ventricles into circulation. From the right ventricle blood is forced into the lungs (to take on oxygen) and then into the left atrium where the process is repeated.

Physicians and scientists use a vector system to examine the heart biopotentials from a system of surface electrodes. In the classical ECG system, the physician will record differential signals that use the right leg (RL) as the common reference point (see Fig. 11-4A). Signal vectors are picked up at points such as the chest (six different points), left arm (LA), right arm (RA), and left leg (LL). In addition, certain modern recording systems look at an xyz coordinate system that looks at vectors that run from head to toe, center of chest to center of back, and left flank to right flank.

Figure 11-4B shows the classical vector system used in ordinary ECG recording (disregarding

the xyz system and chest leads V1-V6). This illustration is a frontal view showing the axis and directions of the six major nonchest vectors. In medical terminology, these vectors are called "leads." In a moment I will discuss the details of acquiring these signals electronically. But first, let's look at a typical Lead I ECG signal in order to appreciate a little about what is going on.

Figure 11-4C shows the Lead I ECG signal with the various signal designations used in medicine to denote the segments. Interestingly, the timing of the ECG signal within each waveform is essentially constant (by point of reference the QRS "spike" complex takes about 40 milliseconds), and only the number of waveforms per minute changes with heartrate. The P wave corresponds to the firing of the atria, which begins pumping blood

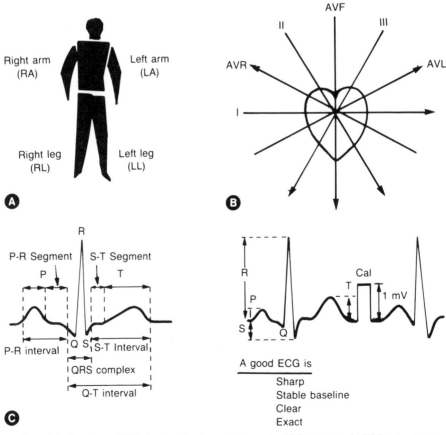

Fig. 11-4. A) Human ECG standard lead connections, B) ECG vectors, C) ECG waveforms.

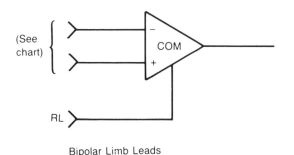

Bipolar Limb Leads

Lead	+ input	− input
I	LA	RA
II	LL	RA
III	LL	LA

Fig. 11-5. Bipolar limb leads.

through the heart. The QRS segment is the major feature of the waveform because it represents the deep contraction of the ventricles as they force blood into the circulatory system. The T wave corresponds to a refractory period when the cells of the heart muscle repolarize to be able to fire again. The trained physician is able to diagnose the type, site, and extent of heart disease from evaluation of the ECG and other clinical factors. In one of the waveforms of Fig. 11-4C, we see a 1 mV calibration pulse for comparison purposes.

As a person interested in things electrical, you may have heard from time to time that short duration electrical shocks, such as discharge from a high voltage capacitor, will kill some victims and not others. A principal reason for this difference is that an electrical shock that occurs during the T-wave segment, when the heart is repolarizing, often causes fatal ventricular fibrillation, while

shocks at other times do not — the difference is a matter of a few milliseconds and bad luck.

Figures 11-5 through 11-7 show the three major recording configurations for ECG signals. The set of leads (I, II, and III) shown in Fig. 11-5 comprise what medical people call *Einthoven's triangle*. These bipolar limb leads are recorded as follows (in all cases RL is the reference electrode):

Lead I. LA is connected to the amplifier's noninverting (+) input, while the RA is connected to the amplifier's inverting (−) input.

Lead II. LL is connected to the amplifier's noninverting (+) input, while the RA is connected to the inverting (−) input.

Lead III. LL is connected to the amplifier's noninverting (+) input, while the LA is connected to the inverting (−) input.

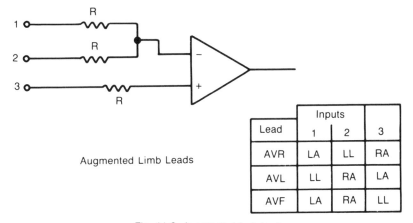

Augmented Limb Leads

Lead	Inputs		
	1	2	3
AVR	LA	LL	RA
AVL	LL	RA	LA
AVF	LA	RA	LL

Fig. 11-6. Augmented limb leads.

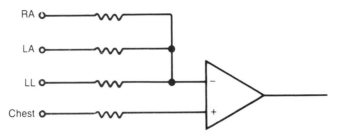

Fig. 11-7. Unipolar chest leads.

In the system above the RL serves as the reference point, and the amplifier measures a differential voltage across two limbs (the unused limb electrode in each case is shorted to the RL).

Figure 11-6 shows the *augmented limb leads,* also called the unipolar limb leads. In this type of recording, the noninverting (+) input of the biophysical amplifier is independent, while the inverting (−) input sees the summation of two the other two limb signals. The chart in Fig. 11-6 shows the connection scheme. The augmented limb leads are designated AVR, AVL, and AVF:

Lead AVR. RA is connected to the noninverting (+) input of the amplifier, while LA and LL are summed at the inverting (−) input.
Lead AVF. LL is connected to the noninverting (+) input of the amplifier, while LA and RA are summed at the inverting (−) input.
Lead AVL. LA is connected to the noninverting (+) input of the amplifier, while LL and RA are summed at the inverting (−) input of the amplifier.

The unipolar chest leads are measured with an electrode on any six locations on the chest applied to the noninverting (+) input (see Fig. 11-7), and the other limb electrode signals summed at the inverting (−) input. The summed signals form what clinicians are pleased to call the *indifferent electrode.*

BIOELECTRIC AMPLIFIERS

Bioelectrical signals acquisition requires amplifiers with special specifications. For example, the input circuitry of an ECG amplifier consists of

the high impedance of the internal amplifier's input terminals (typically well over 1 megohm in even low-quality equipment), a lead selector switch (if required), a 1-mV calibration source, and a means for protecting the amplifier against high-voltage electrical shock. This latter requirement applies especially to clinical equipment where the patient may require a high-voltage "jump start" from a defibrillator during resuscitation attempts.

In clinical medicine, as well as those research applications involving human subjects, the biopotentials amplifier is usually electrically isolated. The subject of isolation amplifiers is covered in Chapter 15. The reason for the isolation amplifier is that certain patients are at risk from minute electrical currents that would pass unnoticed in normal situations. But when the body is invaded, as in surgery or certain other medical procedures, currents as low as 20 microamperes are thought to be harmful — thus leakage from the ac power lines can be dangerous.

The standard −3 dB frequency response of the amplifier used in making diagnostic-grade recordings of ECG signals is 0.05 Hz to 100 Hz, while monitoring instruments are often bandwidth limited to 0.05 to 45 Hz or so. The reason for this difference is that in monitoring situations the higher frequency response makes muscle artifact more severe — ruining the recording or making it difficult to read. By filtering out the higher frequencies you obtain a smoother and more useful waveform, but only at the cost of certain diagnostic features. For example, a notch in the descending portion of the R-S segment that sometimes indicates heart block can be all but obliterated by the filtering effect of a monitoring-grade instrument.

Other biophysical signals will have somewhat different frequency-response requirements. For example, the EEG (brainwave) signal is mostly sinusoidal. We retain the low-frequency response, but can cut the high-frequency response to 30 or 40 Hz. The EMG (skeletal muscle) signals, on the other hand, often have higher frequency components, so a 1000-Hz amplifier is needed. It is not generally a good idea to make the frequency response of any data-acquisition system wider than needed because noise introduced can create recording artifacts.

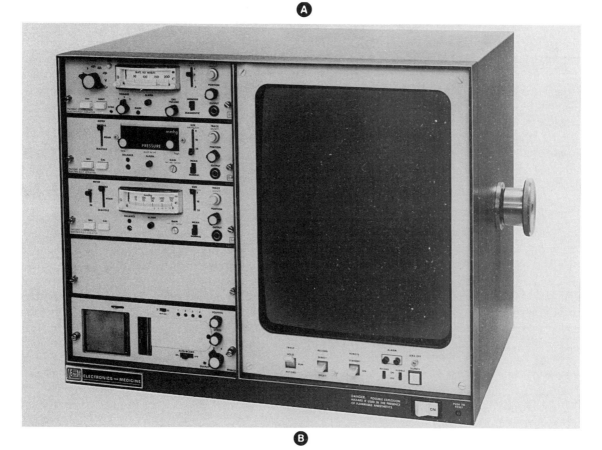

Fig. 11-8. A) Plug-in ECG amplifier, B) bedside monitor.

Figure 11-8A shows a typical ECG/biophysical amplifier used in both clinical and research applications. This Electronics for Medicine, Inc., unit plugs into a bedside or laboratory mainframe such as Fig. 11-8B. Note in Fig. 11-8A the various controls. The input connector here is a six-pin female jack that accepts the four limb electrodes, a chest electrode, and a shield. The LEAD switch selects the recording configuration per the system of Figs. 11-5 through 11-7. This amplifier has upper and lower -3-dB filter switches — controls that are typically missing in most clinical amplifiers, except for those used in catheterization laboratories and other specialized applications. The SIZE control is merely a display control and is not related to the ECG signal. This amplifier also has a high-level (1-volt) input for use in other applications.

The bedside unit of Fig. 11-8B shows a collection of biophysical data acquisition modules along with two display devices: a large oscillosope and a strip-chart recorder (lower left). This particular setup is customized to include (from top to bottom): an ECG preamplifier, a digital readout blood pressure monitor, and an analog readout pressure monitor. Blood-pressure measurements are often made by placing an in-dwelling fluid-filled catheter into an artery in the arm (blood pressure measurement is covered along with other pressure systems in Chapter 7.

Two types of oscilloscopes are widely used in biophysical measurements work. Simple analog, or "bouncing ball," instruments use a long-persistence phosphor that allows most of the wave complex to be on the screen at one time. A better solution is to use a digital storage oscilloscope that holds everything in memory and displays on the analog screen as many as five minutes of immediate past waveforms. These oscilloscopes are called "nonfade" models, and allow the medical staff to review the immediate history of the waveforms. Typically, the digitized waveform is stored in a recirculated shift register from which the analog screen is refreshed via a D/A converter. In general, 256 data points per second are taken to satisfy Nyquist's criterion for sampled waveforms, and refresh occurs every $\frac{1}{64}$ second or faster.

RESPIRATION SIGNALS

There are several ways to measure respiration signals on humans. Most of them were covered in Chapter 9, and are capable of rendering quantitative measures of flow volume and rate. In this section, however, I will discuss a simple method that uses a bioelectric parameter to extract only respiration rate information. This system is used in clinical medicine to monitor patients, especially infants, under situations where it might be difficult or inconvenient to use visual techniques.

Figure 11-9A shows a schematic view of the impedance pneumograph, while an equivalent circuit is shown in Fig. 11-9B. The method is based on the fact that the ac impedance across the chest varies as the lungs inflate and deflate during breathing. The change of impedance, ΔR, is tiny' but sufficient for our purposes.

A low-voltage (100 mV p-p or less) 50 to 500 kHz (200 kHz being common) excitation is applied to the chest through a pair of 100-kilohm, matched-precision resistors. A differential ac amplifier (stabilized and optimized for the excitation frequency) receives the signal from chest electrodes. In most cases the same electrodes as used in Lead I ECG recordings do double duty as respiration-monitoring sites. The output of the ac amplifier, the signal which contains the respiration signal, and a sample of the 200-kHz oscillator are fed to a synchronous detector that extracts the respiration waveform. Because of the distortions in the system, no data other than respiration rate (breaths per second) is retrievable from this method.

You can see from Fig. 11-9B how this instrument works. R1 and R4 are the isolation resistors (each 100 kohms); resistor R is the chest impedance at rest (lungs empty), and ΔR is the change of value of R as the patient inspires air. The output voltage, V_o, is determined by either of the following methods:

$$V_o = I(R + \Delta R)$$

or,

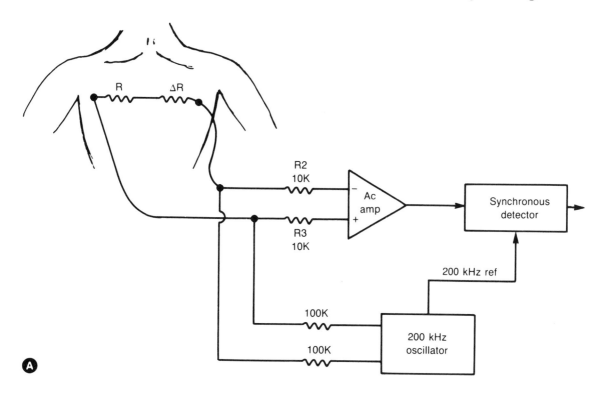

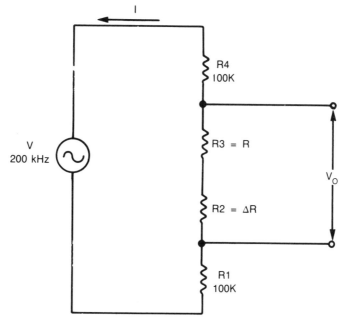

Fig. 11-9. A) Impedance pneumonography, B) equivalent circuit.

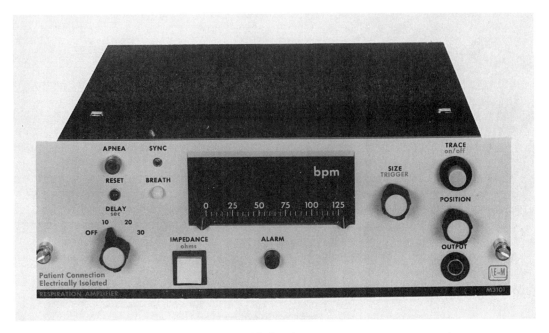

Fig. 11-10. Respirometer.

$$V_o = \frac{(V)(R + \Delta R)}{(R1 + R4 + R + \Delta R)}$$

Assuming an amplifier input impedance high enough to make R2 and R3 too small to need consideration, either of the above expressions will yield the respiration signal. If this signal is integrated using pulse-counting techniques, then you have a respiration monitor, also called an *apnea monitor* (note: although these instruments are called "apnea" monitors, they measure chest movement, not true apnea. The FDA does not prefer the term "apnea monitor" because it conveys a sense of capability that is not actually present. Nonetheless, the respiration-monitoring method shown above is a valuable means for acquiring respiration rate data).

CARDIAC OUTPUT COMPUTER

The signals-acquisition systems problems presented by the biopotentials and respiratory-monitoring methods discussed above are somewhat tame compared with the problems presented in measuring cardiac output. The principal difference is that cardiac output is usually measured using an invasive technique. This type of measurement is presented in order to demonstrate indirect data acquisition technique (most practical cardiac output measurements are inferred from another parameter).

Cardiac output (CO) is defined as the rate of blood pumping volume. The question being asked of the CO measurement is "how much blood is this person pumping per unit of time." Cardiac output is measured in units of liters of blood per minute of time (1/min). In adults, CO reaches a value between 3 and 5 1/min for healthy individuals.

A quantitative measure of cardiac output is the product of the stroke volume and the heart rate. The stroke volume is merely the volume of blood expelled from the ventricle during a single contraction of the heart. Cardiac output is calculated from:

$$C.O. = V \times R$$

where

C.O. is the cardiac output in liters per minute (1/min)

V is the stroke volume in liters per beat (1/beat)

R is the heart rate in beats per minute (beat/min)

Example. Calculate the cardiac output for a patient with a heart rate of 70 beats per minute and a stroke volume of 40 ml/beat.

$$C.O. = \frac{40 \text{ ml}}{\text{beat}} \times \frac{1 \text{ 1}}{1000 \text{ ml}} \times \frac{70 \text{ beats}}{\text{min.}}$$

$$C.O. = \frac{[(40) \times (1 \text{ 1}) \times (70)]}{1000 \text{ min}}$$

$$C.O. = 2.8 \text{ 1/min}$$

It is difficult, and usually impossible, (except on animals in laboratory settings) to measure cardiac output using any technique based on the above equation. The main problem is obtaining good stroke volume data without excessive risk to the patient.

One method for obtaining cardiac output data is to measure blood flow rate, and then integrating the measurement to obtain the time-averaged cardiac output. A problem with this measurement, however, is that it only applies to open-heart surgical situations bcause the transducer must be applied directly over the blood vessel being measured. It is simply not accurate to obtain data from one of the downstream arteries; only the pulmonary artery (between right ventrical and lung) or aorta (main artery out of the left ventrical) can be used with any degree of accuracy.

The only really useful practical methods for measuring cardiac output are the dye dilution methods. Four methods are used 1) Fick's method (oxygen dilution), 2) radioactive dye dilution, 3) optical dye dilution, and 4) thermal dilution. Of these, the thermal dilution method presents the best measurement, but the others will be described as well.

In a dilution measurement of cardiac output (C.O.), a known concentration of some tracer material is injected into the bloodstream just before the blood reaches the heart. The diluted concentration is measured downstream (past the heart), yielding the indirect measure of C.O. From the dilution curve the C.O. can be inferred. Dilution-method C.O. measurements are done routinely in cardiac catheterization laboratories, special procedures laboratories, and, in some cases, at bedside in intensive or coronary-care-unit settings. At least one manufacturer builds a C.O. computer designed for operation by registered nurses.

All dilution methods depend upon knowing the injectate concentration, and the ability to measure the output concentration. The C.O. is determined from:

$$C.O. = \frac{\text{Injection rate (mg/min)}}{\text{Distal concentration (mg/l)}} \text{ (1/min)}$$

Fick Method. This method was one of the earliest techniques used to measure cardiac output, but is still nonetheless used. It is used especially in physiological research where healthy subjects are being studied, rather than in clinical settings where they are being evaluated for coronary disease. The Fick method uses normal room air oxygen inhaled during normal respiration as the injectate. The oxygen is injected into the bloodstream by the normal operation of the subject's lungs.

The infusion rate data is determined by the measurement of respiratory gases. It is necessary to subtract the oxygen concentration of the patient's exhaled air from room air concentrations, usually taken (quite accurately) to be 21 percent.

Concentration is determined by measuring the oxygen concentration of arterial blood as it leaves the lungs. But this term does contain at least some error because returning venous blood is not oxygen free, even though most of the oxygen is given up in the circulatory system. This amount is significant and so must be determined and subtracted from the arterial blood oxygen concentration. Repeated measurements made over a period of several minutes are averaged for a final result.

The Fick method normally requires a substantial amount of external equipment, although at least one company offers an automated version in a smaller package. In clinical use the Fick method has been all but supplanted by other methods.

Dye Dilution Methods. There are two dye dilution methods. In one, a radioactive substance is injected into the bloodstream so that the blood "lights" up to a gamma camera scanning the heart. Such a method is called a "nuclear medicine" procedure, and is found only in the best-equipped hospitals. A problem with this method (other than the use of radioactive tracers) is that the camera is too large for bedside use, and is very expensive compared with other methods that yield results that are just as good.

Optical dye dilution uses an opague (to light) dye such as indocyanine green, and an optical densitometer to measure the concentration. Figure 11-11 shows the concentration curve at a point distal to the injection site. Shortly after the injec-

tion the concentration at the measurement site rises abruptly to a maximum value and then falls off exponentially as the injectate bolus passes the point. This exponential decay is due to the fact that the injectate bolus is spread out (i.e., diluted) over a longer path as the blood flows.

Cardiac output is then measured by integrating the curve. This information is used in an equation of the general form:

$$\text{B.F.R.} = \frac{k \times M}{C \, dt} \, (\text{ml/min})$$

where

B.F.R. is the blood flow rate in milliliters per minute (ml/min).

k is a constant between 20 and 200, depending upon injectate.

M is the volume of injectate in milliliters (ml).

C is the concentration of injectate (measured) in mg/ml.

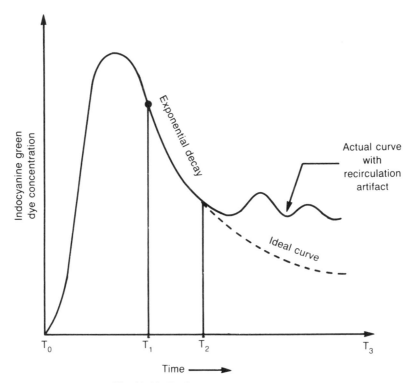

Fig. 11-11. Cardiac output dilution curve.

There is a problem with this method, however. Note time T2 in Fig. 11-11. The normal path for an exponential decay follows the curve marked "Ideal Curve," but the measured concentration follows the "Actual Curve," which represents a recirculation artifact in the data. Some of the blood is recirculated back through the heart before the measurement is completed. The artifact changes the shape of the curve (and increases its integral) between times T2 and T3.

In early instruments the doctor recorded the dye dilution curve on a paper chart recorder using an analog pen on graph paper. Graphical means were then used to extrapolate the area under the curve. In some cases, the person reading the chart would use a French Curve with an exponential side to continue the known exponential curve before the recirculation artifact distorted the data. The curve could then be integrated either using a mechanical planometric integrator, or simply by counting the squares under the curve on the paper. When I discuss the thermodilution method below I will return to the issue of integrating a distorted curve such as Fig. 11-11. There is a method called *geometric integration* that permits automated integration despite the recirculation artifact.

Thermodilution. This method has become the most common method for measuring cardiac output in clinical settings, and is also popular among laboratory scientists. Thermodilution technique forms the basis for most clinical and research-grade cardiac-output computers now on the market. One reason why thermodilution is preferred is that no strange injectates are used (as they are in radio or optical dye dilution), only ordinary intravenous (IV) solutions such as normal saline or 5 percent dextrose in water ("D_5W").

Most thermodilution cardiac-output computers operate on a version of the following equation:

$$C.O. = \frac{k \times G(b) \times G(i) \times V(i) \times [T(b) - T(i)]}{U(b) \times U(i) \quad T(b)' \, dt} \quad (1/min)$$

where

C.O. is the cardiac output in liter per minute (l/min)

k is a constant (approximately 20 to 150, depending upon the type of injection catheter used — value supplied by manufacturer).

G(b) is the density of human blood in kg/m³

G(i) is the density of injectate in kg/m³

V(i) is the injectate volume in liters (l)

T(b) is the blood temperature in degrees Celsius (C)

T(b)' is the post-injection site temperature of the blood, as measured distal to the heart.

T(i) is the injectate temperature (note: usually either iced saline at 0 degrees or room temperature saline at 25 degrees is used).

U(i) is the heat energy of the injectate in joules (J)

U(b) is the heat energy of the blood in joules (J)

The equation above looks like a bear, but is easily simplified once it is recognized that most of the terms are constants, or easily standardized or assumed. For example, one C.O. computer manufacturer calculates C.O. based on the following simplified version of the generic equation:

$$C.O. = \frac{(64.8) \times C(t) \times V(i) \times [T(b) - T(i)]}{T(b)' \, dt}$$

where

C.O. is the cardiac output in liters per minute (l/min)

64.8 is a collection of other constants and the conversion factor from seconds to minutes.

C(t) is a constant that is supplied with the injectate catheter that accounts for the temperature rise in the portion of the outside of the patient's body.

All other terms are as defined above.

A special cardiac output computer test fixture enters a temperature signal that simulates a temperature change of 10 degrees Celsius for a period of 10 seconds.

Example. Find the expected reading during a test of the instrument if the following front panel settings are entered: $C(t) = 49.6$, $T(b) = 37$ degrees, injectate temperature $T(i)$ is 25 degrees, and injectate volume $V(i)$ is 10 ml.

$$C.O. =$$

$$\frac{(64.8) \times C(t) \times V(i) \times [T(b) - T(i)]}{T(b)' \, dt} \, (l/min)$$

$$C.O. = \frac{\begin{array}{c}(64.8) \times (49.6) \times (10ml \times 1l/1000ml) \\ \times [(37) - (25)]\end{array}}{(10 \text{ degrees}) (10 \text{ sec})}$$

$$C.O. = \frac{3214 \times 0.01 \times 12}{100}$$

$$C.O. = 385.68/100 = 3.9 \, l/min$$

The thermodilution measurement of cardiac output is made using a special catheter that is inserted into a vein, usually on the patient's right arm (the brachial vein is popular). The catheter is multilumened, and one of the lumens has its output hole several centimeters from the catheter tip. This proximal lumen is situated so that it is outside the heart (close to the input valve on the right atrium) when the tip is all the way through the heart, resting in the pulmonary artery. There are other lumens in the catheter output at the tip, so are used to measure pressures in the pulmonary artery in other procedures. A thermistor in the tip registers a resistance change with changes in temperature.

The thermistor is usually connected in a Wheatstone-bridge circuit (see Chapter 5). The dc excitation of the bridge is critical. Either the short-term stability of this voltage must be very high, or a ratiometric method must be used to cancel excitation potential drift. In addition, it is also necessary to limit the bridge excitation potential to above 200 mV for reasons of safety to the patient (electrical leakage is especially dangerous because the thermistor is inside the heart or pulmonary artery). This value is also consistent with thermistor stability from self-heating problems.

Figure 11-12 shows a simplified schematic of a cardiac-output computer front-end circuit. The thermistor is in a Wheatstone bridge circuit consisting also of R1 through R3, with potentiometer R5 serving to balance the bridge. In some modern machines, an auto-balancing or zeroing method is used. These circuits use a digital-to-analog converter (DAC) to inject a current into one node of the bridge, and that current nulls the bridge circuit to zero. The doctor waits for the thermistor to equilibrate with blood temperature (usually it is in this condition by the time it is threaded through the venous system to the pulmonary artery). The

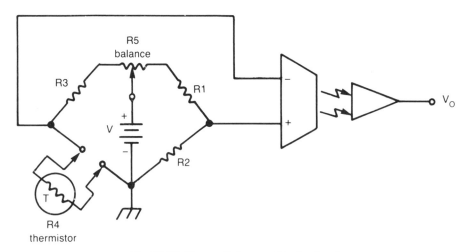

Fig. 11-12. Thermodilution amplifier front-end.

output of the bridge, depending upon the design, is typically 1.5 to 2.2 millivolts per degree Celsius, with 1.8 mV/° C being quite common. This signal is amplified approximately 1000 times to 1 volt/° C by the preamplifier. This preamplifier is an isolated amplifier (see Chapter 15) for reasons of patient safety. The output of this circuit, V_o, is used in the denominator of the equation as above.

The block diagram for a sample analog cardiac output computer is shown in Fig. 11-13. The front-end circuitry from Fig. 11-12 is in the blocks marked "Bridge" and "Pre-amp." The isolator circuit is merely a buffer amplifier that permits V_o to be output to an analog paper chart recorder. Analysis of the waveshape reveals errors of technique — and thus explains odd readings that are not supported by other clinical facts. The temperature signal (V_o) is integrated and then sent to an analog divider where it is combined with the temperature difference signal $[T(b) - T(i)]$ and the constants (all represented by a voltage). The low-pass filtered output of the analog multiplier is a measure of the cardiac output and is displayed on a digital voltmeter.

Research and catheterization laboratory models of C.O. computers are flexible as regards all factors in the measurement. For example, the doctor can set the injectate temperature $T(i)$, injectate volume (5 to 25 ml), blood temperature $T(b)$, and so forth. But in ICU/CCU settings this same flexibility results in erroneous readings that are due to what might be euphemistically called "cockpit problems" (i.e., the doctor or nurse fouled up). The patient may either receive inappropriate treatment, or the readings will be disbelieved so no changes in treatment are ordered.

By design, however, the data-acquisition engineer can alleviate some of these problems. For example, only permit one injectate volume (10 ml is usually selected), and use additional thermistors to measure injectate temperatures. A rectal thermistor or pn-junction temperature sensor can be used to measure blood temperature. One manufacturer only permits room-temperature saline to be used, even though (in controlled settings) the higher $[T(b) - T(i)]$ differential of iced saline yields more accurate readings. The reason for this

change is that iced saline changes temperature more quickly than room-temperature saline, and that makes the operator's technique important. I have seen clinical situations in which one doctor consistently obtained poorer results than his colleagues. The problem was the length of time he held the saline injectate syringe after it was removed from the ice bath. Body heat from his hands caused a tremendous "delta-T" factor that placed the $T(i)$ term in considerable error. The problem is eliminated by using either room-temperature saline (delta-T goes to nearly zero) or a syringe pump for iced saline.

The C.O. computer shown in Fig. 11-13 is representative of most analog forms. In Chapter 29 I will again present this topic under the rubric of digital computer instruments and a digital version will be discussed at that time.

The recirculation artifact (mentioned earlier in conjunction with the dye dilution method) is small in thermodilution computers; so small, in fact, that some machines neglect it altogether. The best, most accurate, machines have a means for compensating for the artifact. This compensation comes in the form of geometric integration of the thermal curve used in the denominator of the equation. Figure 11-14 shows how geometric integration works (also see computer program number 14 in Appendix A).

The thermodilution curve is known from empirical observations to be exponentially decaying from the point where the concentration of injectate drops off to 85 percent of its peak value (point C_2 in Fig. 11-14). The data acquired is clean prior to time T_2 when the curve reaches this exponentially decaying region, so from initiation at time T_1 until time T_2 you can simply integrate concentration C using the normal technique. For this region no special method is needed, but because the recirculation artifact falls within the exponential decay portion of the curve, you can use geometrical integration.

In geometric integration of an exponentially decaying curve, you can depend on the fact that a rectangle can be constructed that approximates the area under the exponential portion of the curve, including the area distorted by the artifact.

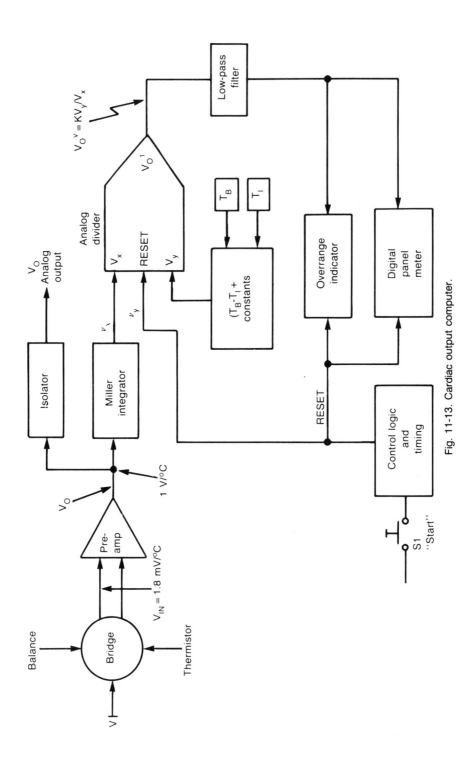

Fig. 11-13. Cardiac output computer.

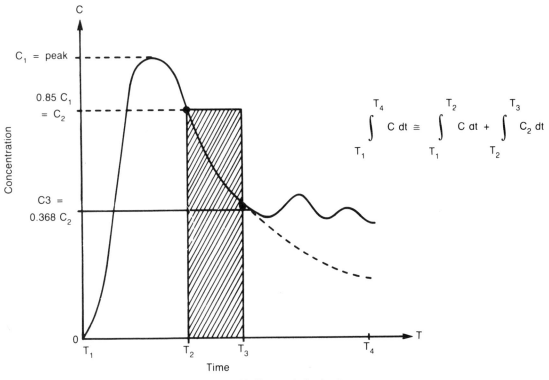

Fig. 11-14. Geometric integration.

This rectangle takes one set of coordinates from the instant when the exponential portion is entered, (C_2, T_2), and the other as the time when the curve passes through the one time-constant point $e^{-1 \text{ unit}}$, or 36.7879 percent of the initial value. This point is (C_2, T_3), with T_3 being the time at which $C = 0.367879 \times C_2$. The calculation of the area of this rectangle is simple:

$$\text{Area} = C_1 \times (T_3 - T_2)$$

Thus, you can write an expression for the integration of the entire dilution curve as follows:

$$\int_{T_1}^{T_4} C \, dt = \int_{T_1}^{T_2} C \, dt + \int_{T_2}^{T_4} C \, dt$$

Or, substituting the geometric method for the second integral yields the final result.

$$\int_{T_1}^{T_4} C \, dt = \int_{T_1}^{T_2} C \, dt + \frac{C_2}{(T_3 - T_2)}$$

CONCLUSION

In this chapter we have hopefully given the data acquisition engineer or other concerned professional some indication of the types of problems associated with biophysical signals and their acquisition. These signals are critical to instrumentation, control devices and data acquisitions efforts in both research and clinical applications.

Chapter 12

Electrodes for Biomedical Applications

I N THE LAST CHAPTER YOU LEARNED THAT A SIZable portion of the signals-acquisition efforts in biology, physiology, and medicine are electrical signals or biopotentials. Most biopotentials are acquired from one of three types of electrode: surface macroelectrodes, in-dwelling macroelectrodes, and microelectrodes. Of these, the first two are generally used *in vivo,* while the latter are used *in vitro.* In this chapter I will elaborate on the acquisition of biopotentials by dealing with the types of electrodes commonly used in biomedical instrumentation.

SURFACE ELECTRODES

Surface electrodes are those that are placed in contact with the skin of the subject. Also in this category are certain needle electrodes of such a size as to prevent their being inserted inside of a single cell (this criterion defines a microelectrode). There is some case for including needle electrodes under the rubric of in-dwelling electrodes, but that is not generally the practice in biomedical engineering.

Surface electrodes (other than needle electrodes) vary in diameter from 0.3 to 5 centimeters (cm) in diameter, with most being in the 1-cm

range. Human skin tends to have a very high impedance compared to other "voltage sources" that you might be familiar with. Typically, normal skin impedance as seen by the electrode varies from 1 kilohm for sweaty skin surfaces to 20 kilohms for dry surfaces. Problem skin, especially dry, scaley, or diseased skin, may reach impedances in the 500 kilohms range. In any event, one must regard surface electrodes as a high-impedance source—a fact which tends to influence the design of biopotential-amplifier input circuitry. In most cases the rule of thumb for a voltage amplifier is to make the input impedance of the amplifier at least ten times the source impedance. For biopotentials amplifiers this requirement means 5 megohms or greater input impedance—a value easily achieved these days.

Another problem regarding surface electrodes results from the fact that there is an exchange of ions between the metallic electrode and the skin, which is an electrolytic substance. The resulting *half-cell voltage* ranges from −3 to +3 volts, with −2 to +2 volts being the most common. Although a range is given for these potentials in the various electrochemistry textbooks, the value for any one metallurgical mixture is relatively stable.

Besides its initial materials dependency, the actual potential exhibited by any given electrode may change slowly with time. Some candidate materials may look good initially, but experience such a large change with time and chemical environment (biological subjects are a hostile environment!) that they are rendered almost useless in practical applications. For these reasons materials like gold, silver-silver chloride (Ag-AgCl) and platinum-platinum black are used. In general use for surface recording is the Ag-AgCl electrode. Unless otherwise specified, you can generally assume this material is used. By international scientific agreement, the zero reference point is the hydrogen-hydrogen (H-H) electrode, which is given a half-cell potential of zero volts by convention. All other electrode half-cell potentials are measured against the H-H zero reference.

The electrode half-cell potential becomes a big problem in bioelectric signals acquisition because of the tremendous difference between these dc potentials and the biopotentials. A typical half-cell potential for a biomedical electrode is 1.5 volts, while biopotentials are more than 1000 times less! The surface manifestation of the ECG signal (see Chapter 11) is 1 to 2 millivolts (mV), while EEG scalp potentials are on the order of 50 microvolts (μV). Thus, the half-cell electrode voltage is 1500 times greater than the peak ECG potential, and 30,000 times greater than the EEG signal.

The instrument designer must provide a strategy for overcoming the effects of the massive half-cell potential offset. Because the half-cell potential forms a large dc component for the minute signal voltage, you will find the appropriate strategy a combination of the following approaches:

First, you could use a differential amplifier to acquire the signal. If the electrodes are identical, then the half-cell potentials should be the same. Theoretically, at least, the equal potentials would be seen as a single common-mode potential, and thus would cancel in the output. A limitation of this approach is that the gains required to process low-level signals also act on tiny differences between the two half-cell potentials. A difference of 1 mV —only 0.1 percent of the total—looks like any

other 1 mV dc signal to the gain of 1000 ECG amplifier.

Second, you could design the signals-acquisition circuit to provide a counter-offset voltage to cancel the half-cell potential of the electrode. While this approach has certain appeal, it is limited by the fact that the half-cell potential changes with time and the relative motion between skin and electrode.

Third, you could ac-couple the input amplifier. This approach can permit removal of the signal component from the dc offset. This option is, perhaps, the most appealing—especially where variations of the dc offset are substantially in lower frequency than the signal frequency components. In that case the normal −3-dB frequency-response limit can be useful to tailor the attenuation of variations in the dc offset.

In some biomedical applications, however, signal components are near-dc. For example, the frequency content of the ECG signal is 0.05 to 100 Hz. In medical ECG equipment, therefore, one can expect the baseline to shift every time the patient moves around in bed.

In most cases the first and third options are selected for biopotentials amplifiers. The user will require an ac-coupled, differential-input amplifier for signals acquisition.

ELECTRODE MODEL

Figure 12-1 shows a circuit model of a biomedical surface electrode. This model more or less matches the equivalent circuit of ECG and EEG electrodes. In this circuit a differential amplifier is used for signals processing, which will cancel the effects of electrode half-cell potentials, V1 and V2. Resistance R3 represents the internal resistances of the body, which are typically quite low. The biopotentials signal is represented as a differential voltage, V3. The other resistances in the circuit represent the resistances at the electrode-skin contact interface.

The surprising aspect of Fig. 12-1 are the values usually associated with capacitors C1 and C2. While some capacitance is normally expected, it usually surprises people to learn that these con-

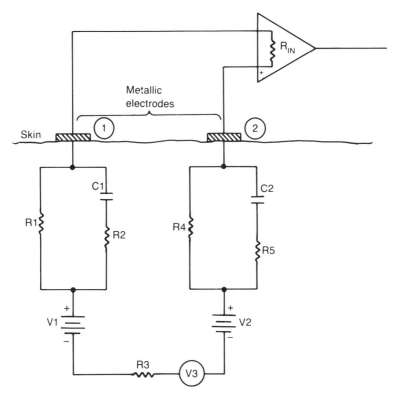

Fig. 12-1. Skin electrode equivalent circuits.

tact capacitances can attain values of several microfarads.

TYPICAL SURFACE ELECTRODES

A variety of electrodes has been designed for surface acquisition of biomedical signals. Perhaps the oldest form of ECG electrodes still in clinical use are the strap-on variety. These electrodes are 1 to 2 square inch brass plates that are held in place by rubber straps. A conductive gel or paste is used to reduce the impedance between the electrode and skin.

A related form of ECG electrode is the suction cup electrode shown in Fig. 12-2. This device is used as a chest electrode in short-term ECG recording. For longer term recording or monitoring, such as continuous monitoring of a hospitalized patient in a coronary or intensive-care unit, the paste-on column electrodes of Fig. 12-3 are used instead.

The column electrodes are shown in Fig.

12-3A, while a sections schematic is shown in Fig. 12-3B. The electrode consists of an Ag-AgCl button at the top of the gel-filled column. This assembly is held in place by an adhesive-coated, foam-rubber disk.

The use of a gel-filled column that holds the actual metallic electrode off the surface reduces movement artifact. For this reason (among several others), the electrode of Fig. 12-3 is preferred for monitoring hospitalized patients.

There are, however, several problems associated with this type of electrode. One of the problems is the inability of the adhesive to stick for long on sweaty or "clammy" skin surfaces. The user also must avoid placing the electrode over boney prominences. Usually, the fleshy portions of the chest and abdomen are selected as electrode sites. Various hospitals have different protocols for changing the electrodes, but in general, the electrode is changed at least every 24 hours—and often more frequently as few last as long as 24 hours. In some hospitals the electrode sites are moved — and electrodes changed — once every eight-hour nursing shift in order to avoid ischemia of the skin at the site.

The surface electrodes that I have discussed thus far are noninvasive types. That is, they adhere to the skin without puncturing it. Figure 12-4 shows a needle electrode. This type of ECG electrode is inserted into the tissue immediately beneath the skin by puncturing the skin at a large oblique angle (i.e., close to the horizontal with respect to skin surface). The needle electrode is only used for exceptionally poor skin, especially on anesthetized patients, and in veterinary situations. Of course, infection is an issue in these cases, so needle electrodes are typically sterilized in ethylene oxide gas.

IN-DWELLING ELECTRODES

In-dwelling electrodes are intended to be inserted into the body. These are not to be confused with needle electrodes (described above), which are intended for insertion into the layers beneath the skin. The in-dwelling electrode is typically a tiny, exposed metallic contact at the end of a long, insulated catheter. In one application the electrode is threaded through the patient's veins (usually in the right arm) to the right side of the heart in order to measure the intracardiac ECG waveform. Certain low-amplitude, high-frequency features (such as the Bundle of His element) become visible only when an in-dwelling electrode is used.

MICROELECTRODES

The "microelectrode" is an ultra-fine device that is used to measure biopotentials at the cellular level (Fig. 12-5). In practice, the microelectrode penetrates a cell that is immersed in an "infinite" fluid (such as physiological saline), that is in turn connected to a reference electrode. Although several kinds of microelectrode exist, most of them are of one of two basic types: metallic contact or

Fig. 12-2. Suction cup electrode.

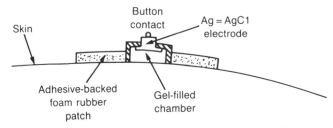

Fig. 12-3. Paste-on electrodes A) photo, B) schematic.

fluid filled. In both cases an exposed contact surface of about 1 to 2 micrometers (1 $\mu M = 10^{-6}$ M) is in contact with the cell. As might be expected, this fact makes microelectrodes very high impedance devices.

Figure 12-6 shows the construction of a typical glass-metal microelectrode. A very fine platinum or tungsten wire is slip-fitted through a 1.5 to 2 millimeter (mm) glass pipette. The tip is etched, and then fire formed into the shallow-angle taper

shown. The electrode can then be connected to one input of the signals amplifier.

There are two sub-categories of this type of electrode. In one type the metallic tip is flush with the end of the pipette taper, while in the other there is a thin layer of glass covering the metal point. This glass layer is so thin as to require measurement in Angstroms, and drastically increases the impedance of the device.

A fluid-filled microelectrode is shown in Fig.

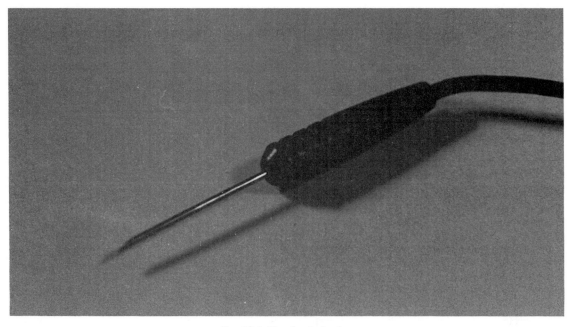

Fig. 12-4. Needle electrode.

12-7. In this type the glass pipette is filled with a 3 molar solution of potassium chloride (KCL), and the large end capped with an Ag-AgCl plug. The small end need not be capped because the 1 μM opening is small enough to contain the fluid.

The reference electrode is likewise filled with 3 molar KCL, but is very much larger than the microelectrode. A platinum plug contains fluid on the interface end, while an Ag-AgCl plug caps the other end.

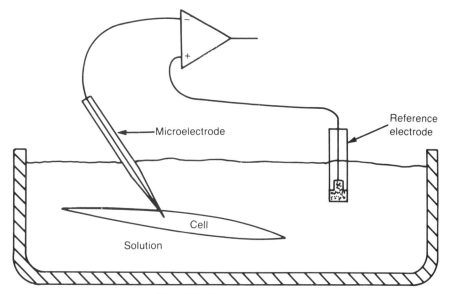

Fig. 12-5. Microelectrode use.

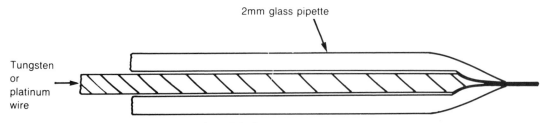

Fig. 12-6. Glass microelectrode.

Figure 12-8 shows a simplified equivalent circuit for the microelectrode (disregarding the contribution of the reference electrode). Analysis of this circuit reveals the signals acquisition problem due to the RC components. Resistor R1 and capacitor C1 are due to the effects at the electrode-cell interface, and are (surprisingly) frequency dependent. These values fall off to a negligible point at a rate of $1/(2\Delta F)^2$, and are generally considerably lower than R_s and C2.

Resistance R_s in Figure 12-8 is the spreading resistance of the electrode, and is a function of the tip diameter. The value of R_s in metallic microelectrodes without the glass coating is approximated by:

$$R_s = \frac{P}{4\Delta r}$$

where

R_s is the resistance in ohms

P is the resistivity of the infinite solution outside of the electrode (e.g., 70 ohm-cm for physiological saline).

r is the tip radius (typically 0.5 μM for a 1 μM electrode)

Example. Assuming the typical values given above, calculate the tip spreading resistance of a 1 μM microelectrode.

$$R_s = P/4\Delta r$$

$$R_s = \frac{70 \text{ ohm-cm}}{4\Delta \ 0.5 \ \mu M \times \dfrac{10^{-4}}{\mu M}}$$

$$R_s = 111,000 \text{ ohms}$$

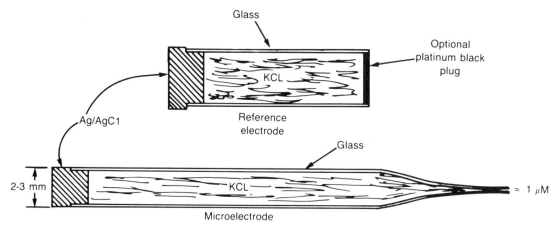

Fig. 12-7. KCL microelectrode.

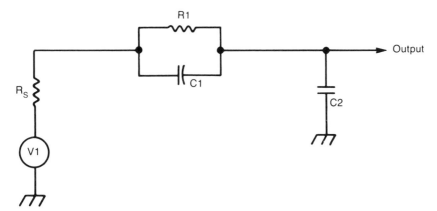

Fig. 12-8. Equivalent circuit.

The impedance of glass coated metallic microelectrodes is at least one or two orders of magnitude higher than this figure.

For fluid-filled KCL microelectrodes with small taper angles ($\Theta/180$ radians), the series resistance is approximated by:

$$R_s = \frac{2P}{\Theta\, r\, a}$$

where

R$_s$ is the resistance in ohms
P is the resistivity (typically 3.7 ohm-cm for 3M KCL)
r is the tip radius (typically 0.1 μM)
a is the taper angle (typically $\Theta/180$)

Example. Find the series impedance of a KCL microelectrode using the values shown above:

$$R_s = \frac{2P}{\Theta\, r\, a}$$

$$R_s = \frac{(2)(3.7 \text{ ohm-cm})}{10^{-4}\,(3.14)}$$

$$(3.14)\ 0.1\ \mu M \times \frac{}{\mu M\ \ 180}$$

$$R_s = 13.5 \text{ megohms}$$

The capacitance of the microelectrode is given by:

$$C2 = \frac{0.55e}{nl(R/r)} \text{ pF/cm}$$

where

e is the dielectric constant of glass (typically 4)
R is the outside tip radius
r is the inside tip radius (r and R in same units)

Example. Find the capacitance (C2 in Fig. 12-8) if the pipette radius is 0.2 μM and the inside tip radius is 0.15 μM.

$$C2 = \frac{0.55e}{nl(R/r)} \text{ pF/cm}$$

$$C2 = \frac{(0.55)(4)}{nl(0.2\ \mu M/0.15\ \mu M)} \text{ pF/cm}$$

$$C2 = 7.7 \text{ pF/cm}$$

How do these values affect performance of the microelectrode? Resistance R$_s$ and capacitor C2 operate together as an RC low-pass filter. For example, a KCL microelectrode immersed in 3 cm of physiological saline has a capacitance of approximately 23 pF. Suppose it is connected to the ampli-

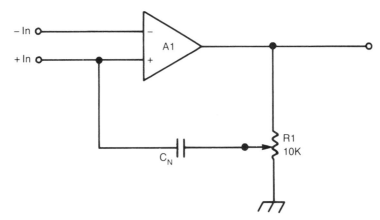

Fig. 12-9. Capacitance compensation circuit.

fier input (15 pF) through 3 feet of small-diameter coaxial cable (27 pF/ft, or 81 pF). The total capacitance is $(23 + 15 + 81)$ pF $= 119$ pF. Given a 13.5 megohm resistance, the frequency response (at the -3 dB point) is:

$$F = \frac{1}{2 \pi R C}$$

where

F is the -3 dB point in hertz (Hz)
R is the resistance in ohms
C is the capacitance in farads

Example. For $C = 119$ pF $(1.19 \times 10^{-10}$ farads) and $R = 1.35 \times 10^{7}$ ohms, find the upper -3 dB frequency-response point.

$$F = \frac{1}{(2)(3.14)(1.35 \times 10^{7})(1.19 \times 10^{-10})}$$

$$F = 99 \text{ Hz} \quad 100 \text{ Hz}$$

Clearly, a 100-Hz frequency response, with a -6 dB/octave characteristic above 100 Hz, results in severe rounding of the fast rise-time action potentials (electronics engineers, who speak of nanosecond rise-times, will please refrain from smirking at the term "fast" in this context). A strategy must be devised in the instrument design to overcome the effects of capacitance in high-impedance electrodes.

Neutralizing Microelectrode Capacitance

Figure 12-9 shows the standard method for neutralizing the capacitance of the microelectrode and associated circuitry. A neutralization capacitance, C_n, is in the positive feedback path along with a potentiometer voltage divider. The value of this capacitance is:

$$C_n = \frac{C}{A - 1}$$

where

C_n is the neutralization capacitance
C is the total input capacitance
A is the gain of the amplifier

Example. A microelectrode and its cabling exhibit a total capacitance of 100 pF. Find the value of neutralization capacitance (Fig. 12-9) required for a gain-of-10 amplifier.

$$C_n = C/(A - 1)$$
$$C_n = (100 \text{ pF})/(10 - 1)$$
$$C_n = 100 \text{ pF}/9 = 11 \text{ pF}$$

Chapter 13

Chemistry/Laboratory Transducers

I N THIS CHAPTER I WILL DISCUSS BRIEFLY WITH transducers used in chemical analysis, although be aware that this is merely an elementary survey and you should study one of the entire books that are completely dedicated to this topic. A heavy emphasis is placed on medical laboratory instruments, although the basic principles are broadly applicable.

PHOTOCOLORIMETRY

One of the most basic forms of instrument is also both the oldest and most commonly used: photocolorimetry. These devices are used to measure oxygen content in blood, CO_2 content of air, water vapor content, blood electrolyte (Na and K) levels, and a host of other measurements in the medical laboratory.

Figure 13-1 shows the basic circuit of the most elementary form of colorimeter. Although the circuit is very basic, this is the very circuit of a widely used blood oxygen meter. The circuit is a Wheatstone bridge that uses a pair of photoresistors (see Chapter 10) as the transducing elements. Potentiometer R5 is used as a balance control, and it is adjusted for zero output ($V_o = 0$) when the same light shines on both photoresistors. Recall from the Chapter 10 discussion that output voltage V_o will be zero when the two legs of the bridge are balanced. In other words, V_o is zero when R1/R2 = R3/R4. It is not necessary for the resistor elements to be equal (although that is usually the case), only that their ratios be equal. Thus, a 500k/50k ratio for R1/R2 will produce zero output voltage when R3/R4 = 100k/10k.

The photoresistors are arranged so that light from a calibrated source shines on both equally and fully, except when an intervening filter or sample is present. Thus, the bridge can be nulled to zero using potentiometer R5 under this zero condition. In most instruments a translucent sample is placed between the light source and one of the photocells. The amount of transmission to light allowed by the sample is a measure of its density, and is thus transducible. Let's look at a couple of different types of instrument to see how this principle is applied.

Blood O_2 Level. An almost classical (but still used) method for measuring blood oxygen level is the colorimeter of Fig. 13-1. It works on the fact that the redness of blood is a measure of its oxygenation. This instrument is nulled with neither standard filter cell nor blood in the light path. A

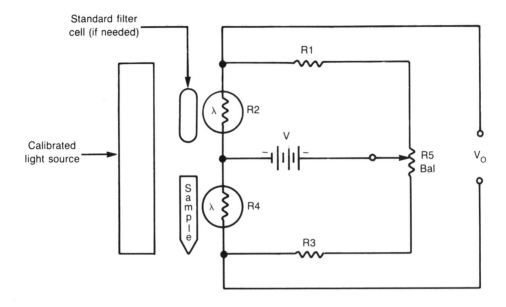

$$V_O = V \left[\frac{R2}{R1 + R2} - \frac{R4}{R3 + R4} \right]$$

Fig. 13-1. Basic colorimeter.

standard filter cell is introduced between the light source and R2, and a blood sample is placed in a standardized tube between the light source and R4. The degree of blood saturation in the sample is thus reflected by the difference in the bridge reading. On one model a separate resistor across the R1/R2 arm is used to bring the bridge back into the null condition, and the dial for that resistor is calibrated in percent of O_2. More modern instruments based on digital computer techniques provide the measurement in a more automatic manner.

Respiratory CO_2 Level. The exhaled air from humans is roughly 2 to 5 percent carbon dioxide, while the percentage of CO_2 in room air is negligible. A popular form of end CO_2 meter is based on the fact that CO_2 absorbs infrared (IR) waves. The "light source" in this type of instrument is actually either an IR LED or a Cal Rod (identical to the one that heats your coffee pot!); the photocells are selected for IR response. In this type of instrument, room air is passed through a cuvette placed between R2 and the heat source,

while patient expiratory air is passed through the same type of cuvette that is placed between the heat source and R4. The difference in IR transmission is a function of the percentage of CO_2 in the sample circuit.

The associated electronics (not shown) will allow zero and maximum span (i.e., gain) adjustment. The zero point is adjusted with room air in both cuvettes, while the maximum scale (usually 5 percent CO_2) is adjusted with the sample cuvette purged of room air and replaced with a calibration gas (usually 5 percent CO_2, 95 percent nitrogen). This calibration gas must be obtained from a local supplier and be specified as a calibration gas. Otherwise, the quantities are only approximate. Also, be sure of the type of measurement: calibration gases are available by either weight or volume.

Blood Electrolytes. Blood chemistry includes levels of sodium (Na) and potassium (K). An instrument commonly used for these forms of measurement is the *flame photometer* (Fig. 13-2). This form of colorimeter replaces the light source with a flame produced in a gas carburetor. The sample is injected into the carburetor, and burned along with the gas/air mixture. The colors emitted on burning are proportional to the concentrations of Na and K ions in the sample. A special gas is used to burn cleanly with a blue flame when no sample is present. In medical applications a specified size sample of patient's blood is mixed with an indium calibrating solution (also a predetermined amount). The solution is well-mixed, and then applied to the carburetor. By comparing the intensities of the colors generated by burning Na and K ions with the intensity of the calibration color, the instrument can infer the concentration of the two elements.

A warning is in order regarding these flame photometers. During the time that I was in the biomedical engineering profession I saw a number of flame photometers give terribly flawed readings. In all cases it was not the fault of the instrument, but rather the operator. The carburetor and surrounding glass structures must be cleaned frequently, or the buildup of material from past tests will unduly bias the results of the present test. In addition, pure carbonization of the associated glass windows will obscure the flame and may create both a lack of sensitivity and an erroneous reading (especially if both windows are not carboned the same amount).

The uses of colorimeters do not end with

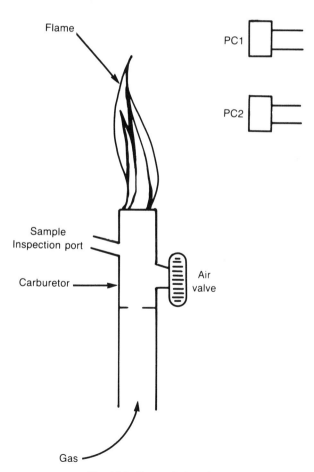

Fig. 13-2. Flame photometer.

the medical laboratory. The clue to looking for a transducible event is detection of either a density change or an absorption differential to one or two colors that is a function of the parameter being measured; for example, IR is absorbed by CO_2, and a certain O_2 saturation passes light at 800 nM wavelength.

In the past we had to rely only on those phenomena that were either linear or changed in an easily discerned manner (e.g., logrithmically or exponentially). Today, however, the modern programmable digital microcomputer can be trained to unravel quite nonlinear phenomena and present us with intelligible information. We can either program the computer to solve the complicated mathematical equation that describes the phenomena, or create a look-up table from purely empiricle data. The latter method is the one usually used. Alternatively, with certain types of curve, we can

create a reverse or inverse curve to fit the data, and output that to the human operator.

PHOTOSPECTROMETER

This class of instruments depends upon the fact that certain chemical compositions absorb wavelengths of light to different degrees. If you are familiar with the spectrum of sunlight created when the light passes through a prism or is reflected off of a diffraction grating, then you are familiar with the basis for the spectrophotometer (see Fig. 13-3). You must move either the light beam or the photodetector to examine each of the light wavelengths in turn. In most cases the sample and the photodetector are kept stable while a diffraction grating or prism is rotated to permit all colors of the spectrum to fall on the same photo detector. By comparing the amplitudes of the light at different angles, you can infer the transmission

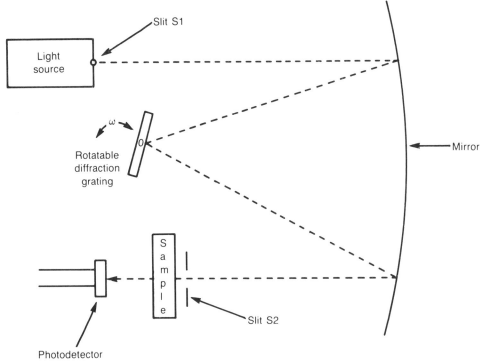

Fig. 13-3. Photospectometry.

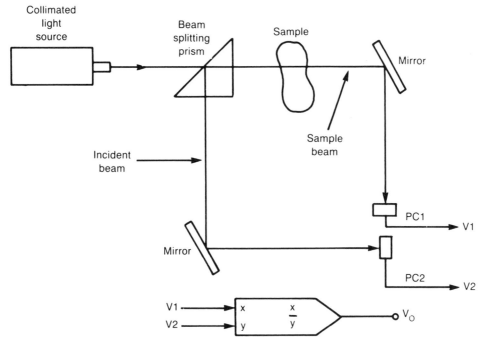

Fig. 13-4. Ratiometric measurements

of the respective colors and create a chart that shows amplitude vs wavelength.

Ratiometric Measurements

A significant problem with photometric measurements is keeping the light source constant in both intensity and color. The simple fact is that all types of light emitters change, and that change introduces artifact in the acquired data — if it occurs between calibration and measurement. The answer to many of these problems is the use of ratiometric measurement technique (Fig. 13-4).

In a *ratiometric system* the collimated light beam is passed through a 50-percent beam-splitting prism. This optical device splits the beam into two equal amplitude beams at right angles to each other. The incident beam is passed through an optical path of length L to photosensor PC2; the sample beam also passes through an optical path of length L, but passes through the sample being measured as well. If the optical properties of the two paths are the same and they are of the same length, then the light intensities arriving at the two colorimeter photosensors will be the same, except for any lost in the sample. Thus, the Wheatstone concept can be used. But first you must do a little signal processing on the signal. You must take the ratio of V_1 and V_2, the outputs of PC1 and PC2, respectively. Either in the computer, or in an analog multiplier, you must take the ratio V_1/V_2. Thus, a change in the light level affects both photosensors equally, so the only difference between the two is the properties of the sample.

Chapter 14

Radiations and X-Ray Signals

X-RAYS ARE SIMILAR TO LIGHT WAVES IN THAT they are electromagnetic waves with wavelengths shorter than either visible light or ultraviolet radiation. X-rays are long familiar to most readers because they find widespread use in medicine, science, industry, and at airport security checkpoints. X-rays are usually generated by a phenomenon called *bremstrahlung* (see Fig. 14-1). When an incident electron with energy E_i is smashed into a target containing heavy nucleii, then a strange thing happens. As the electron is deflected around the nucleus, it loses some energy and assumes a new energy level, E_d. The difference between incident and deflected electron energy levels must, according to the Law of Conservation of Energy, go somewhere, so it becomes a photon of X-ray energy.

Figure 14-2 shows a basic X-ray generator tube. It is a vacuum tube containing an electron-emitting cathode (a heated thermionic filament) and a target anode. The materials of the target anode are selected to make X-ray generation easier. Besides the type of materials, the applied high-voltage potential (V+) determines the kinetic energy of the accelerated electrons, and hence the frequency and wavelength of the emitted X-rays.

In a few very old TV sets, there was once a scare of X-ray emission caused by a new type of high-voltage regulator tube that operated at higher than usual potentials. Medical, scientific, and industrial *bremstrahlung* generators are somewhat more tightly designed than TV tubes, however.

A variation on the basic tube uses a rotating anode. The heat created by the kinetic energy of electrons smashing into the target is tremendous, and overheating is a major cause of lost X-ray tubes. The rotating anode spreads the heat energy over a larger volume of metal, and incidentally produces a more narrowly focused beam.

Many of the sensors used in light-operated devices will also work at X-ray wavelengths. Certain phototubes, photomultiplier, and photodiodes (and transistors) will also work in X-ray measurements (see Chapter 10). In this chapter you will look at two sensors that are unique to radiation measurements: the Geiger-Mueller tube and the scintillation cell.

GEIGER-MUELLER TUBES

The Geiger-Mueller tube (Fig. 14-3) is a glass or metallic cylinder that has been evacuated of air and then refilled to a less-than-atmospheric pres-

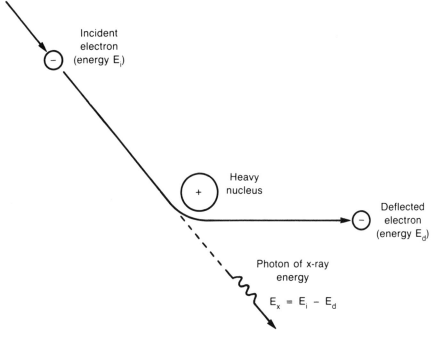

Fig. 14-1. Bremstrahlung phenomenon.

sure with an ionizing gas (usually argon with a touch of bromine, at 100-Torr pressure). When radiation impinges on such a gas, its energy forces the gas into ionization, thereby altering the electrical characteristics of the G-M tube. These electrical characteristics are our transducible event.

The general circuit for the G-M tube is shown in Fig. 14-3. In most applications the external circuitry consists of a power supply and a series current-limiting or load resistor. There are three modes of operation for the G-M tube, and these modes affect both the subsequent circuitry and the

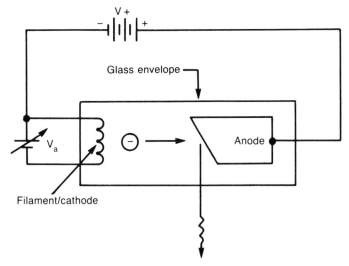

Fig. 14-2. Simple X-ray generator tube.

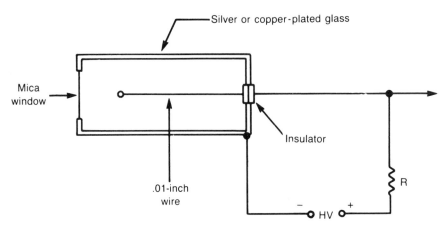

Fig. 14-3. Geiger-Mueller tube.

voltage level required. These regions are shown in the curves in Fig. 14-4 and are designated as follows: A) ionization chamber mode, B) proportional counter mode, and C) Geiger counter mode; these modes occur at points A, B, and C in Fig. 14-4.

In the ionization chamber mode, a weak electric potential is applied across the G-M tube, so only a few electrons are generated by the radiation. Nearly all of these electrons are collected by the electode. Therefore, you can assume that the cur-

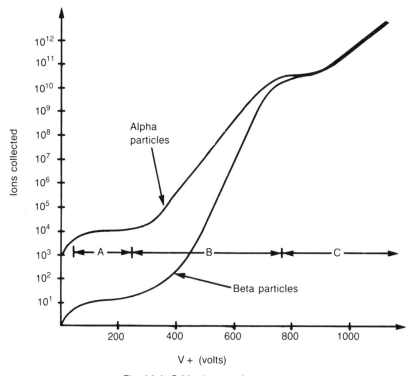

Fig. 14-4. G-M tube transfer curves.

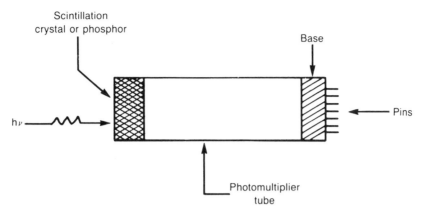

Fig. 14-5. Scintillation counter.

rent in the external load resistor is a measure of the amount of radiation.

In the proportional counter mode, the electric field is stronger, so the electrons generated by the ionizing gas reach sufficient kinetic energy to release additional electrons by kinetic collisions with gas molecules. The output across the load resistor will be a spike with an amplitude that is proportional to the kinetic energy of the ionizing radiation particle. In this mode the G-M tube can count ionizing particles on a one-for-one basis.

The Geiger-counter mode uses very high potentials, on the order of 800 to 2000 volts. The operation is similar to that in the proportional counter mode, except that the kinetic energies are so high that a multiplication effect takes place; the tube operates in avalanche condition. The output pulses are of approximately the same amplitude ("all or nothing" operation) every time the tube fires. An external counter circuit will tell you the number of pulses per unit of time and hence the level of radiation.

SCINTILLATION COUNTERS

An example of a scintillation cell or counter is shown in Fig. 14-5. The word *scintillation* is used to denote a process similar to that which generates light on the screen of a cathode-ray tube. When a radiation particle strikes an atom of certain phosphorous materials, its kinetic energy may be added to the energy of the orbital electrons. When the electrons are thus excited, they jump to a higher energy state, which is unstable. When they fall back to their ground state, the energy absorbed goes off in the form of a photon of light. Certain crystal materials and minerals possess this property.

Figure 14-5 shows a scintillation device in which a scintillation crystal window is attached to the light input window of a photomultiplier tube. Radiation causes the crystal to scintillate, and the light thus produced is picked up and amplified by the P-M tube. The current at the output of the P-M tube is proportional to the light, and hence the radiation level.

Chapter 15

Basics of Operational Amplifiers

I N THIS CHAPTER I DISCUSS THE BASICS OF OPERA-
tional amplifiers. The role of these devices in
data acquisition is so great that an elementary review is in order. Topics to be considered include
operational amplifiers plus certain other forms of
amplifier that are based on integrated-circuit operational amplifiers.

OPERATIONAL AMPLIFIERS

One of the early texts on operational amplifiers tells us that the IC operational amplifier has
made ". . . the contriving of contrivances a game
for all." There was a physics professor who was
struggling with the design of a transistor amplifier
for his undergraduate students to use in a
classroom experiment. It seemed very difficult to
build a circuit that 1) had a gain of precisely 100
without the need for adjustment, and 2) was repli-
cable for the thirty or so amplifiers required for the
class (using "on-hand" components of mixed and
doubtful ancestry). After he learned about the 741
IC operational amplifier his problems were all but
over! The little 741, although today not even considered for many jobs because of the newer types
now on the market, then cost just over a dollar (yet
obeyed very simple design equations). You can de-

sign op-amp voltage amplifiers by using only the
ratio of two resistors.

The operational amplifier was originally designed to perform mathematical operations in analog computers (hence the name "operational" amplifiers). The first op amps were vacuum-tube
models and only approximated the behavior of the
ideal mathematical model of the device. Later,
transistor amplifiers became available, and finally
integrated-circuit types came on the market. One
of the first linear ICs made commercially was the
μA-709 operational amplifier. Although now considered primitive, this device sold for $109 in one
mid-sixties catalog (the current price is about five
for a dollar — where they are still available).

Figure 15-1A shows the usual circuit symbol
of the operational amplifier. There is an alternate
symbol shown in Fig. 15-1B that is preferred by
some people. This symbol is technically the correct
symbol to use. That version uses a curved back to
which the input leads are attached. The version
shown in Fig. 15-1A is used almost universally,
however, even though it technically denotes any
amplifier — including but not limited to operational
amplifiers. I will maintain the *de facto* standard of
Fig. 15-1A.

Note the pin-outs for the amplifiers in Fig. 15-1. The pin numbers given are for the 741 device and have become something of an industry standard. There are two input connections, two power-supply connections, and one output connection. There is no ground or common connection. The ground and signal common are taken from the power supply common line (more on this later).

The two power supply connections are V+ and V−. The V+ supply is positive with respect to common, while the V− is negative with respect to common. The range for these voltages is typically ±4 volts to ±18 volts, although a number of examples exist with wider (or slightly different) voltage ranges. An RCA CA-3140 BiMOS device, for example, operates at potentials up to ±22 volts for V− and V+, while a certain "low-power" operational amplifiers operate down to ±1.5 volts dc.

In addition to the absolute voltage limits, there are also sometimes relative limitations. For example, the 741 device has a 30-volt limit for the voltage defined by the expression [(V+) − (V−)], even though each V− and V+ can be as high as 18 volts. Thus, if V+ is +18 volts, then V− must be not greater than −12 volts in order that the differential not be greater than 30 volts [(+18) − (−12) = +30 volts].

The selection of power-supply voltages might also depend somewhat on the maximum anticipated output voltage. If the amplifier is being designed

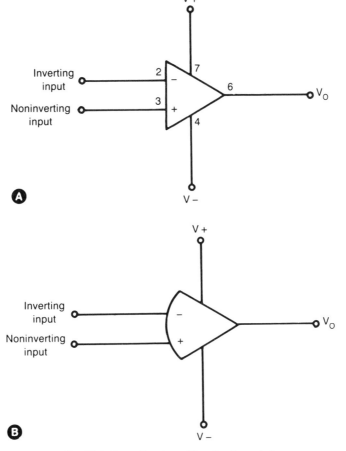

Fig. 15-1. Operational-amplifier circuit symbols.

for use with an analog-to-digital converter that has a range of -10 to $+10$ volts input, then you would certainly want the output of the amplifier to be capable of achieving these limits. But there is a limit on how high the output voltage can go, and that limit is a function of the power-supply voltage. In general, the limitation is based on the number of pn junctions between the output terminal on the IC and the two power-supply terminals. Each pn junction has a 0.7-volt drop that must be accounted for. If there are four pn junctions between the output terminal and the V+ power-supply terminal, for example, the maximum allowable output voltage will be $[(V+) - (4 \times 0.7)]$ volts, or 2.8 volts lower than V+. Thus, when you want the output terminal to swing to +10 volts, the absolute minimum V+ power supply voltage will be $10 + 2.8$ volts, or +12.8 volts dc. Obviously, a +12-volt dc power supply will not work in this case. In general, ordinary bipolar-transistor operational amplifiers require a supply voltage of 2 to 4.5 volts higher than the maximum required output voltage, yet must remain within the V+ and V− constraints of the device. Some BiMOS and BiFET devices are available in which the maximum output signal voltage can be as close as 0.5 volts to the dc power supply potential.

The two inputs for the operational amplifier form a so-called "differential pair" because they operate 180 degrees out of phase with each other. The inverting input produces a 180-degree phase shift between input signal and output signal (in other words, a positive-going input signal produces a negative-going output signal, and vice versa). The noninverting input produces a 0 degree phase shift in the output signal. Since one input produces an in-phase output, and the other produces an out-of-phase output, application of the same voltage to both inputs simultaneously produces a zero net output potential. I will use this fact in a later section to create the immensely useful differential amplifier. The two inputs on the operational amplifier have a very high impedance (which is infinite in the ideal model). Thus, they are an essentially perfect voltage amplifier input.

The output of the operational amplifier is also suited for use as a voltage-amplifier circuit. The output impedance of the typical op amp is usually quite low (50 to 200 ohms), so forms a nearly perfect voltage source.

OPERATIONAL-AMPLIFIER DC POWER SUPPLIES

Figure 15-2 shows a model of the typical operational-amplifier power supply. Although batteries are shown here for sake of simplicity, electronic power supplies operated from the ac power lines are commonly used. Recall that there are two different voltages in the op-amp power supply, V+ and V−. Voltage V+ is supplied by battery B1, while V− is supplied by battery B2. The common (or ground) connection is the junction between the

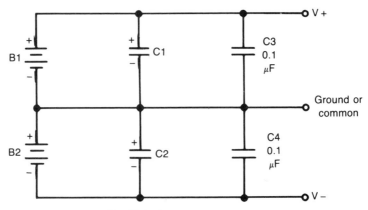

Fig. 15-2. Op-amp power supplies.

two batteries. Normally, B1 and B2 will have the same voltage rating, although that is not strictly a requirement unless other circuit considerations apply.

The capacitors shown in Fig. 15-2 are used for decoupling, especially when multiple stages are fed from the same power supply. Capacitors C1 and C2 are normally 50 to 200 μF and are used for decoupling low-frequency signals. Capacitors C3 and C4 are used for decoupling higher frequency signals. You cannot normally use the higher value C1/C2 for high-frequency signals because these are normally electrolytic capacitors, which are ineffective at high frequencies.

The signal common or "ground" connection is used as the zero reference for input and output signals on the operational amplifier. Whether it is actually grounded or not depends upon circuit design considerations. In most cases, however, it is grounded for the sake of simplicity.

In most applications electronic power supplies used for B1 and B2 must be voltage regulated. Although there are certainly numerous applications where voltage-regulated dc power supplies are not strictly required, they are almost always "good engineering practice."

THE IDEAL OPERATIONAL AMPLIFIER

Before getting further into operational-amplifier circuits, let's set the stage for our simplistic circuit analysis by discussing the properties of the so-called *ideal operational amplifier*. This ideal device has the following properties:

1. Infinite open-loop gain
2. Zero output impedance
3. Infinite input impedance
4. Zero noise contribution
5. Infinite bandwidth
6. Differential inputs "stick together"

What do these statements mean, and how do they comport with real, practical IC operational amplifiers?

Infinite Open-Loop Gain. This property means that the voltage gain of the ideal operational amplifier in the open-loop (i.e., no feedback) configuration is infinite. Real operational amplifiers do not even approach the ideal but are still good enough approximations to make the device behave properly. The ability of practical operational amplifiers to approach the ideal is dependent upon having an extremely high open-loop gain (otherwise, the equations behave poorly). In practical devices you find that the open-loop voltage gain, A_{VOL}, will be 20,000 in low-cost devices, and well over 1,000,000 in premium devices.

Zero Output Impedance. The operational amplifier is supposed to be a perfect voltage amplifier and so it requires a zero output impedance. Real devices have output impedances of 50 to 200 ohms, with most being under 100 ohms.

Infinite Input Impedance. This parameter means that the input will neither sink nor source electrical current. Recall that input impedance is $Z_{IN} = V_{IN}/I_{IN}$, so for the input impedance to be infinite, I_{IN} must be zero. In real operational amplifiers, of course, you find this value is non-zero and is one of the primary differences between premium and low-cost devices. Low-cost amplifiers have input (transistor) bias currents to contend with of up to 1 or 2 milliamperes, but certain other devices measure the input current in picoamperes or nanoamperes. The RCA BiMOS operational amplifiers (CA-3140, etc.) use MOSFET input transistors to produce an input impedance of more than 10^{12} ohms.

Zero Noise Contribution. The noise referred to is the internal noise of the amplifier added to the signal. This is one primary difference between the low-cost and premium devices. The low-cost amplifiers add considerable "hiss" noise and so are unusable on low-signal applications.

Infinite Bandwidth. This parameter means that there is no limit to the operating frequency of the device, which in real operational amplifiers is obviously unrealistic. Unconditionally stable, frequency-compensated devices like the 741 may have an upper frequency limit of only a few kilohertz, while other operational amplifiers operate to several megahertz. Only a few high HF or low VHF devices are available, and they are usually labeled "video operational amplifiers," or something similar.

Differential Inputs Stick Together. This property is essential to the simplified circuit analysis used here. The importance of this property is that a voltage applied to one input will also appear on the other input. Both inputs must be treated mathematically alike in this regard. Thus, if you apply a voltage to the noninverting input, then you must treat the inverting input as if it also sees that voltage. This is not just some theoretical trick used to make equations work. If you apply a real voltage to a real noninverting input, and then use a real voltmeter at the inverting input, you will measure a real voltage at that point.

These properties will be important in the circuit descriptions that follow.

STANDARD OPERATIONAL-AMPLIFIER PARAMETERS

Designing circuits with operational amplifiers requires understanding the various parameters given in the specification sheets. The list below is not intended to be exhaustive, but it represents the most commonly needed parameters.

Open-Loop Voltage Gain (A_{VOL}). Voltage gain is defined as the ratio of output voltage to input signal voltages (V_O/V_{IN}), which is a dimensionless quantity. The open-loop voltage gain is the gain of the circuit without feedback (i.e., with the feedback loop open). In an ideal operational amplifier A_{VOL} is infinite, but in practical devices it will range from about 20,000 for low-cost devices to over 1,000,000 in premium devices.

Large-Signal Voltage Gain. This gain figure is defined as the ratio of the maximum allowable output voltage swing (usually several volts less

than V− and V+) to the input signal required to produce a swing of ±10 volts (or some other standard).

Slew Rate. This parameter specifies the ability of the amplifier to change from one output voltage extreme to the other extreme while delivering full, rated output current to the external load. The slew rate is measured in terms of voltage change per unit of time. The 741 operational amplifier, for example, is rated at 0.5 volts per microsecond (0.5 V/μS). Slew rate is usually measured in the unity-gain noninverting follower configuration (more on this later).

Common-Mode Rejection Ratio (CMRR). A common-mode voltage is one that is presented simultaneously to both inverting and noninverting inputs (voltage V3 in Fig. 15-3). In an ideal operational amplifier, the output resulting from the common-mode voltage is zero, but in real devices it is non-zero.

The common-mode rejection ratio (CMRR) is the measure of the device's ability to reject common-mode signals and is expressed as the ratio of the differential gain to the common mode gain. The CMRR is usually expressed in decibels, with common devices having ratings between 60 dB and 120 dB (the higher the number, the better the device).

Power-Supply Rejection Ratio (PSRR). Also called *power-supply sensitivity,* the PSRR is a measure of the operational amplifier's insensitivity to changes in the power-supply potentials. The PSRR is defined as the change of the input offset voltage (see below) for a 1-volt change in one power-supply potential (while the other is held constant). Typical values are in microvolts or millivolts per volt of power-supply potential change.

Input Offset Voltage. The voltage required at the input to force the output voltage to zero

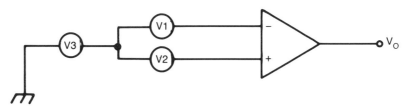

Fig. 15-3. Common mode and differential-mode signals.

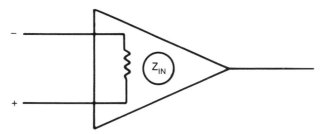

Fig. 15-4. Input equivalent circuit.

when the signal voltage is zero. The output voltage of an ideal operational amplifier is zero when V_{IN} is zero.

Input Bias Current. This current is the current flowing into or out of the operational-amplifier inputs. In some sources this current is defined as the average difference between currents flowing in the inverting and noninverting inputs.

Input Offset (Bias) Current. The difference between inverting and noninverting input bias current when the output voltage is held at zero.

Input Signal Voltage Range. The range of permissible input voltages as measured in the common mode configuration, i.e., the maximum allowable value of V3 in Fig. 15-3.

Input Impedance. The resistance between the inverting and noninverting inputs (Z_{IN} in Fig. 15-4). This value is typically very high: 1 megohm in low-cost bipolar operational amplifiers and over 10^{12} ohms in premium BiMOS devices.

Output Impedance. This parameter refers to the "resistance looking back" (to borrow from Thevenin's theorem) into the amplifier's output terminal, and is usually modeled as a resistance in series with the output signal and ground (despite the lack of a ground terminal on the device). Typically the output impedance is less than 100 ohms.

Output Short-Circuit Current. The current that will flow in the output terminal when the output load resistance external to the amplifier is zero ohms (i.e., a short to ground).

Channel Separation. This parameter is used on multiple-operational-amplifier integrated circuits, i.e., two or more operational amplifiers sharing the same package with common power-supply terminals. The separation specification tells us something of the isolation between the op amps inside the same package and is measured in decibels (dB). The 747 dual operational amplifier, for example, offers 120 dB of channel separation. From this specification you may infer that a 1 microvolt change will occur in the output of one of the amplifiers when the other amplifier output changes by 1 volt (1 V/1μV = 120 dB).

MINIMUM AND MAXIMUM PARAMETER RATINGS

Operational amplifiers, like all electronic components, are subject to certain maximum ratings. If these ratings are exceeded, then the user can expect premature — often immediate — failure, or unpredictable operation. The ratings mentioned below are the most commonly used.

Maximum Supply Voltage. This potential is the maximum that can be applied to the operational amplifier without damaging the device. The operational amplifier uses V+ and V− dc power supplies that are typically −/+ 15 volts dc, although some exist with much higher maximum potentials. The maximum rating for either V− or V+ often depends upon the value of the other (see below).

Maximum Differential Supply Voltage. This potential is the algebraic sum of V− and V+, namely [(V+) − (V−)]. It is often the case that this rating is not the same as the sum of the maximum supply-voltage ratings. For example, one 741 operational amplifier specification sheet lists V− and V+ at 15 volts each, but the maximum differential supply voltage is only 28 volts. Thus, when both V− and V+ are at maximum (i.e., 15 volts dc each), the actual differential supply voltage is [(+15 V) − (−15 V)] = 30, which is 2 volts over the maximum rating. Therefore, when either V− or V+ is at maximum value, the other must be proportionally lower. For example, when V+ is +15 volts, the maximum allowable value of V− is [28 V − 15 V] = 13 volts.

Power Dissipation, P_d. This rating is the maximum power dissipation of the operational amplifier in the normal ambient temperature range (80 degrees Celsius in commercial devices, and 125 degrees Celsius in military-grade devices). A typical rating is 500 milliwatts (0.5 watts).

Maximum Power Consumption. The maximum power required, usually under output short-circuit conditions, that the device will survive. This rating includes both internal power dissipation and device power requirements.

Maximum Input Voltage. This potential is the maximum that can be applied simultaneously to both inputs. Thus, it is also the maximum common-mode voltage. In most bipolar operational amplifiers, the maximum input voltage is equal to the power-supply voltage, or nearly so. There is also a maximum input voltage that can be applied to either input when the other input is grounded.

Differential Input Voltage. This input voltage rating is the maximum differential-mode voltage that can be applied across the inverting (−) and noninverting (+) inputs.

Maximum Operating Temperature. The maximum temperature is the highest ambient temperature at which the device will operate according to specifications with reasonable reliability. The usual rating for commercial devices is 70 or 80 degrees Celsius, while military components must operate to 125 degrees Celsius.

Minimum Operating Temperature. There is a minimum operating temperature, i.e., the lowest temperature at which the device operates within specifications. Commercial devices operate down to either 0 or −10 degrees Celsius, while military components operate down to −55 degrees Celsius.

Output Short-Circuit Duration. The length of time the operational amplifier will safely sustain a short circuit of the output terminal. Many modern operational amplifiers are rated for indefinite-output short-circuit duration.

Maximum Output Voltage. The output potential of the operational amplifier is related to the dc power-supply voltages. Most operational amplifiers have several bipolar pn junctions between the output terminal and either V− or V+ terminals, and the voltage drop across these junctions reduces the maximum output voltage. For example, if there are three pn junctions between the output and power-supply terminals, then the maximum output voltage is [(V+) − (3 × 0.7)], or [(V+) − 2.1] volts. If the maximum V+ voltage permitted is 15 volts, then the maximum allowable output voltage is [(15 V) − (2.1 V)], or 12.9 volts. It is not always true that the maximum negative output voltage is equal to the maximum positive output voltage. A related rating is the maximum output voltage swing, which is the absolute value of the voltage swing from maximum negative to maximum positive.

INVERTING-FOLLOWER CIRCUITS

Figure 15-5 shows the inverting-follower circuit. In this circuit the noninverting input is grounded, so you must treat the inverting input as if it were also grounded (recall ideal property number 6). This fact gives rise to a somewhat confusing concept known as *virtual ground*. The inverting input is not actually grounded, but since it is at zero potential because the other input is grounded, we say that it is "virtually grounded." The concept is simple, only the words are confusing.

Let's consider the currents appearing in node "A" of Fig. 15-5. You know from property no. 3 that I3, the input bias current, is zero. You also know from Kirchoff's Current Law (KCL) that all currents into and out of a junction algebraically sum to zero. Thus,

$$I_1 = -I_2$$

You also know from Ohm's law that

$$I_1 = V_{IN}/R_f$$

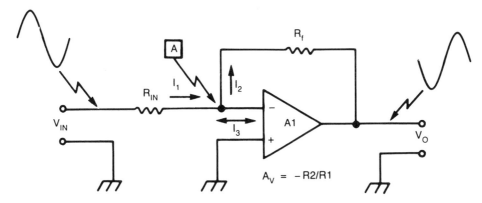

Fig. 15-5. Inverting-follower circuit.

and,

$$I_2 = V_O/R_{IN}$$

Thus, when you substitute these two equations into KCL above:

$$(V_{IN}/R_{IN} = -(V_O/R_f)$$

You know that a voltage amplifier's transfer function is V_O/V_{IN}, so solving the above equation for the transfer function yields:

$$V_O/V_{IN} = -R_f/R_{IN}$$

Thus, the voltage gain A_V of the inverting follower is given by the ratio of two resistors:

$$A_V = -R_f/R_{IN}$$

We may, then, design the inverting amplifier simply by manipulating the values of these two resistors (see Fig. 15-5). There is sometimes a constraint on the minimum allowable value of R_{IN}. Point "A" is essentially grounded, so the input impedance of this circuit is simply the resistance of R_{IN}. There is a rule of thumb in voltage-amplifier design that says the input impedance of a stage must be five times (some prefer ten times) the source impedance of the signal source.

Example. Design a gain-of-100 inverting amplifier that has an input impedance of 10 kilohms or more.

Solution. Because the input impedance must be 10 kilohms, R_{IN} must be 10 kilohms or more. Let's set it at 10 kilohms. Solving the gain equation for R_f yields:

$$R_f = A_V \times R_{IN}$$
$$R_f = (100) \times (10,000 \text{ ohms})$$
$$R_f = 1,000,000 \text{ ohms}$$

Thus, a 10-kilohm input resistor and a 1-megohm feedback resistor yields a gain of 100. Because this is an inverting follower, the gain is actually -100 (note: the "$-$" sign indicates a 180-degree phase reversal).

NONINVERTING FOLLOWERS

The noninverting follower applies a signal to the noninverting input. There are two basic configurations for the noninverting follower:

1. Unity-gain noninverting follower
2. Noninverting follower with gain

Figure 15-6A shows the unity-gain noninverting follower. The output terminal is connected to the inverting input, producing 100 percent feedback. The output voltage is equal to the input voltage. So what use is a unity gain (i.e., gain of 1) voltage amplifier? There are three uses: buffering, impedance transformation, and power amplification. *Buffering* means using an amplifier to isolate a circuit from its load. Some transducers and most

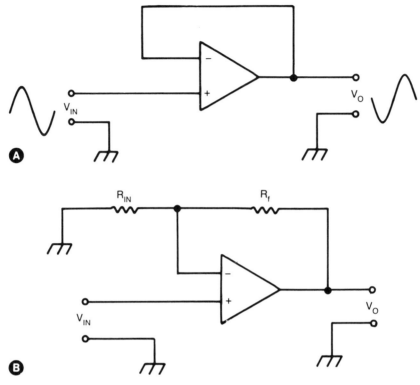

Fig. 15-6. Noninverting followers A) unity gain, B) gain.

oscillators or astable multivibrators will change frequency if the load impedance changes, so a unity-gain noninverting follower will help "buffer" the circuit.

Impedance transformation occurs because the input impedance is very high, while the output impedance is very low. You can use this circuit, for example, when acquiring a signal from biological or chemical sources where the impedance is high. A pH electrode, for example, has source impedances ranging from 10 megohms to 100 megohms.

The power amplification is small, but the voltage remains constant ($V_O = V_{IN}$) even though the impedances are unequal. Obviously, because power is defined by $P = V^2/R$, reducing R while keeping V the same results in higher power (P).

The noninverting follower with gain circuit of Fig. 15-6B retains the properties of the unity-gain circuit but produces voltage gain as well. Keeping in mind that the inverting input (point "A") is at

V_{IN}, analysis similar to the previously used method produces a voltage gain of:

$$A_V = (R_f/R_{IN}) + 1$$

The noninverting follower circuits are used wherever extremely high input impedance is needed, or where no phase reversal can be tolerated.

OPERATION FROM A SINGLE POWER SUPPLY

Operational amplifiers are normally powered from a bipolar dc power supply. Such a power supply has V+ and V− voltages that are each referenced to ground or common. This system essentially requires two semi-independent dc supplies. In some cases, either ultimate use or other design constraints force the use of a single, monopolar dc power supply. In this section I discuss simple methods for operating the amplifier from a single dc power supply.

Some schemes exist for creating a split power supply from a monopolar supply in order to mimic bipolar power supply operation. One such scheme connects two zener diodes in series across the single supply, along with the necessary current-limiting resistors. The junction between the two zener diodes becomes the signal common. A limitation of this method is that the dc supply cannot be chassis referenced.

Another scheme is to use the regular monopolar dc power supply for V+, and then use a dc-to-dc converter circuit for V−. Such a circuit is little more than an ac oscillator in the 20 to 500-kHz range, with its output signal rectified and filtered to produce the V− voltage.

Figure 15-7 shows the method for biasing the operational amplifier inputs to permit single supply operation. This technique is based on the simple resistor voltage-divider circuit of Fig. 15-7A. The output voltage (V_1) is given by the standard voltage-divider equation:

$$V_1 = \frac{R2 \times (V+)}{R1 + R2}$$

In most cases the value of V_1 will be one-half V+, so that the operational amplifier has a quiescent output point that is midway between extremes. This bias level is achieved by making R1 = R2. The value of R1 and R2 is usually selected such that it falls between 1k and 100k.

The capacitor-shunting resistor, R2, is used to decouple ac variations. The value of capacitance is selected for a reactance value of one-tenth R2 (i.e., R2/10) at the lowest frequency of operation. For example, suppose R2 = 10k, and the lowest frequency of operation is 10 Hz. If R2 is 10k, then the capacitive reactance of the shunt capacitor should be: R2/10 = (10k)/10 = 1k. Solving the usual capacitive reactance equation for C give us:

$$C_{\mu F} = \frac{10^6}{2f \; X_C}$$

$$C_{\mu F} = \frac{1,000,000}{(2)(3.14)(10 \text{ Hz})(1000 \text{ ohms})}$$

$$C_{\mu F} = \frac{1,000,000}{62,800}$$

$$C_{\mu F} \quad 15.9 \; \mu F$$

The value, 15.9 μF, is not standard, so a 20 or 22 μF unit would be selected.

Figure 15-7B shows the method for biasing an operational amplifier in the inverting-follower configuration. In the bipolar supply version of this circuit (Fig. 15-5), the noninverting input is grounded (i.e., set to zero volts). But in single supply operation you apply bias voltage V1 to the noninverting input. This voltage (V_1) will also appear on the inverting input (hence the need for dc-blocking capacitor C1). The output terminal will be biased up according to the value of V_1, so it may require a dc-blocking capacitor also (shown in Fig. 15-7C) if such a voltage adversely affects the following stage.

The value of capacitor C1 is selected to have a low impedance at the lowest frequency of operation, using a protocol similar to that discussed above for the voltage-divider shunt capacitor. A general rule of thumb is to regard $R_{IN}C1$ as a high-pass filter with a cut-off frequency, f_C, equal to $1/(2 \pi R_{IN}C1)$. The object is to select a value of C1, given a value for R_{IN}, that results in a value of F_C lower than the lowest operating frequency.

The circuit configuration for noninverting follower circuits is shown in Fig. 15-7C. This circuit is the same as for inverting followers except for resistor R3. The purpose of R3 is to maintain a high input impedance to signals applied to the noninverting input. The minimum value of R3 is at least ten times the output resistance of the driving stage. In practical cases, however, the source impedance is usually low enough that it is possible to set R3 to 100 or 1000 times the source impedance. Typical values range from 10k to 1 megohm, with 100k predominating.

The value of C1 is selected so that the cut-off

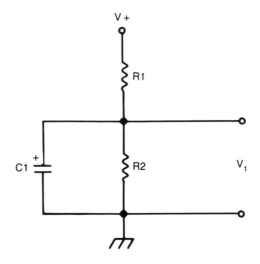

$$\text{①} \quad V_1 = \frac{R2 \times (V+)}{R1 + R2}$$

② $X_{C1} \leq R2/10$ at lowest frequency of operation.

③ $C1 \geq C\ \mu F \geq \dfrac{10^6}{2\ \pi F\ X_C}$

A

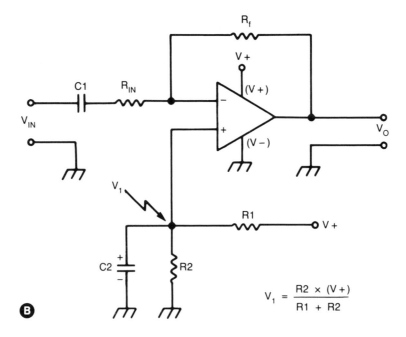

$$V_1 = \frac{R2 \times (V+)}{R1 + R2}$$

B

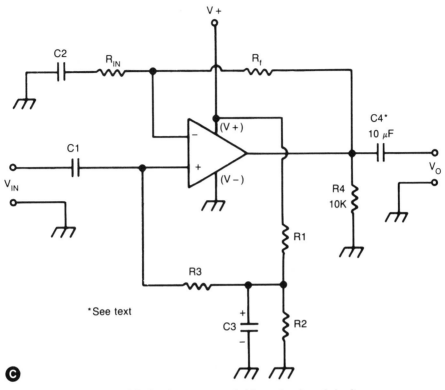

Fig. 15-7. Single power supply bias network and circuits.

frequency of the filter formed by C1R3 is lower than the lowest operating frequency. The same equation applies here as above.

Capacitor C4 and resistor R4 are used when the (V+)/2 bias on the output terminal will adversely affect a following stage or instrument. Again, the "lowest frequency of operation" rule is invoked when setting the value of C1, with the resistance being the input resistance of the stage following.

OPERATIONAL AMPLIFIER PROBLEMS AND THEIR SOLUTIONS

Earlier in this chapter I discussed the concept of the ideal operational amplifier. While such a hypothetical device is a useful learning tool and makes our analysis easier, it doesn't exist and cannot be purchased and used in practical circuits. All real operational amplifiers depart somewhat from the ideal, and the quality of the device is sometimes measured by the degree of that departure. We find, for example, that open-loop gain is not really infinite, but rather ranges from 20,000 to over 1,000,000. Similarly, real operational amplifiers don't have infinite bandwidth, and in fact some are heavily bandwidth limited. This is especially true of "unconditionally stable" or "frequency compensated" operational amplifiers such as the 741 device. While the stability is highly desirable, it is obtained at the expense of frequency response. In this section I will deal with some of the more common problems in real devices and their solutions.

Offset Cancelation

The ideal operational amplifier will produce a zero output voltage when the two inputs are at the same potential. But real operational amplifiers often produce an offset potential at the output. These voltages are those which exist on the output terminal when it should be zero. There are several sources of these voltages. One is the input bias currents used to operate the transistors in the input stage of the operational amplifier. Typically, these currents are non-zero. When the current flows into or out of the inverting input, it will create a voltage drop equal to the product of the current and the parallel combination of R_{IN} and R_f. How do you deal with this problem? The voltage drop produced by the input bias currents is amplified and appears at the output as an offset potential.

Figure 15-8 shows the use of a compensation resistor, R_C. This resistor has a value equal to the parallel combination of the other two resistors. Because the same bias current flows in both inputs, this resistor will produce the same voltage drop at the noninverting input as at the inverting input. Because the inputs are differential, the net output voltage is zero.

Figure 15-9 shows two methods for nulling output offsets, regardless of the source. Figure 15-9A shows the use of offset null terminals that are found on some operational amplifiers. A potentiometer is placed between the terminals, while the wiper is connected to the V-power supply. This potentiometer is adjusted to produce the null required. The input terminals are shorted together, and the pot is adjusted to produce zero volts output.

Figure 15-9B shows a circuit that can be used on any operational amplifier, inverting or noninverting, except the unity-gain noninverting follower. A counter-current (I3) is injected into the summing junction (point "A") of a magnitude and polarity to cancel the output offset voltage. The voltage at point "B" is set to produce a null offset at the output. The output voltage component due to this voltage is:

$$V_O' = -V_B \times (R_f/R2)$$

If greater (i.e., finer) control over the output offset is required, then one of the two circuits of Fig. 15-10 can be used to replace the potentiometer of Fig. 15-9B.

Power-Supply Decoupling

Operational amplifiers, like all other active electronic components, are somewhat affected by variations in power-supply voltage and noise signals riding on the power-supply potentials. It is also possible that signal from one operational amplifier is coupled to others over the power-supply lines. The cure for these problems, and certain stability problems which I will discuss in due course, is the decoupling scheme shown in Fig. 15-11. The V− and V+ power supply lines are each decoupled with two capacitors. C1 and C2 are 0.1 μF capacitors and are used for high-frequency signals; C3

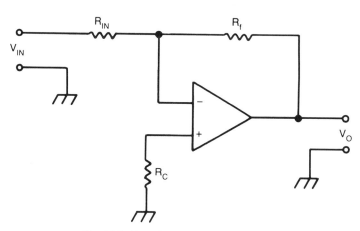

Fig. 15-8. Use of a compensation resistor.

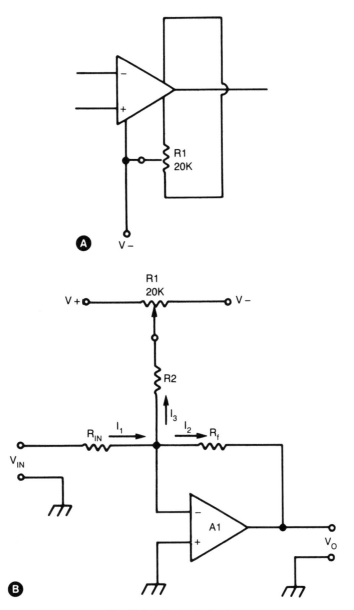

Fig. 15-9. Offset null circuits.

and C4, on the other hand, are higher valued and are for low-frequency signals. The reason for using two capacitors is that electrolytic capacitors used for C3 and C4 are ineffective at high frequencies.

The decoupling capacitors should be mounted as close as possible to the body of the operational amplifier. This constraint is especially true when the operational amplifier is not frequency compensated, and has a high gain-bandwidth product.

Frequency Stability

Operational amplifiers that are not internally frequency compensated are susceptible to oscillations. Figure 15-12 shows a plot of the open-loop

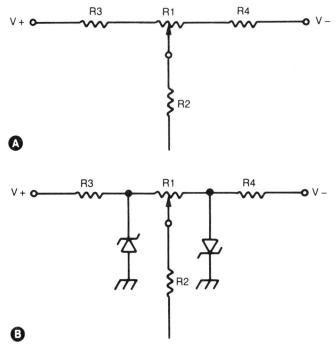

Fig. 15-10. High resolution offset null circuits.

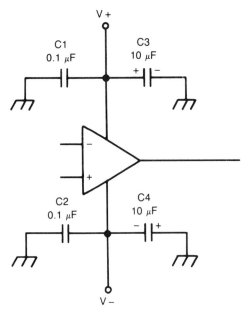

Fig. 15-11. Proper power-supply decoupling.

phase shift versus frequency for a typical operational amplifier. From dc to a certain frequency there is essentially zero phase-shift error, but above that breakpoint the phase error increases rapidly. This change is due to the internal resistances and capacitances of the amplifier acting as a phase-shift network. At some frequency, f, the phase-shift error reaches 180 degrees, which when added to the 180-degree inversion normal to inverting-follower amplifiers, adds up to the 360-degree phase shift that satisfies Barkhausen's criteria for oscillation. At this frequency the amplifier will become an oscillator.

The use of power-supply decoupling helps somewhat for this problem, and it is considered poor engineering practice to use an uncompensated operational amplifier without these decoupling capacitors. In other cases you may need to use a variant of the methods shown in Fig. 15-13. Figure 15-13A shows lead compensation. If the

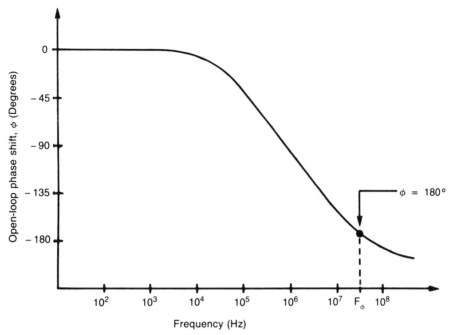

Fig. 15-12. Phase-shift vs Frequency.

operational amplifier is equipped with compensation terminals (usually pins 1 and 8 on "standard" packages), then connect a small value capacitor (20 to 100 pF) as shown. An alternate scheme is to connect the capacitor from a compensation terminal to the output terminal.

The recommended capacitance in manufacturer's specification sheets is for the unity-gain noninverting-follower configuration. For a gain follower the capacitance is reduced by the feedback factor, B:

$$C = C_m B$$

where

C is the required capacitance
C_m is the recommended unity gain capacitance
B is the feedback factor $R_{IN}/(R_{IN} + R_f)$

Lag compensation is shown in Fig. 15-13B. In this case you connect either a single capacitor or a resistor-capacitor network from the compensation

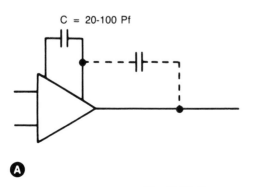

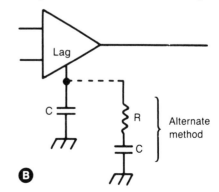

Fig. 15-13. Frequency compensation methods.

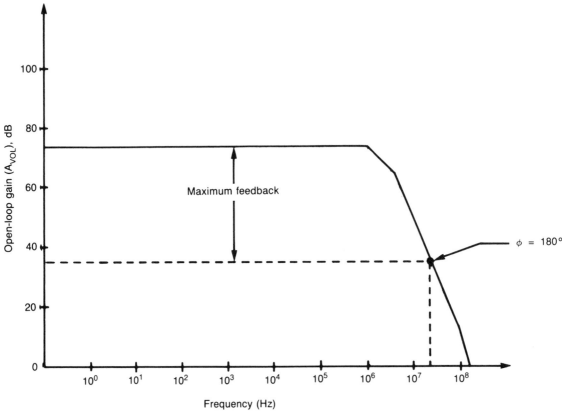

Fig. 15-14. Frequency response.

terminal to ground. A related method places the resistor-capacitor series network between the inverting and noninverting-input terminals.

The object of these methods (see Fig. 15-14) is to reduce the high-frequency loop gain of the circuit to a point where the total loop gain is less than unity at the frequency where the 180-degree phase shift occurs. The amount of compensation required to accomplish this goal determines the maximum amount of feedback that can be used without violating the stability requirement.

DC DIFFERENTIAL AMPLIFIER

A differential amplifier is one that produces an output voltage that is the product of the gain and the difference between the two input voltages. Figure 15-15 shows a simple dc differential amplifier that is based on a single operational amplifier. The gain is:

$$A_{VD} = R3/R1$$

or,

$$A_{VD} = R4/R2$$

provided that R1 = R2, and R3 = R4.

The output voltage from the dc differential amplifier is the product of the differential voltage gain (above) and the difference between the two input voltages (V2-V1). This difference potential is known as the *differential mode signal,* and the differential amplifier ideally responds only to such signals.

A common mode signal is one that is applied to both inputs simultaneously. In the ideal differential amplifier, the common-mode gain is zero, so the common-mode signal causes no change in the output signal. Instrumentation applications, including

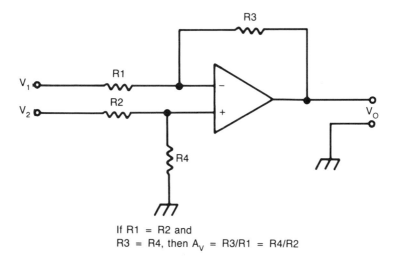

If R1 = R2 and
R3 = R4, then A_V = R3/R1 = R4/R2

Fig. 15-15. Dc differential amplifier.

many in the data-acquisition field, are ideal for differential amplifiers. Consider a case where a low-level signal must pass over wires in an environment that is saturated with 60-Hz ac fields from the power lines. In that case you would normally expect the signal to be obscured by 60-Hz noise. If you use a differential amplifier, however, you can make the design such that the signal voltage is a differential mode potential. The lines from the signal source can be made equal length and pass

through the same environment. In this case both lines would receive the same 60-Hz signal levels, in phase, and that noise signal is therefore common mode. The differential amplifier will reject the interfering signal.

Good common-mode rejection is possible in the circuit of Fig. 15-15, but only if the resistors are matched (R1 = R2 and R3 = R4 to a close tolerance). An improved circuit is shown in Fig. 15-16. Here resistor R4 is replaced with a series

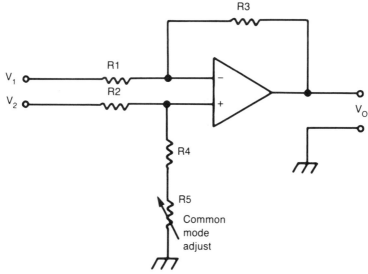

Fig. 15-16. CMRR adjustment circuit.

combination of a fixed resistor and a potentiometer (or just a potentiometer, in some cases). Potentiometer R5 becomes a common-mode adjust control. It is adjusted with $V_1 = V_2$ (which means shorted together and applied to a single signal source) for minimum output voltage.

The basic differential amplifier presented in Figs. 15-15 and 15-16 suffers from the same defects as the inverting-follower circuit. If you want to make a differential amplifier with a very high input impedance, then you must use the three-device Instrumentation Amplifier circuit of Fig. 15-17.

Where would such a circuit be used? Wherever there is a requirement for extremely high impedances, of course. Such applications might include biopotentials amplifiers (EEG, ECG, etc), chemical electrodes (pH, oxygen, CO_2, etc), and certain physics instruments.

The circuit of Fig. 15-17 consists of three operational amplifiers. It is preferable for amplifiers A1 and A2 to be in the same IC package to improve thermal stability, but they can be separate if needed (in a moment I will discuss the IC instrumentation amplifier).

Operational amplifier A3 is used in the same dc differential-amplifier circuit as seen earlier, but its inputs are driven from a balanced circuit consisting of two noninverting followers with gain (A1 and A2) which share a common "input" resistor, R1. The overall voltage gain of this amplifier is given by:

$$A_v = ((2R2/R1) + 1)(R6/R5)$$

assuming R2 = R3, R4 = R5, R6 = R7.

In some cases you will want to make a variable gain control for the differential amplifier. This neat job is accomplished by making R1 adjustable, although you must be cautious in this regard because the term "R1" appears in the denominator of the gain equation. If you place a potentiometer in this

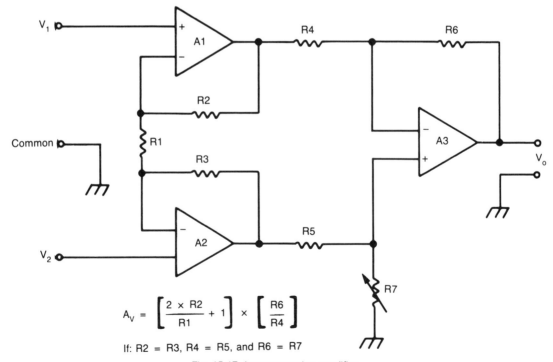

$$A_V = \left[\frac{2 \times R2}{R1} + 1 \right] \times \left[\frac{R6}{R4} \right]$$

If: R2 = R3, R4 = R5, and R6 = R7

Fig. 15-17. Instrumentation amplifier.

slot, it will become zero at one end of its range and will makes the gain go very high — perhaps higher than the circuit can tolerate. In these cases it is prudent to place a fixed resistor in series with the potentiometer which has a value that produces the highest gain desired.

Resistor R7 is used to adjust the common-mode rejection ration (CMRR) of the differential amplifier. Common-mode signals are those which affect both inputs equally. In an ideal differential amplifier, these signals will null each other to zero, so no output occurs. In a practical amplifier, however, there will be some small differential gain error, so some signal will appear on the output in response to common-mode signals. One source of differential gain errors is the normal tolerance differences of the resistors. Minimizing these errors (e.g., using 1 percent or better, low-temperature coefficient resistors) helps. If you make R7 adjustable, then it is possible to adjust out the error.

The CMRR adjustment is made by shorting the two inputs together and then connecting them to a signal source (1-volt @100-Hz, for example). Monitor the output of the amplifier on an oscilloscope, and adjust R7 for minimum output. You will have to continuously adjust the gain of the oscilloscope to more sensitive levels in order to "fine tune" the CMRR adjustment.

The three-device instrumentation amplifier (IA) has been made available in integrated circuit (IC) form. There are several advantages to making the instrumentation amplifier in integrated form (let's call such a device an "ICIA"). For one thing, drift is better controlled because all three operational amplifiers and all of the resistors except R1 share a common silicon substrate, and hence a common thermal environment. All stages drift in the same manner, and often cancel each other's effects. Drift is not zero, but with proper design the ICIA drift is reduced significantly compared with all but the best discrete circuit designs.

A second advantage is that of component density. An IA made from discrete components will require at least 2 to 3 square inches of printed-circuit-board space. In ICIA form, however, the same amplifier might require less than 1 square inch!

Still another advantage for many users is cost. While the unit cost of the ICIA, especially in the higher grades, is often large compared with simple discrete op amps, the aggregate cost of the final product is often lower because the design is simpler, assembly easier, printed-circuit-board layout simpler, and the general overall "hassle factor" for production people is less. The advantages of the ICIA are sufficiently interesting that these chips will someday eclipse, if not replace, the conventional operational amplifier.

The size advantages of the ICIA are not always of primary importance. In some cases, however, size is a concern, or it can make an existing product better. In the case of a transducer or electrode in a noisy environment, especially, it becomes possible to install the amplifier on or in the transducer. This design allows you to send a larger signal down the line to the receiving instrument or data-acquisition system. In one example a pressure transducer had a built-in $\times 100$ ICIA so that it could produce an output of 100 mV to 1 volt, instead of 100 to 1000 microvolts. Obviously, if you expect to pick up some 60-Hz line noise, you are better off sending a 1000-millivolt signal than a 1-millivolt signal. In the following section, I will examine several commercial ICIA devices.

Commercial ICIA Devices

The typical ICIA device will contain a circuit similar to the one shown in Fig. 15-17. One exception is that resistor R1 will be external to the IC package. This external resistor is sometimes designated "R_g", and the pins that connect it to the ICIA are labelled "gain set" or something similar.

Typical gain equations for the ICIA devices will be of the form:

$$A_V = (50k/R_g) + 1$$

Which, of course, means that for a circuit such as Fig. 15-17, resistors R2 and R3 are 25 kilohms each, and R4 = R5 = R6 = R7.

As is common in too many texts, the form of the equation given is not what is actually needed. For most of us, the gain is known or otherwise determinable, and what we need to know is the resistor that will yield that required gain. Thus, it will be helpful to avoid the minor algebra needed to convert the above equation to find the resistance required:

$$R_g = 50k/(A_V - 1)$$

Example. Select a resistor that will produce a gain of ×100.

$$R_g = 50k/(A_V - 1)$$
$$R_g = 50k/(100 - 1)$$
$$R_g = 50k/99$$
$$R_g = 0.51 \text{ kilohms} = 510 \text{ ohms}$$

The ICIA may come in any one of several packages. There are several examples in 8- or 10-pin round metal cans (the original form of IC package!), while others are available in 8-, 14-, or 16-pin DIP packages. There are also several hybrid instrumentation amplifiers that are similar enough in function to the ICIA to be considered under the same rubric.

The National Semiconductor LM-363-xx family of devices includes three ICIA devices in round metal can IC packages. A related device, the LM-363-AD, offers three switch-selectable gains in a 16-pin DIP package. The three LM-363-xx devices have fixed gains determined by the "-xx" suffix.

Model	Fixed Voltage Gain
LM-363-10	×10
LM-363-100	×100
LM-363-500	×500

The purpose of these devices is to provide a fixed, standard gain in small packages. The gains selected (×10, ×100 and ×500) are among the most commonly encountered in electronic circuits, and so are well chosen. Consider an example where a fixed-gain LM-363 device might be used. Suppose you have a low-output-level, force-displacement transducer operating in a noisy environment. If you place an ×10 or ×100 amplifier at the

transducer, then you improve the signal-to-noise ratio at the input of the data-acquisition system. I once built such a transducer amplifier around a 741 operational amplifier, and it required several resistors and a potentiometer (see TAB Book No. 1012, *How To Design and Build Electronic Instrumentation*)]. That ×10 amplifier was built in a small die-cast Pomonoa box that was mounted on a Grass FT-3 transducer. That gain-of-10 preamplifier can now be built inside of the ITT-Cannon connector used on the Grass transducer — should someone feel a compelling need to do so!

Figure 15-18 shows the LM-363-xx fixed-gain ICIA device, while Fig. 15-19 shows a practical circuit. The amplifier is built inside of an 8-pin metal can. The LM-363 series has a gain-bandwidth product of nearly 30 MHz. Pin no. 8 of the LM-363-xx is used for frequency compensation. The values of R and C connected to this pin will custom tailor the ICIA to your own needs. Direct current power is applied to pin numbers 1 and 4 of the LM-363-xx so that V+ is applied to pin no. 1, and V− is applied to pin no. 4.

The decoupling capacitors that are attached to these pins are needed, as in regular operational amplifiers, to prevent oscillation or instability of the amplifier in a practical circuit. Two capacitors are used on each power supply line because of the wide bandwidth of the amplifier. The 4.7 μF capacitors are used to decouple the lower frequencies and are usually tantalum electrolytics. The 0.1 μF units are intended to decouple high-frequency signals and should be either disc ceramic, mylar, or other high-frequency capacitors. The reason why low and high value capacitors are put in parallel is that high value electrolytics, which are needed for low-frequency signals, do not work very well at high frequencies. The decoupling capacitors, especially the high frequency units, must be mounted as close as possible to the body of the ICIA.

The other version of the LM-363 family is a 16-pin miniDIP (Figure 15-20) that will provide step-selectable fixed gains of ×10, ×100 or ×1000. The gain is selected by shorting together specific pins (2, 3 and 4). The protocol is shown below:

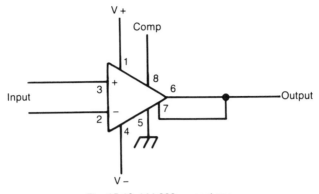

Fig. 15-18. LM-363-xx package.

Gain	Short Together
10	(all pins left open)
100	3 & 4
1000	2 & 4

The circuit of Fig. 15-21 uses guard shield driver terminals on the LM-363-AD. Their use is optional, but highly recommended when low-level signals are anticipated—especially at higher gains.

The Burr-Brown INA-101 ICIA is shown in Fig. 15-22. The internal circuit is similar to the IA circuit of Fig. 15-15, with resistor R1 brought out to the external gain set terminals, and the feedback resistors (R2/R3) set to 20 kilohms. The voltage gain available from the INA-101 is set by:

$$A_V = (40 \text{ k}\Omega/R_g) + 1)$$

where R_g is the external gain resistor in kilohms

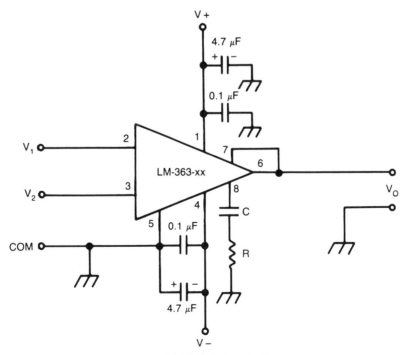

Fig. 15-19. LM-363-xx circuit.

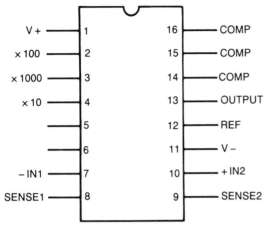

Fig. 15-20. LM-363-AD package.

The INA-101 device is a low-noise amplifier that sports a minimum 60 Hz CMRR of 106 dB, and an input impedance of 10^{10} ohms. The drift is 25 $\mu V/°C$. The INA-101 will operate from dc power supplies of $\pm$ 5 volts to $\pm$ 20 volts, with $\pm$ 15 volts being the intended operating potentials. The current packaging of the INA-101 is the 10-pin metal can that is usually described as "similar to TO-5."

Figure 15-23A shows the Precision Monolithics, Inc. AMP-01 device. The AMP-01 is housed in an 18-pin DIP package. The basic circuit for the AMP-01 is shown in Fig. 15-23B. Notice how simple the circuit is! There are few connections: differential inputs, dc power supplies (V−

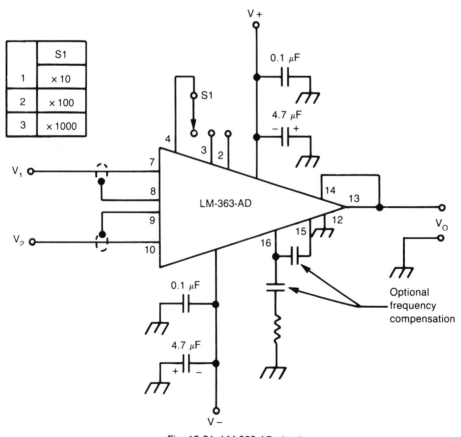

Fig. 15-21. LM-363-AD circuit.

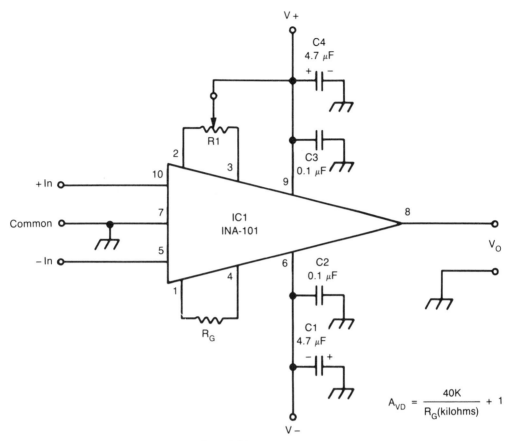

Fig. 15-22. INA-101 circuit.

and V+), output, ground, and two gain-setting resistors. The voltage gain of this circuit is given by:

$$A_{VD} = 20 \ R_s/R_g$$

Suppose we want to make a differential voltage amplifier with a gain of ×1000. We need to make a resistor ratio of 1000/20, or 50:1. Thus, if R_s is set to 100 kilohms, and R_g is 2 kilohms, we will have the required gain of 1000. The permissible gain range is 0.1 to 10,000.

The dc power supply voltages are up to −/+ 18 volts dc. Note in Fig. 15-23B that the dc power supply lines are heavily bypassed. The 0.1-μF units are used to bypass high frequencies, while the 1-μF units are for low frequencies. The 0.1-μF units must be mounted as close as possible to the body of the amplifier.

The maximum operating frequency depends upon the gain. At a gain of 1, the maximum small-signal input frequency is 570 kHz, while at a gain of 1000 it is reduced to 26 kHz.

ISOLATION AMPLIFIERS

An isolation amplifier is one in which the impedance between the input terminals and the power-supply terminals is maximized. A typical isolation amplifier symbol is shown in Fig. 15-24. The input side of the amplifier has its own power-supply terminals, separate from the main amplifier power-supply terminals. While the main amplifier power is derived from a dc power supply connected to the ac power mains, the input-side dc power is derived either from a special electronic (oscillator type) dc power supply, or batteries. The isolation from the

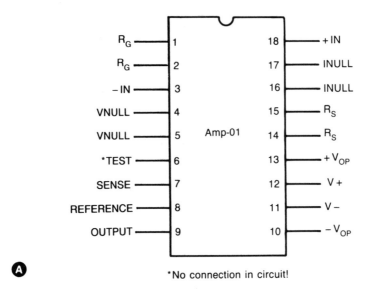

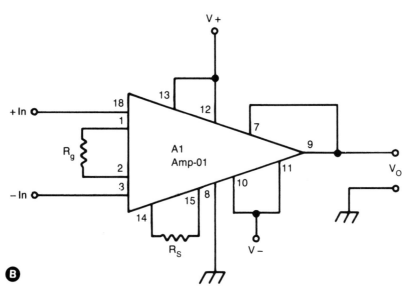

Fig. 15-23. AMP-01 package and circuit.

ac power mains approaches 10^{12} ohms in some cases.

But why use an isolation amplifier? There are many applications for instrumentation amplifiers that are dangerous for either the circuit or the user. In biomedical applications the issue is patient safety. There are numerous signals acquisition needs in biomedical instrumentation where the patient is at risk. Even the simple ECG machine, which measures and records the heart's electrical activity, was once implicated in patient safety problems. Another problem area in biomedical applications is catheterization instruments. There are several tests where the doctors insert an electrode or transducer into the body, and then measure the resulting signal: the intracardiac ECG places an electrode inside the heart by way of a blood vein. The cardiac output computer uses a signal from a thermistor inside a catheter placed in the heart (also through a vein), and simple electronic blood pressure monitors use a transducer that connects to an artery. In all of these cases we do not want the patient exposed to small differences of potential due to current leakage from the 60-Hz ac power lines. The solution is use of an isolation amplifier.

Another application is signals acquisition in high-voltage circuits. You do not want to mix high voltage sources with low voltage electronics because you don't want the low-voltage circuits to blow out. Again, the solution is the isolation amplifier.

Figure 15-24 shows the basic symbol for the isolation amplifier. The break in the triangle used to represent any amplifier denotes the fact that there is an extremely high impedance (typically 10^{12} ohms) between the inputs and output terminal of the isolation amplifier.

Note that there are two sets of dc power-supply terminals. The V− and V+ terminals are the same as found on all ICIA or op-amp devices. These dc power supply terminals are connected to the regular dc supply of the equipment where the

device is used. Such a power supply derives its dc potentials from the ac power mains by way of a 60-Hz transformer. The isolated dc power supply inputs (V_I- and V_I+) are used to power the input amplifier stages, and must be isolated from the main dc power supply of the equipment. The V_I- and V_I+ terminals are usually either battery powered or powered from a dc-to-dc converter that produces a dc output from the main power supply by using a high-frequency (50 to 500 kHz) oscillator. The high-frequency "power-supply" transformer does not pass 60-Hz signals well, so the isolation is maintained.

Figure 15-25 shows the circuit of an isolation amplifier based on the Burr-Brown 3652 device.

The dc power for both the isolated and nonisolated sections of the 3652 is provided by the 722 dual dc-to-dc converter. This device produces two independent −/+ 15 vdc supplies that are each isolated from the 60-Hz ac power mains and from each other. The 722 device is powered from a +12 vdc source that is derived from the ac power mains. In some cases the nonisolated section (which is connected to the output terminal) is powered from a bipolar dc power supply that is derived

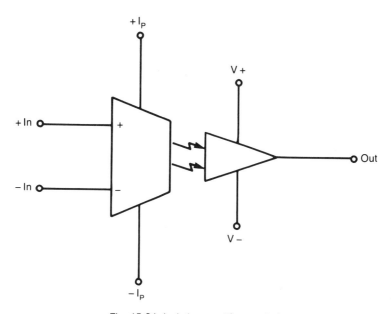

Fig. 15-24. Isolation amplifier symbol.

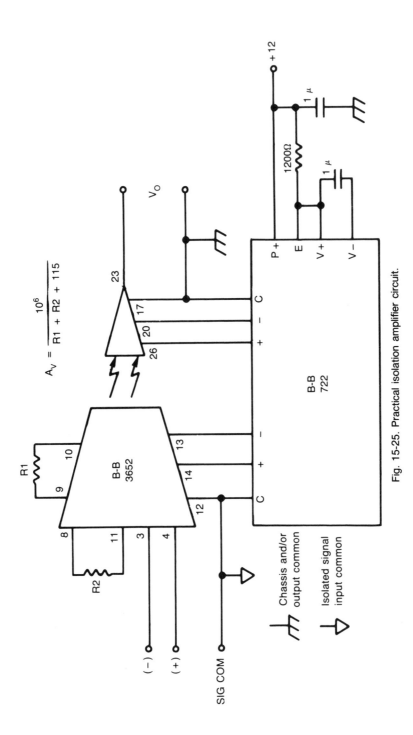

$$A_V = \frac{10^6}{R1 + R2 + 115}$$

Fig. 15-25. Practical isolation amplifier circuit.

from the 60-Hz ac mains, such as a $-/+$ 12 vdc or $-/+$ 15 vdc supply. In no instance, however, should the isolated dc power supplies be derived from the ac power mains.

There are two separate ground systems in this circuit, symbolized by the small triangle and the regular three-bar "chassis" ground symbol. The isolated ground is not connected to either the dc power supply ground/common, or the chassis ground. It is kept floating at all times and becomes the signal common for the input signal source.

The gain of the circuit is approximately:

$$Gain = \frac{1,000,000}{R1 + R2 + 115}$$

In most design cases the issue is the unknown values of the gain setting resistors. You can rearrange the equation above to solve for (R1 + R2):

$$(R1 + R2) = \frac{1,000,000 - (115 \times Gain)}{Gain}$$

where

R1 and R2 are in ohms
Gain is the voltage gain desired

Let's work an example. Suppose you need a differential voltage gain of 1000. What combination of R1 and R2 will provide that gain figure?

If Gain = 1000

$$(R1 + R2) = \frac{1,000,000 - (115 \times 1000)}{1000}$$

$$(R1 + R2) = \frac{1,000,000 - (115,000)}{1000}$$

$$(R1 + R2) = \frac{885,000}{1000}$$

$$(R1 + R2) = 885 \text{ ohms}$$

In this case, you need some combination of R1 and R2 that adds to 885 ohms. The value 440 ohms is "standard," and will result in only a tiny gain error if used.

CONCLUSION

In this chapter you have studied the basic operational amplifier. In the next chapter I will follow up with a discussion of operational amplifier applications that are interesting in a data-acquisition context.

Chapter 16

Linear Amplifier Applications

I N CHAPTER 15 I DESCRIBED THE OPERATIONAL amplifier, some basic operational-amplifier circuits, and the integrated-circuit instrumentation amplifier (ICIA). In this chapter I further develop this topic by considering some circuit applications for these devices. Special attention is given to those that are most useful in the data-acquisition field, even though many of the circuits have general utility.

Unless otherwise specified, you may assume that all circuits presented here operate from bipolar, regulated dc power supplies regardless of whether or not the dc power-supply terminals are shown in the drawing. In general, V− and V+ are both 12 to 15 volts dc, regulated. Also assume that both V− and V+ are bypassed to ground with a 0.1 μF capacitor and a 4.7 μF tantalum electrolytic capacitor (see Chapter 15).

SUMMING CIRCUIT

A summing circuit examines two or more input signal voltages and produces an output voltage that is proportional to their algebraic sum and the circuit gain for that particular input. An operational-amplifier summing circuit is shown in Fig. 16-1.

The basis for the summing amplifier of Fig. 16-1 is the ordinary inverting-follower amplifier circuit. Recall from Chapter 15 that the voltage gain of an inverting follower is the ratio of the feedback and input resistors:

$$A_V = -R_f/R_{IN} \qquad \textbf{Eq. 16-1}$$

where

A_V is the voltage gain (dimensionless)
R_f is the feedback resistance
R_{IN} is the input resistance (both R_{IN} and R_f in same units)

The negative sign indicates that a 180-degree phase reversal takes place between input and output signals. In Fig. 16-1 there is a single feedback resistor, but each input has its own independent input resistance. The transfer function of Fig. 16-1 can be written as:

$$V_O = R_f \frac{V1}{R1} + \frac{V2}{R2} + \frac{V3}{R3} + \ldots + \frac{V_n}{R_n} \quad \textbf{Eq. 16-2}$$

Equation 16-2 contains terms for only three inputs

as shown in Fig. 16-1, but can be extended to N inputs. Consider the following practical example.

Example. A circuit such as Fig. 16-1 has the following resistor values: $R_F = 100k$, $R1 = 100k$, $R2 = 50k$ and $R3 = 50k$. Find the output voltage if the following dc signal voltages are applied: $V1 = 100$ mV, $V2 = 100$ mV, and $V3 = 400$ mV.

$$V_0 = R_f\frac{V1}{R1} + \frac{V2}{R2} + \frac{V3}{R3}$$

$$V_0 = (100K)\frac{(100\text{ mV})}{(100k)} + \frac{(200\text{ mV})}{(50k)} + \frac{(400\text{ mV})}{(50k)}$$

$V_0 = 100$ mV $+ (2 \times 200$ mv$) + (2 \times 400$ mV$)$
$V_0 = 100$ mV $+ 400$ mV $+ 800$ mV
$V_0 = 1300$ mV $= 1.3$ volts

There are a number of applications for the summing circuit. In analog instrumentation, for example, a derived value may require the summation of two voltage analogs of physical parameter values. In digital data-acquisition systems, you may be sharply limited on the processing time line or there may be a limit to the number of available analog inputs. In these cases you might want to sum the values prior to input to the computer.

The summing circuit can also be used as a linear mixed for ac signals such as audio. In this case you might want to place dc blocking capacitors in series with each input resistor (more on this later). Each input resistor may be used as a channel gain control, while the feedback resistor forms a master gain control.

Still another application is adding a dc component to a varying dc or ac signal. Here you would apply the ac signal to one input and the dc reference bias to the other input.

AC AMPLIFIERS

Amplifiers for ac signals fall into two distinct categories. First, there are those that merely tailor the frequency response at one or both ends of the passband, but may also be dc coupled. Second, there are those that are designed to pass only ac signals while blocking dc signals. Many amplifiers are designed to fit both categories, as you shall see later in this section.

LOW-PASS AMPLIFIER

Figure 16-2 shows an inverting amplifier (Fig. 16-2A) and its frequency response plot (Fig. 16-2B). At low frequencies the voltage gain of this circuit is the same as for dc signals: $-R_f/R_{IN}$. At a frequency higher than F_h, the response falls off at a rate of -6 dB/octave (-20 dB/decade). In other words, above F_h the gain drops off 6 dB for every 2:1 change in excitation frequency. The breakpoint frequency, F_h, is -3 dB down from the low frequency and dc gain, ($-R_f/R_{IN}$, and is defined as:

$$F_h = 1/(2\pi R_f C1) \qquad \textbf{Eq. 16-3}$$

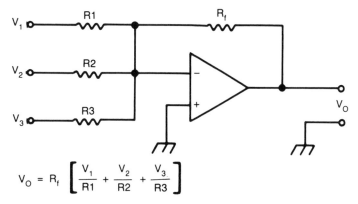

$$V_O = R_f \left[\frac{V_1}{R1} + \frac{V_2}{R2} + \frac{V_3}{R3} \right]$$

Fig. 16-1. Three-input dc mixer/adder.

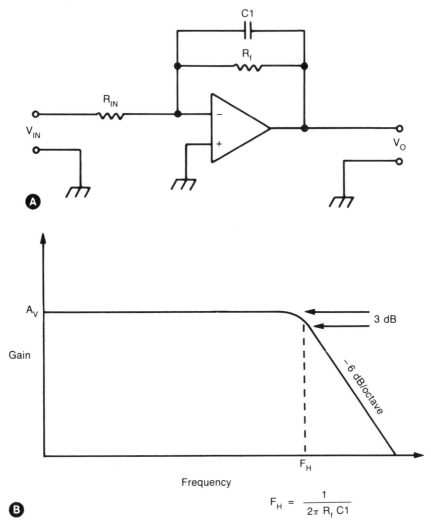

Fig. 16-2. Integrator A) circuit, B) response.

The circuit of Fig. 16-2A acts like a first-order low-pass filter with a cut-off frequency equal to F_h. This circuit is probably the simplest form of active filter, so simple that it is only occasionally called by the august name "filter."

Setting component values is done the same as for any inverting-follower amplifier. The input impedance is governed by R_{IN} because the inverting input is grounded. A rule of thumb for the minimum value of input resistance holds that it should have a value not less than ten times the output impedance of the driving source. A common prac-

tice is to set R_{IN} at 10 kilohms or more, and then select a value of R_f that yields the required voltage gain ($R_f = ABS|A_v| \times R_{IN}$).

Once the value of the feedback resistor is determined, you can use a rewritten form of Eq. 16-3 to select the value of capacitor C3:

$$C1 = 1/(2 \pi f_h R_f) \qquad \text{Eq. 16-4}$$

where

R_f is the feedback resistance in ohms

f_h is the desired frequency breakpoint in hertz (Hz)

C1 is the capacitance in farads

Example. Find the capacitance required to make a −3 dB breakpoint at 3000 Hz, when the feedback resistance is 100 kilohms.

$C = \frac{1}{2} \pi f R$

$C = 1/[(2)(3.14)(3000 \text{ Hz})(100{,}000 \text{ ohms})]$

$C = 1/1.88 \times 10^9$

$C = 5.31 \times 10^{-10} \text{ farads} = 0.00053 \ \mu F$

Circuits such as Fig. 16-2 are used to roll off high-frequency response. A typical application might be decreasing high frequencies in order to reduce noise without sacrificing signal.

HIGH-PASS AMPLIFIER

A high-pass amplifier (Fig. 16-3) is used to roll off frequencies below a certain −3 dB cut-off breakpoint. A typical application for data acquisition is the biopotentials amplifier used in medicine. Consider an electrocardiographic (ECG) amplifier,

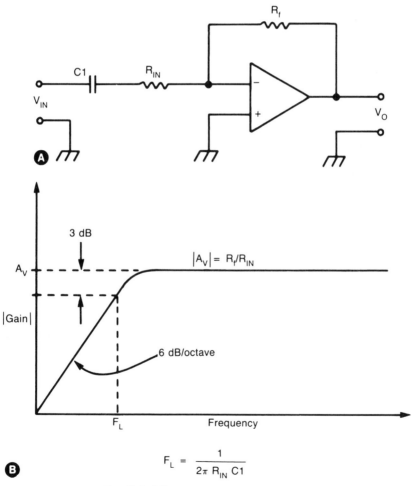

Fig. 16-3. Differentiator A) circuit, B) response.

for example. The ECG amplifier uses a high-input-impedance differential amplifier to acquire signals from metallic electrodes on the patient's skin (see Chapters 11 and 12).

There are two signal artifacts that obscure the 1-millivolt (mV) ECG signal. One is the half-cell potential (a dc offset bias) caused by the interface between the metallic electrode and the electrolytic skin. The half-cell potential can be 1000 or more times the peak amplitude of the ECG signal. The other artifact is low-frequency drift caused by changes in the electrode interface (an impedance-reducing gel or paste dries out over a period of several hours). Both of these artifacts are reduced by the ac coupling of the differential-amplifier inputs.

The example above illustrates using the ac-coupled amplifier to block a dc component in the signal. In this case, you must be certain that the low-frequency response is adequate for the signal being acquired. The human ECG signal, for example, has a diagnostically significant frequency spectrum of 0.05 to 100 Hz.

The circuit of Fig. 16-3A has a frequency response similar to that shown in Fig. 16-3B. The absolute value of the circuit voltage gain at all frequencies above the -3 dB breakpoint, f_L, is given by the familiar expression, R_f/R_{IN}. At frequencies lower than F_L, the response drops off at a -6 dB/octave (-20 dB/decade) rate. Frequency f_L in Fig. 16-3B is found from:

$$f_L = 1/(2 \pi R_{IN} C1) \qquad \textbf{Eq. 16-5}$$

where

    f_L is the -3 dB frequency in Hertz (Hz)
    R_{IN} is the input resistance in ohms
    C1 is the capacitance in farads (F)

In most practical applications you will know f_L as a specification parameter based on the application and will set resistance R_{IN} independently of Eq. 16-5. You will, therefore, prefer a rewritten form of Eq. 16-5 to find the required capacitance:

$$C' = 10^6/(2 \pi f_L R_{IN}) \qquad \textbf{Eq. 16-6}$$

where C' is the capacitance in microfarads (μF), and all other parameters are as specified in Eq. 16-5.

Example. Find the capacitance required for a -3 dB point of 30 Hz when the input resistance is 10 kilohms

$C' = 10^6/(2 \pi f_L R_{IN})$
$C' = 10^6/[(2)(3.14)(30 \text{ Hz})(10,000 \text{ ohms})]$
$C' = 10^6/1.884 \times 10^6$
$C' = 0.53 \ \mu\text{F}$

BANDPASS AMPLIFIERS

The bandpass amplifier (Fig. 16-4) combines the properties of both low-pass and high-pass amplifiers. The circuit of Fig. 16-4A produces a frequency response such as Fig. 16-4B. Like the other circuits, this one is an inverting follower with a mid-band voltage gain defined by the ratio of the feedback and input resistors ($-R_f/R_{IN}$).

The frequency responses at high and low ends of the passband are tailored by capacitors C1 and C2 acting with their respective resistors. Below frequency f_L and above f_h, the response drops off from the mid-band gain at a rate of -6 dB/octave, or -20dB/decade. The -3 dB breakpoints are defined as follows:

$$f_L = 10^6/(2 \pi R_{IN} C1) \qquad \textbf{Eq. 16-7}$$

and,

$$f_h = 10^6/(2 \pi R_f C2) \qquad \textbf{Eq. 16-8}$$

Or, written to find the capacitances:

$$C1 = 10^6/(2 \pi R_{IN} f_L) \qquad \textbf{Eq. 16-9}$$

and,

$$C2 = 10^6/(2 \pi R_f f_L) \qquad \textbf{Eq. 16-10}$$

where

    C1 is in microfarads (μF)
    C2 is in microfarafs (μF)

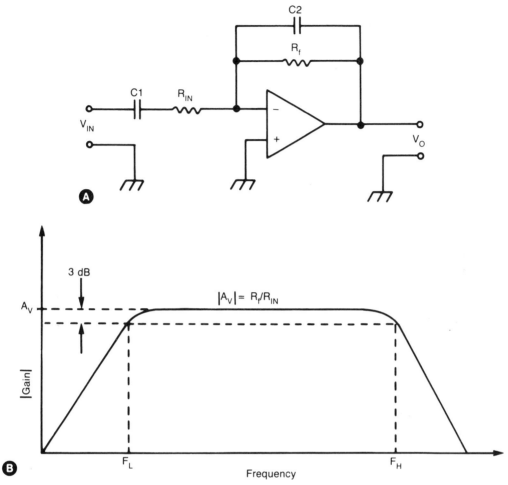

Fig. 16-4. Practical differentiator A) circuit, B) response.

R_{IN} is the input resistance in ohms
R_f is the feedback resistance in ohms
f_L is the low-end -3 dB point in hertz (Hz)
f_h is the high-end -3 dB point in hertz (Hz)

Example. Design an inverting passband amplifier with a mid-band gain of 100 and an input resistance of 10 kilohms or more. Make the -3 dB point 10 Hz and the upper -3 dB point 1000 Hz.

Discussion. The higher the value of resistance, the lower the required capacitance value for any given frequency. Thus, set R_{IN} at 100 kilohms in order to get lower values of capacitance. The gain $= 100 = R_f/100k$. So $R_f = (100k)(100) = 10$ megohms Therefore, $R_{IN} = 100$ kohms, and $R_f = 100$ megohms. Thus,

$C1 = 10^6/(2 \pi R_{IN} f_L)$
$C1 = 10^6/[(2)(3.14)(10 \text{ Hz})(100{,}000)]$
$C1 = 10^6/(6.28 \times 10^6)$
$C1 = 0.159 \ \mu F$

And the other capacitor is:

$C2 = 10^6/(2 \pi f_h R_f)$
$C2 = 10^6/[(2)(3.14)(1000 \text{ Hz})(10^7)]$
$C2 = 10^6/6.28 \times 10^{10}$
$C2 = 0.0000159 \ \mu F = 15.9 \ pF$

AC-COUPLED OUTPUT CIRCUIT

If a signal has a dc offset that may adversely affect a following stage, then you can use the

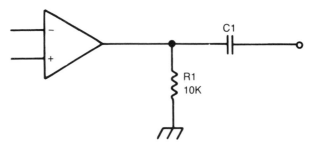

Fig. 16-5. Ac-coupling.

method of Fig. 16-5 to ac-couple the op-amp output. Resistor R1 is used to provide a load for the operational amplifier and to prevent any dc component of the signal from charging coupling-capacitor C1.

The value of C1 is set by the lowest operating frequency and the input resistance of the following stage. In most cases C1 will have a very low value when the input impedance of the following stage is very high (for example, when a noninverting follower is used).

AC-COUPLED NONINVERTING AMPLIFIER

Thus far our examples have been AC-coupled inverting amplifiers, so in this section I will consider the AC-coupled noninverting amplifier (see Fig. 16-6). This circuit is basically the same as the dc-coupled version, except for R1C1.

The circuit of Fig. 16-6 has a high-pass characteristic much like Fig. 16-3B. The gain falls off at −6 dB/octave (−20 dB/decade) at frequencies below the −3 dB point, f_L. At frequencies above

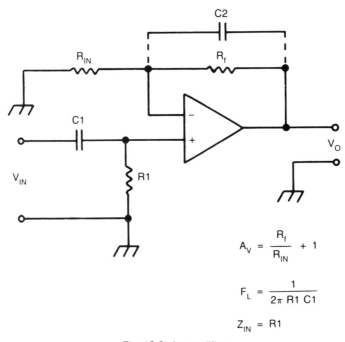

$$A_V = \frac{R_f}{R_{IN}} + 1$$

$$F_L = \frac{1}{2\pi \, R1 \, C1}$$

$$Z_{IN} = R1$$

Fig. 16-6. Ac amplifier.

the −3 dB point, the gain is defined by the usual expression for dc gain:

$$A_V = \frac{R_f}{R_{IN}} + 1 \qquad \textbf{Eq. 16-11}$$

Resistor R1 sets the input impedance of the circuit. The purpose of R1 is to prevent the charging of C1 by the input bias currents of the op amp, so in some very high input impedance premium models R1 is deleted. The value of input resistor R1 may be set almost arbitrarily, provided that it is at least ten times the source resistance. General practice calls for R1 to be as high as possible, despite the "10×" rule. When low-cost devices are used for the op amp, there will be a substantial input bias current, so R1 will be on the order of 100 kilohms to keep the current from changing C1. Most operational amplifiers can accept values of R1 in the 1 megohm to 10-megohm range. A few models feature extremely high input impedance (which implies picoampere-level input-bias currents), and in these cases R1 may be deleted altogether.

The value of the coupling capacitor is set according to the rule:

$$C1 \ 10^6/(2 \ \pi \ R1 \ f_L) \qquad \textbf{Eq. 16-12}$$

where

C1 is in microfarads
R1 is in ohms
f_L is in hertz (Hz)

Example. Find the value of C1 when R1 is 1 megohm, and the lower −3 dB point is 20 Hz.

$C1 = 10^6/(2 \ \pi \ R1 \ f_L)$
$C1 = 10^6/[(2)(3.14)(10^6 \ ohms)(20 \ Hz)]$
$C1 = 10^6/(1.26 \times 10^8)$
$C1 = 0.008 \ \mu F$

In cases where a single dc power supply is used (see Chapter 15), add a capacitor between the cold end of R_{IN} and ground. The value of this capacitor will set the low-end frequency response. The capacitor value is set as above.

AC-COUPLED DIFFERENTIAL AMPLIFIERS

There are many data-acquisitions applications that require a differential amplifier. Such an amplifier (see Chapter 15) delivers an output voltage that is proportional to the gain and the difference between two input voltages.

Figure 16-7 shows a simple, ac-coupled differential amplifier. This circuit is similar to the dc

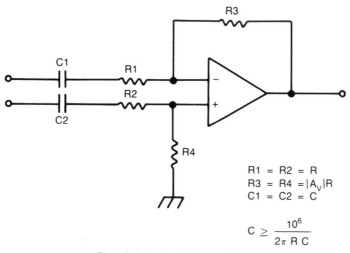

$$R1 \ = \ R2 \ = \ R$$
$$R3 \ = \ R4 \ = |A_V|R$$
$$C1 \ = \ C2 \ = \ C$$

$$C \geq \frac{10^6}{2\pi \ R \ C}$$

Fig. 16-7. Differential ac amplifier.

version and has a high-pass characteristic. The input capacitors are identical and have a value set by:

$$C = 10^6/(2 \pi R f_L) \quad \textbf{Eq. 16-13}$$

if $C1 = 2 = C$, $R1 = R2 = R$, and
$$R3 = R4 = A_V R, \text{ and}$$

where

    C is in microfarads (μF)
    f_L is the low-end -3 dB point in hertz (Hz), and
    all resistances are in ohms

It is essential that balance be maintained in the circuit of Fig. 16-7. This requirement means that $R1 = R2$, $R3 = R4$, and $C1 = C2$. Further-more, C1 and C2 should be physically identical (i.e., of the same type) as well as the same value.

The gain at all frequencies higher than the -3 dB point (and lower than the natural frequency limitation of the op amp) is given by either of the following expressions:

$$A_V = R3/R1 \qquad \textbf{Eq. 16-14}$$

$$A_V = R4/R2 \qquad \textbf{Eq. 16-15}$$

The simple design of Fig. 16-7 suffers from the same input impedance and gain limitations that proved to be a problem in the inverting-follower design. These problems are solved substantially by using the instrumentation amplifier (IA) circuit of Fig. 16-8.

The IA is made from three operational amplifiers, and is essentially similar to the dc-coupled IA

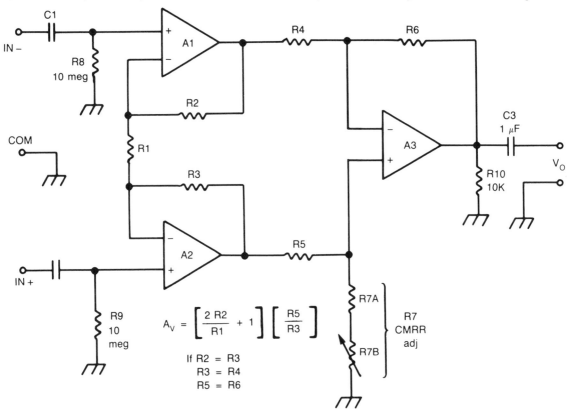

$$A_V = \left[\frac{2\,R2}{R1} + 1 \right] \left[\frac{R5}{R3} \right]$$

If R2 = R3
R3 = R4
R5 = R6

Fig. 16-8. Ac-coupled instrumentation amplifier.

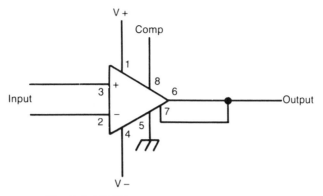

Fig. 16-9. LM-363-xx (repeat of Fig. 15-18).

of Chapter 15. The IA consists of two stages. The input stage is a differential amplifier consisting of A1 and A2. In most cases A1 and A2 should be a single-package, dual op amp so that the two halves of the input amplifier track with each other thermally. The other stage of the IA is the output amplifier, A3. This stage takes the differential output from A1-A2 and converts it to a ground referenced "single-ended" output potential, V_O.

The gain of the IA circuit is the product of the stage gains. In other terms:

$$A_v = \frac{2 \times R2}{R1} + 1 \frac{R5}{R3} \quad \textbf{Eq. 16-16}$$

if:

R2 = R3
R4 = R5
R6 = R7

Resistance R7 forms a common-mode rejection ratio adjustment (CMRR ADJ). A general rule of thumb is to make R7A = 0.9 × R7, and R7B = 0.2 × R7. This ratio offers the advantage that the point at which R7 = R6 will be approximately mid-scale on R7B, allowing ample room to adjust for component tolerances.

The input circuitry is ac-coupled in a manner similar to the noninverting amplifier case cosidered earlier. The input impedance is set by resistors R8 and R9. These resistors must be equal, and preservation of the CMRR is best achieved if they are matched. The purpose of R8 annd R9 is to prevent bias current from the amplifier inputs from reaching capacitors C1 and C2.

Capacitors C1 and C2 should likewise be matched. The value of these capacitors sets the low-frequency response breakpoint and is found from:

$$C = 10^6/(2 \pi R f_L) \quad \textbf{Eq. 16-17}$$

if:

R8 = R9 = R, and C1 = C2 = C

where

R is in ohms

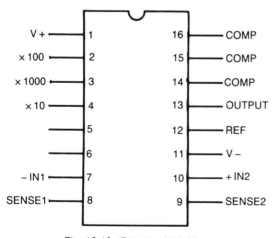

Fig. 16-10. (Repeat of 15-20.)

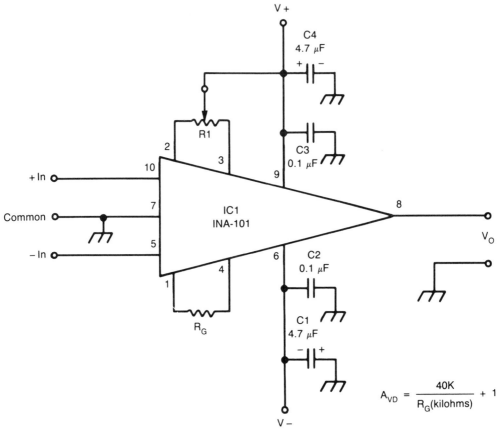

Fig. 16-11. (Repeat of 15-22.)

$$A_{VD} = \frac{40K}{R_G(\text{kilohms})} + 1$$

C is in microfarads (μF)

f_L is in hertz (Hz)

The use of the output network C3-R10 depends upon whether or not a dc component exists, and can be ignored in almost all cases except where a single dc power supply is used to power A1, A2, and A3.

A UNIVERSAL "REAR END" FOR ACQUISITIONS AMPLIFIERS

One of the major failings of locally designed projects, in addition to not a few professionally designed instruments, is inadequate attention to the rear-end circuitry. In the majority of the instruments the most effort and thought go into the front end, and the rear-end is merely the output section. But proper design of the rear (or "output") end of an electronic instrument/control circuit can make the difference between a mediocre and really useful device.

First, though, let's get a little terminology straight. What do "rear end" and "front end" mean, if not human anatomy? The *front end* of a device is the section where the input signals are received. Examples include rf amplifier in a communications receiver, the transducer amplifier in a temperature monitor, the differential biopotentials amplifier in an ECG amplifier, and the Wheatstone-bridge amplifier in a strain-gage transducer circuit. Almost all of the signals processing or shaping is done in the front end of the circuit. The rear end, on the other hand, is less glamorous but nonetheless important. The *rear end* of the circuit includes the output stages.

Figure 16-12 shows a semi-universal rear-end circuit that I have used in any number of cases. During the years when I worked in a medical school/hospital environment, I built this circuit into several physiological (biopotentials) amplifiers, quite a few transducer amplifiers, and numerous other projects. The circuit provides several features:

1. Gain (if R2 > 10 kohms), otherwise unity gain
    2. Gain control (0 to 1, or 0 to A_V)
    3. Dc Balance control/offset null
    4. Position control

The circuit is constructed from two 1458-type dual operational amplifiers. The 1458 contains two 741-family operational amplifiers inside of an 8-pin mini-DIP package. Any other standard operational amplifier will also work; the 1458 is used to reduce the parts count and wiring time. A type 324 would place all of the components in a single IC package.

The dc power supply for this circuit must be bipolar. In other words, it requires both positive-to-ground (V+) and negative-to-ground (V−) dc power supplies, as with any other operational amplifier circuit that does not use special techniques. Potentials between −/+4.5 and −/+15 volts dc will serve nicely.

The circuit of Fig. 16-12 is made up of four inverting followers in cascade, so it serves overall as a noninverting follower. Removing any one stage (U1B recommended) will make the circuit into an inverting follower because there would be an odd number of inverting stages.

The gain of an inverting-follower operational amplifier stage is set by the ratio of the feedback and input resistors. For example, the gain of stage U1A is −R2/R1, where the negative sign indicates that inversion (180-degree phase reversal) is taking place. Similarly, the gain of U1B is −R6/R5; of U2A −R8/R7; and of U2B −R10/R9. The overall gain of the entire circuit is simply the product of all the individual gains:

$$A_{V_T} = A_{V_1} \times A_{V_2} \times A_{V_3} \times A_{V_4} \quad \textbf{Eq. 16-18}$$

Because all gains in the circuit as shown in Fig. 16-12 are unity, i.e., one, the overall gain is one. But you can increase or decrease the overall gain by varying the stage gains. The recommended procedure is to vary the gain of the fixed stage, U1B. In this case you can assume that the gain of the overall circuit will be −R6/R5. Leave R5 equal to 10 kilohms in most cases, unless it is impossible to find a value of R6 that will result in the correct gain without modifying the value of R5. In any event, do not let R5 become less than 1000 ohms. The rules for changing the gain are:

1. Unity gain: leave as is (R5 = R6 = 10 kilohms)
2. Less than unity gain: R5 > R6 (e.g., R5 = 10 kilohms, R6 = 2000 ohms, yields a gain of 2,000/10,000 or 0.20).
3. Greater than unity gain: R5 < R6 (e.g., R5 = 10 kilohms, R6 = 100 kohms, yields 100/10 = 100).

The gain control (R8) is used to vary the overall gain of the circuit from zero to full gain. If the values are as shown in Fig. 16-12, then R8 varies the gain from 0 to 1. This potentiometer is usually a front panel control and is accessible to the user of the project.

The null control is used to cancel the effects of dc offsets created both in this circuit and in previous stages. It also provides the dc balance effect noted earlier. The dc balance control on some instruments is used to cancel the change of output baseline as the sensitivity control is varied — a most disturbing effect to someone making a measurement! Potentiometer R4 is adjusted using a dc output meter at V_O, and is adjusted until there is no shift in dc output when the gain control is varied through its full range. If there are dc offsets present in the input signal (V_{IN}), then such a shift will be noted in the output. The function of R4 is to provide an equal but opposite polarity offset signal to cancel the offset from all other sources. In some cases there might be 10-kilohm resistors (similar to R14 and R15 near R12) between the ends of the potentiometer and the power-supply potentials.

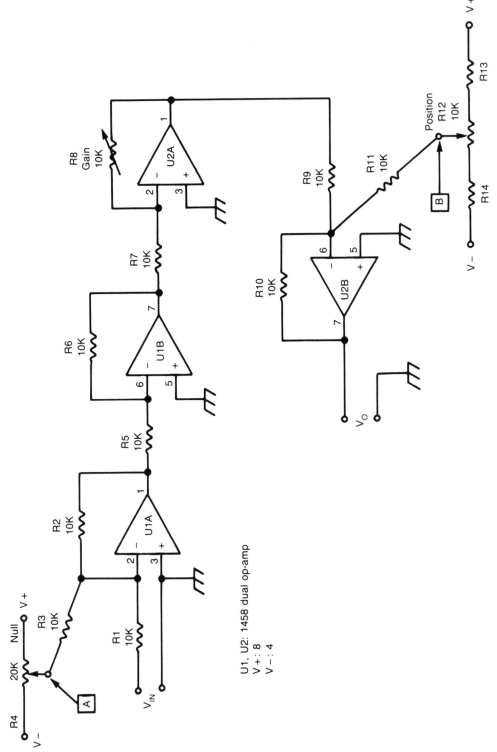

Fig. 16-12. Universal rear end.

These resistors reduce the offset range while increasing the resolution of the adjustment. Use these resistors only if there is a problem homing in on the correct value.

The position control, R12, is optional, and is normally used when the output signal is to be displayed on an analog paper chart recorder, or dc CRT oscilloscope. Potentiometer R12 provides an intentional offset to the final stage (U2B) that is quite independent of the input signal. The effect is to position the output waveform anywhere on the oscilloscope or chart recorder vertical axis that you might desire.

In some cases the range of the potentiometer is too great. Only a little adjustment of the potentiometer will send the trace offscreen. You can counter this problem with the simple expedient of selecting values for R13 and R14 (note: R13 = R14) that allow the trace to just disappear off the top when the potentiometer reaches the limit of its upward travel, and off the bottom when the potentiometer reaches its lower limit.

ADJUSTMENT

The adjustment of this circuit requires either a dc voltmeter or a dc-coupled oscilloscope that has a grid on the screen so that potentials can be read. If the oscilloscope is used, then short the input with the oscilloscope switch marked for that purpose ("GND" in "AC-GND-DC" on some models), and set the trace to exactly the center of the vertical lines on the grid. Select a sweep speed that yields a nonflickering line. Next place the switch into the "dc" position. The vertical deflection factor should be around 0.5 volts/division.

Procedure:

1. Disconnect V_{IN} from the front-end circuit and short this input to ground.
2. Using a dc voltmeter, set the potential at point A to 0.00 volts.
3. Similarly, set the potential at point B to 0.00 volts.
4. Set R8 to maximum resistance (highest gain).

5. Make all adjustments to the front-end circuits as needed, and then return to the rear-end circuit.
6. Adjust R8 through its range from 0 to 10 kilohms several times while watching the output indicator ('scope or meter); if the output potential shifts, then adjust R4 until the shift is cancelled. You will have to continually run R8 through its range while adjusting R4; this adjustment is somewhat interactive, so try it several times, or until no further improvement is attainable.
7. Check the range of the position control.

The univeral rear-end circuit is a simple project that can give your instrument and control projects that final "professional" touch that makes them more useful to all users.

ACTIVE FILTERS

The ac-coupled amplifiers presented earlier offer high or low-pass response characteristics, with roll-off beyond the cut-off frequency being on the order of -6 dB/octave. This slope means that a $2:1$ change in frequency produces a -6 dB change of gain. Such circuits are known as *first-order filters* and produce the same result as simple passive RC filters. When a faster roll-off is needed, then you might opt for an active circuit rather than cascaded passive circuits. In this section, I will discuss certain filter circuits that are capable of giving -12 dB/octave (-40 dB/decade) performance. These are called *second-order filters*. Please note, however, that the subject of active filters is too broad for the present book, and you should consult a text on active operational amplifier filters if your needs transcend what's given here.

Figure 16-13A shows the multiple-feedback-path active low-pass filter. This circuit is based on the ordinary operational amplifier, and has a simplified response characteristic similar to Fig. 16-13B. Note that the attenuation slope below the cut-off frequency falls off at a rate twice that for the simple circuits presented earlier.

The cut-off frequency for this circuit is dependent upon the values of resistor and capacitors selected, and can be stated in simplified form as:

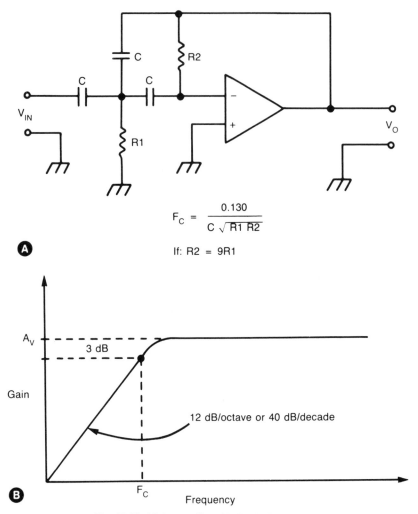

$$F_C = \frac{0.130}{C \sqrt{R1\ R2}}$$

If: R2 = 9R1

Fig. 16-13. High-pass filter A) circuit, B) response.

$$F_c = \frac{0.130}{C\ R1\ R2}$$

if R2 = 2 R1.

By constraining the values of R1 and R2 above you can simplify what would ordinarily be a lot more arithmetic.

You can modify the circuit for high-pass operation as shown in Fig. 16-14A and 16-14B. Here the roles of resistor and capacitor element are reversed, which has the result of changing the simplified frequency cut-off equation to:

$$F_c = \frac{0.210}{R\ C1\ C2}$$

if C1 = 9 C2.

LIMITING AMPLIFIERS

A limiting amplifier is one in which the output-potential excursion is limited in at least one direction. The simplest types of limiting amplifiers are shown in Fig. 16-15A through 16-15D. These circuits use diodes to clip the output potential.

The circuit in Fig. 16-15A is, perhaps, the simplist because it uses a pair of ordinary silicon signal diodes (e.g., 1N4148) to clip the signal to ±0.7 volts or so. When the output signal goes positive, diode D1 will clamp it to +0.7 volts; when it tries to go negative, it is held to −0.7 volts by D2. Between these limits the gain for the circuit is set by the usual expression for the type of amplifier (either inverting or noninverting can be used). The disadvantage of this circuit is the low level of

clamping caused by the junction potentials of D1 and D2 (i.e., 0.6 to 0.7 volts).

A circuit using back-to-back zener diodes is shown in Fig. 16-15B. In this circuit the clamping diodes are series-connected back-to-back zeners, so the output potential is always 0.7 volts higher than the zener rating of the diode for that direction. For example, when V_O is positive, diode D2 is forward biased and so contributes +0.7 volts, and diode D1 is in zener mode. If V1 is the zener

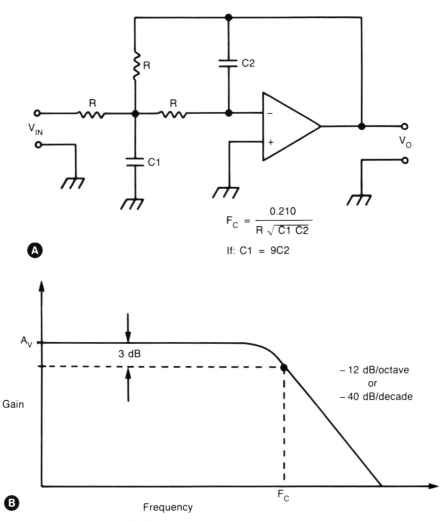

$$F_C = \frac{0.210}{R \sqrt{C1\ C2}}$$

If: C1 = 9C2

Fig. 16-14. Low-pass filter A) circuit, B) response.

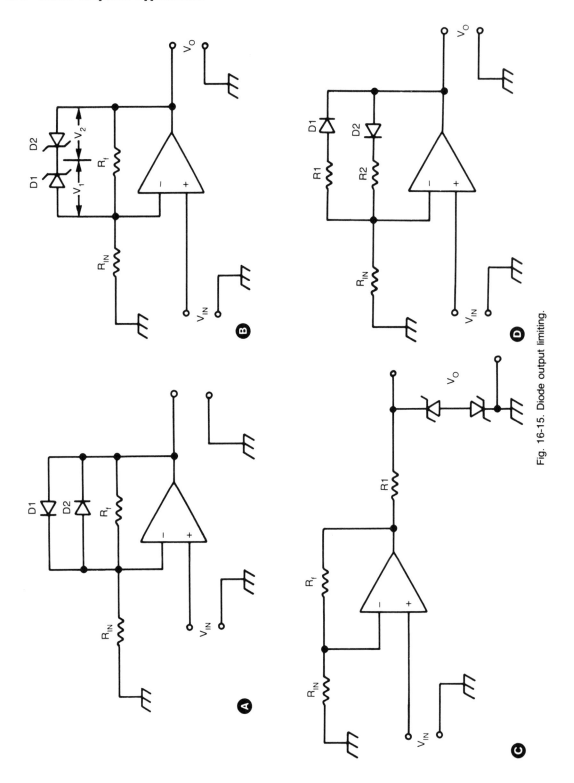

Fig. 16-15. Diode output limiting.

potential of D1, then the output voltage is V1 + 0.7 volts. Similarly, on negative excursions the output potential is held to V2 − 0.7 volts (where V2 is a negative voltage). There is no reason for V1 and V2 to be equal unless you want to maintain output symmetry.

Another zener diode clamping circuit is shown in Fig. 16-15C. This circuit is the one prefered in Jacob Millman's *Microelectronics*, a text with which most EE's are familiar. The output-potential clamping protocol is the same as in the above case. The series resistor is used to limit the current to the zener diodes to a safe level and can be omitted only if the op-amp output is limited by internal circuitry to a value less than that required to stress the diodes.

A slightly different approach is taken in the circuit of Fig. 16-15D. Here we have different gains for the positive and negative voltage excursions. Given that this circuit is a noninverting follower, it follows that the gain for positive input signals will be:

$$A_V = \frac{R2}{R_{IN}} + 1$$

and for negative input signal excursions:

$$A_V = \frac{R1}{R_{IN}} + 1$$

BIOPOTENTIALS AMPLIFIERS

This discussion is basically a continuation of Chapters 11 and 12. I couldn't introduce this material in these chapters because I hadn't yet covered operational and other amplifiers. Before dealing with circuit specifics, lets review the matter of biopotentials.

Biopotentials are electrical signals generated in living things. Most familiar are signals such as the electrocardiogram (ECG, for heart signals), electroencephalogram (EEG, for brain signals), electromyograph (EMG, for skeletal muscle signals) and so forth. These signals are generated at the cell level and are due to differing electrolytic concentrations between inside and outside of the cell. For example, most cells are rich in potassium inside of the cell, and rich in sodium outside the cell (note: both potassium and sodium are found on both sides of the cell wall, but the relative concentration is different). The different ionic potentials of these elements produces a voltage of about −80 millivolts (i.e., the inside is more negative than outside of the cell). When the cell discharges, for example, when the heart contracts, the cell wall selectively breaks down and becomes permeable to the external sodium. The ionic charge on both sides of the wall equilibrates, and the "action potential" snaps to as much as +20 millivolts for a short while. The cell wall then reasserts itself and restores the resting potential (−80 mV) concentrations as before.

You can pick up the potentials from internal muscles like the heart from surface electrodes. Unfortunately, not all is easy in this respect, for there are some severe problems.

One of the first problems is that the signal is very low level. While EMG signals may look like 100 millivolt powerhouses, other signals are not so strong. The typical ECG waveform, for example, is on the order of 1 millivolt, with significant components much less than 1 mV. The EEG signal is a small signal indeed — on the order of 5 to 100 microvolts. The biopotentials amplifier must have a high gain in order to correctly display these signals. If you have an oscilloscope or strip-chart (paper) recorder with a 1-volt full-scale input, then you will need the following gains:

Electrocardiogram (ECG):

1000 millivolts = 1 volt, so

$$A_V = 1000/1$$
$$A_V = 1000$$

Electroencephalogram (EEG):

Assuming 50 microvolt signals, and 1 μV = 1/1,000,000 volts

$$A_V = 1 \text{ volt}/0.000050 \text{ volt}$$
$$A_V = 20,000$$

With gains of 1000 to 20,000 required, you clearly need a high gain amplifier! Of course, not all display devices require a 1-volt input, but the principle is the same: A_V = output/input.

The second problem is that biopotentials signals tend to be subject to intense 60-Hz ac fields from the surrounding ac power lines. These fields can introduce a tremendous ac component into the signal, and either shows up as a 60-Hz baseline component or totally obliterates the signal. The solution is to make the amplifier a differential-input type. Given that the two electrode wires are in the same field, they will be affected similarly by the field and produce equal inputs to the amplifier. The nature of the differential amplifier is that it will cancel equal potential inputs, so the 60-Hz interference goes away — and remains away so long as the balance is maintained.

Of course, it is sometimes difficult to maintain that balance, but perhaps never more so then when the operators of the equipment, typically nurses, physicians, or scientists, subvert the intent of the design. I recall one urgent summons to an intensive care unit in the hospital where I worked. The monitoring technician could not get the 60-Hz interference to go away, and so had summoned me — the repairman. The problem was that the patient tended to sweat a lot, and would not keep the electrodes on. The karaya adhesive used on the back of the electrodes kept coming loose, and the electrode would fall off. The nurse, being quick-witted and resourceful, painted the area the electrode site with a solution called Betadyne — a great disinfectant, but also a good insulator when dry. Hence, she made the observation that the ECG monitor worked fine after she made the change, but that it deteriorated within ten minutes and was now all 60 Hz and no ECG! The problem was that the Betadyne was a conductor when it was wet, but an insulator when it dried. The electrodes were essentially free-floating on the patient's body — caused them to act like an antenna for the 60-Hz field.

You now know about two parameters for the biopotentials amplifier: high gain and differential inputs. Is there anything else?

In Chapter 12 I noted the complex RC equivalent circuit beneath the skin. Researchers generally agree that the electrode-skin contact is actually a complex network, rather than a simple ohmic contact such as between two similar metals. There are actually two problems here: dc offset and impedance. The impedance of the electrode is anything from 1000 ohms to 20,000 ohms, so the input of the amplifier must be a high impedance. Thus, our amplifier must be a high-input impedance, high gain, differential amplifier (it's getting more complicated!)

The second problem is the dc offset potentials, V1 and V2. Because the electrodes are metallic and the skin is electrolytic, there will be a half-cell potential between the electrode and the skin when the two are placed in contact with each other. This potential can be many times higher than the signal potential, and so will easily saturate the high-gain amplifier needed for the signal. Of course, if potentials V1 and V2 are exactly equal, they will cancel out because of the differential inputs of the amplifier. Unfortunately, this situation does not exist very often, especially since a 1-millivolt difference out of a 1-volt potential can cause the amplifier to saturate!

The solution to be offset voltage problem is to ac couple the amplifier (more on this later). Our amplifier must have the following specifications.

1. Ac-coupled
2. High input impedance
3. Differential
4. High gain

The general diagram for a biopotentials amplifier is shown in Fig. 16-16. Here we see what might be required. The input amplifier, A1, is the biopotentials amplifier discussed above. The next stage is for frequency shaping and signal processing. In general, the frequency-shaping function must do any or all of the following: set low-pass frequency limit, set high-pass frequency limit, and set any notch-filter frequencies (e.g., 60 Hz to take out residual 60-Hz signals not handled by the differential input). Signals processing might be noth-

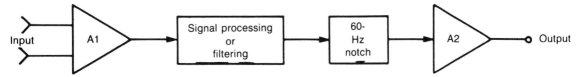

Fig. 16-16. Signal processing circuit.

ing but additional gain (it isn't always wise to put all gain in the same stage), or might include things like differentiation, integration, analog multiplication, and so forth (although many signal-processing functions are now done in software rather than hardware). The "read end" is merely a stage that provides display control over the signal: position, span, dc balance, etc.

Finally, there is the output buffer amplifier. This stage might be part of the rear end section, or provided separately. It has the function of isolating the amplifier from the external world.

AC COUPLING

The dc offset potential of the input electrodes makes it necessary to ac couple the signal amplifier. But there is also a requirement for low-end frequency response. The ECG signal, for example, normally has a frequency component spectrum of 0.05 Hz to 100 Hz. A signal of 0.05 Hz is almost dc, so one might be tempted to use a dc amplifier. Unfortunately, you must provide both 0.05 Hz as a low-end −3 dB point in the amplifier response, but also ac couple it to prevent saturation of the amplifier.

An input circuit solution is shown in Fig. 16-17. This circuit is more widely used. The amplifiers, A1 and A2, are merely the input amplifiers of a differential instrumentation amplifier (or its ICIA version). Resistors R1 and R2 are used to keep the capacitors from charging from the bias currents, and are sometimes deleted when those currents are picoampere in range (e.g., on BiMOS or BiFET operational amplifiers). Do not use less than 10 megohms for this application, as these resistors set the input impedance of the amplifier. Also, make sure to balance them either with hand-selection with an ohmmeter, or, by using 1-percent tolerance resistors.

The capacitors are selected according to the desired lower end −3 dB frequency response limit. The equation is:

$$C = \frac{1,000,000}{6.28\,R\,f}$$

where

C is the capacitance in microfarads
f is the frequency in Hertz
R is the resistance in Ohms

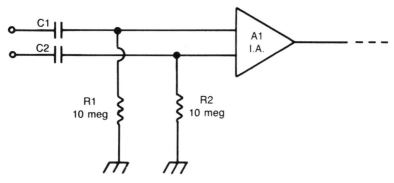

Fig. 16-17. Ac-coupled input.

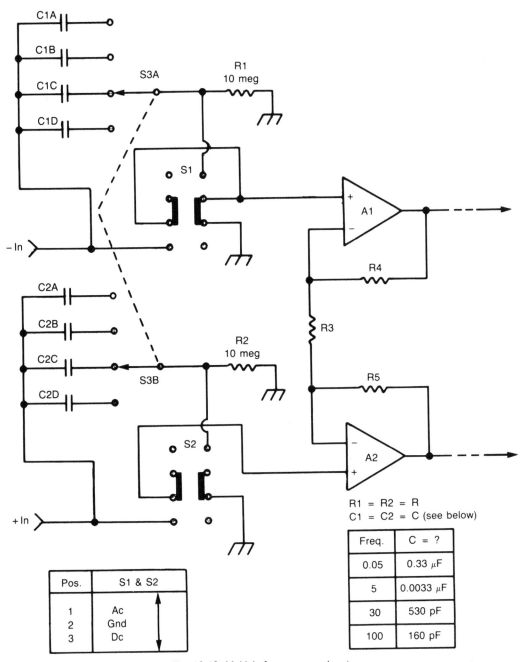

Fig. 16-18. Multiple-frequency ac input.

Pos.	S1 & S2
1	Ac
2	Gnd
3	Dc

R1 = R2 = R
C1 = C2 = C (see below)

Freq.	C = ?
0.05	0.33 μF
5	0.0033 μF
30	530 pF
100	160 pF

Example

$$C = \frac{1,000,000}{6.28 \, R \, f}$$

$$C = \frac{1,000,000}{(6.28)(10,000,000)(0.05 \text{ Hz})}$$

$$C = 0.32 \; \mu F$$

Because 0.33 is the next higher "standard value," one would normally select 0.33 μF for this application: C1 = C2 = 0.33 μF.

The upper -3 dB frequency limit is set by other means. If the operational amplifiers have compensation terminals, then capacitors or RC networks across those terminals are used. Otherwise, select a capacitor to shunt the feedback resistor in any stage. The value of the capacitor is found from the same formula as above.

Figure 16-18 shows an input network that is designed for biopotentials amplifiers used for more than ECG or other single-use clinical measurements. It may be that the user will want to select either of the following features:

1. Single-ended input,
2. Differential input,
3. Ac-coupling,
4. Dc-coupling, and/or
5. Various low-frequency responses

The circuit in Fig. 16-18 shows the input stages of an IC instrumentation amplifier, or the equivalent circuit based on three operational amplifiers (see Chapter 15). The inputs are individually selected for ac-coupling, dc-coupling, or ground by switches S1 and S2. These switches are not ganged together so that the two inputs can be individually selected. The reason for this is that someone might want to make a single-ended amplifier, and can do so by grounding the unwanted input. To make an inverting single-ended amplifier, for example, set switch S2 to the GND position. This setting forces the input of A2 to ground potential, and allows A1 to control the circuit. Similarly, to make a noninverting single-ended amplifier, the user will merely set the other switch (S1) to GND for exactly the same reason.

If the amplifier is to be dc-coupled and differential, then both S1 and S2 are set to position 1. Similarly, if it is to be ac-coupled and differential, you set S1 and S2 to position 3. The ac-dc coupling decision can be made regardless of whether single-ended or differential configuration is selected.

The frequency response is tailored by a ganged pair of single-pole, four-position (SP4P) rotary switches. These switches select the capacitor value used in the input circuit. Values for the common low-frequency responses encountered in biological and medical research are shown in the inset table in Fig. 16-18.

Chapter 17

Laboratory Amplifiers for Data Acquisition

IT IS WELL RECOGNIZED THAT DATA-ACQUISITION chores often require the amplification of analog signals. Even in an era of massive computerization of these types of instruments, there is still need for the analog subsystem (if only to boost an analog signal to the point where it can be input to an A/D converter connected to a computer).

In addition to this simple scaling function, it is also often either possible or desirable to do some of the signal processing in the analog subsystem. While this statement seems like heresy to computer-oriented engineers, it is nonetheless a reasonable trade-off. There may be a situation in which computer hardware or timeline constraints make it less costly to use an analog circuit. "But," protests the computerist, ". . . isn't the computer solution always superior to analog?" While this sentiment is popular, it is also false.

Similarly, it is often stated that the computer solution is "better" than the analog solution. Again, even though this claim may be true most of the time, it is not universally true. Engineers thrive on change: ". . . when the company decides to stick with the present designs in the inventory it is time to find another job" is an old maxim—almost an article of faith in the profession. One wag told me once that the proper Greek letters for a fraternity for engineering students is "delta-delta-delta!" While the yearning of the engineer is for change, and the engineer naturally wants to make every product better, the engineer sometimes has to face the reality that 'better is the enemy of good enough.'

In light of the foregoing, I am going to consider the use of standard laboratory amplifiers for certain data-acquisition chores. In Chapters 15 and 16 I discussed operational and other linear IC amplifiers which you can use to design your own analog subsystem. In this chapter I will discuss certain amplifiers that may be already in existence in your facility.

The term "laboratory amplifiers" describes a wide range of instruments of many and varying capability. Some of these instruments are categorized according to several schemes. For example, we can divide them according to coupling method: dc vs ac. In the case of (especially) ac amplifiers, there is often a frequency-response characteristic that will take some of the burden of filtering in the system. Amplifiers can also be categorized according to gain:

Low gain   1 to 100
Medium gain   100 to 1000
High gain   >1000

Some amplifiers carry names that represent certain special applications. For example, the biopotentials amplifier (see Chapter 11) is used to acquire signals from living things, and because of certain practical problems tend to have very high input impedances and certain other characteristics. Laboratory amplifiers can be either free-standing models, or part of a plug-in mainframe system. An example of the plug-in types are the Hewlett-Packard 8800-series.

In the sections below I will discuss some of the special types of laboratory amplifier that may be useful in certain specific cases.

CHOPPER AMPLIFIERS

One of the unfortunate characteristics of simple dc amplifiers is that they tend to be noisy and possess a certain inherent drift of both gain and dc offset baseline. In low and medium gain applications these problems are less important than in high-gain amplifiers, especially in the lower regions of those gain ranges. As gain increases, however, these problems loom much larger. For example, a drift of 50 μV/°C in an $\times$100 medium gain amplifier produces an output voltage of

$$(50 \ \mu V/°C) \times 100 = 5 \ mV/°C$$

This is tolerable in most cases. But in an $\times$20,000 high-gain amplifier the output voltage would escalate to

$$(50 \ \mu V/°) \times 20,000 = 1-V/°C$$

and that amount will obscure any real signals in a short period of time.

Similarly, noise can be a problem in high-gain applications, where it had been negligible in most low to medium-gain applications. Operational-amplifier noise is usually specified in terms of nano-

volts of noise per square root hertz (i.e., Noise(rms) = nV/(Hz)$^{1/2}$. A typical low-cost operational amplifier has a noise specification of 100 nV/(Hz)$^{1/2}$, so at a bandwidth of 10 kHz the noise amplitude will be

$$Noise(rms) = 100 \ nV \times (10,000)^{1/2}$$
$$Noise(rms) = 100 \ nV \times 100$$
$$Noise(rms) = 10,000 \ nV = 0.00001 \ volts$$

In an $\times$100 amplifier, without low-pass filtering, the output amplitude will be only 1 mV, but in an $\times$100,000 amplifier it will be 1 volt.

A circuit called a *chopper amplifier* will solve both problems, because it makes use of an ac-coupled amplifier in which the advantages of feedback can be optimized.

The drift problem is cured (or at least reduced significantly) because of two properties of ac amplifiers; one is the inability to pass low frequency (i.e., near-dc) changes such as those caused by drift, and the other is the ability to regulate the stage through the use of heavy doses of negative feedback.

But many low-level analog signals are very low frequency, i.e., in the dc-to-30-Hz range (for example, human ECG signals have frequency components down to 0.05 Hz) and will not pass through such an amplifier. The answer to this problem is to chop the signal so that it passes through the ac amplifier, and then to demodulate the amplifier output signal to recover the original waveshape, but at a higher amplitude.

Figure 17-1 shows a block diagram of the basic chopper circuit. The traditional chopper is a vibrator-driven SPDT switch (S1) connected so that it alternately grounds first the input and then the output of the ac amplifier. An example of a chopped waveform is shown in Fig. 17-2. A low-pass filter following the amplifier will filter out any residual chopper "hash" and any miscellaneous noise signals that may be present.

Most of the mechanical choppers used a chop rate of either 60 or 400 hertz, although 100-, 200-, and 500-Hz choppers are also found. The main criterion for the chop rate is that it be twice the highest component frequency that is present in the

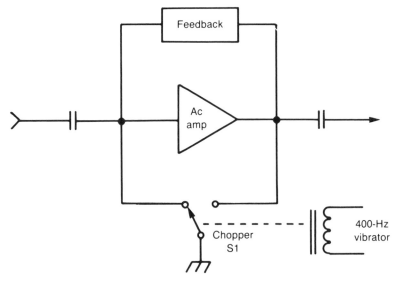

Fig. 17-1. Chopper amplifier.

input waveform. In other words, it must obey Nyquist's sampling criterion.

A differential chopper amplifier is shown in Fig. 17-3. In this circuit an input transformer with a center-tapped primary is used. One input terminal is connected to the transformer primary center-tap, while the other input terminal is switched back and forth between the respective ends of the transformer primary winding.

A synchronous demodulator following the ac amplifier detects the signal and restores the original, but now amplified, waveshape. Again, a low-pass filter smoothes out the signal.

The modern chopper amplifier may not use mechanical vibrator switches as the chopper because a pair of CMOS or JFET electronic switches driven out of phase with each other will perform exactly the same job. Other electronic switches used in commercial chopper amplifiers include PIN diodes, varactors, and optoisolators.

Chopper amplifiers limit the noise because of both the low-pass filtering required and because of the fact that the ac amplifier can be tuned to a narrow passband around the chopper frequency.

At one time the chopper amplifier was the only way to obtain low drift in high-gain situations. Modern IC and hybrid amplifiers, however, have

such improved drift properties that no chopper is needed (especially in the lower end of the "high-gain" range).

CARRIER AMPLIFIERS

A carrier amplifier is any type of signal-processing amplifier in which the signal carrying the desired information is modulated onto another signal, i.e., a "carrier." The chopper amplifier is considered by many to fit this definition, but is usually regarded as a type in its own right. The two principal carrier amplifiers are the dc-excited and ac-excited varieties.

Figure 17-4 shows a dc-excited carrier amplifier. The Wheatstone bridge transducer is excited by a dc potential, V. The output of the transducer, then, is a small dc voltage that varies with the value of the stimulating parameter. The transducer signal is usually of very low amplitude and is noisy. An amplifier builds up the amplitude, and a low-pass filter removes much of the noise. In some models the first stage is actually a composite of these two functions, being a filter with gain.

The signal at the output of the amplifier-filter section is used to amplitude modulate a carrier signal. Typical carrier frequencies range from 400 Hz to 25 KHz, with 1 kHz and 2.5 kHz being very common. The signal frequency response of a carrier amplifier is a function of the carrier frequency, and is usually considered to be one-fourth of that carrier frequency. A carrier frequency of 400 Hz, then, is capable of signal frequency response of 100 Hz, while the 25 kHz carrier will support a frequency response of 6.25 kHz. Further amplification of the signal is provided by an ac amplifier.

The key to the performance of any carrier amplifier worthy of the name is the phase-sensitive

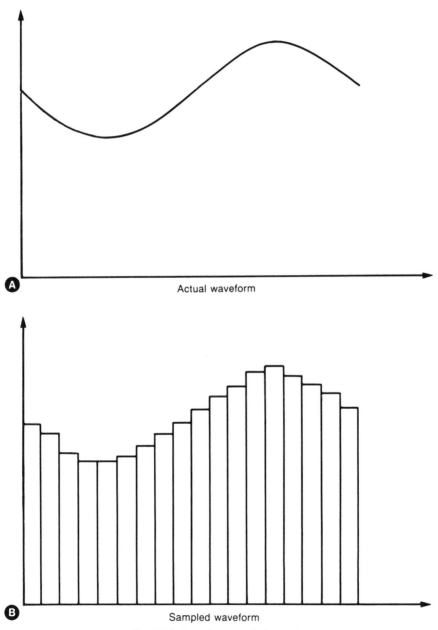

Fig. 17-2. Actual vs sampled waveform.

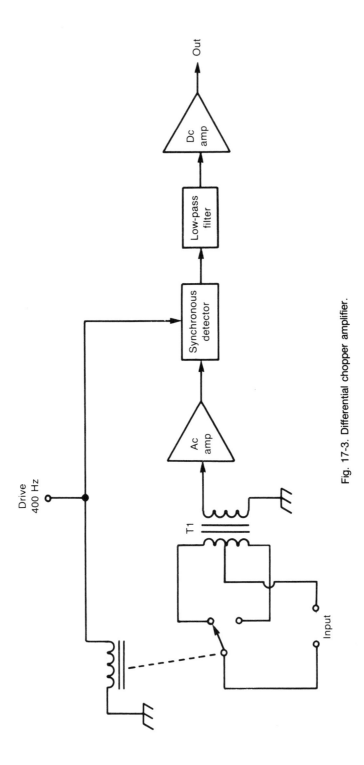

Fig. 17-3. Differential chopper amplifier.

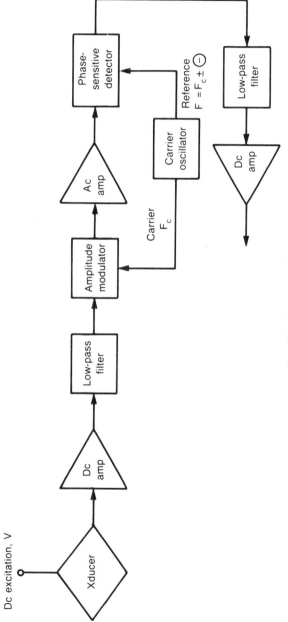

Fig. 17-4. Carrier amplifier.

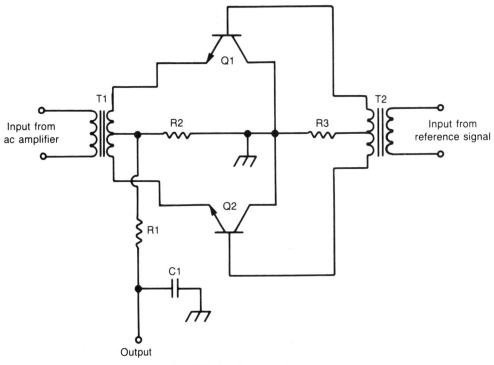

Fig. 17-5. Synchronous detector.

detector (PSD) that demodulates the amplified ac signal. Envelope detectors, while very simple and low in cost, suffer from an inability to discriminate between the real signals and spurious signals.

Figure 17-5 shows a simplified version of one form of PSD circuit. Transistors Q1 and Q2 provide a return path to ground for the opposite ends of the secondary winding of input transformer T1. These transistors are alternately switched into and out of conduction by the reference signal in such a way that Q1 is off when Q2 is on, and vice versa. The output waveform of the PSD is a full-wave rectified version of the input ac signal.

Other electronic switching schemes are also used in PSD design. All systems are designed using the fact that a PSD is essentially an electronic DPDT switch. The digital PSD most often seen uses a CMOS electronic IC switch such as the 4066 (in low-cost systems). These switches are toggled by the reference frequency in such a way that the output is always positive-going, regardless of the phase of the input signal.

The advantages of the PSD include the fact that it rejects signals not of the carrier frequency, and certain signals that are of the carrier frequency. The PSD, for example, will reject even harmonics of the carrier frequency, and also those components that are out of phase with the reference signal. The PSD will, however, respond to odd harmonics of the carrier frequency. Some carrier amplifiers seem to neglect this problem altogether. But in some cases, manufacturers will design the ac-amplifier section to be a bandpass amplifier with a response limited to $F_C -/+ (F_C/4)$. This response will eliminate any third, or higher-order, odd harmonics of the carrier frequency before they reach the PSD. It is then necessary only to assure the purity of the reference signal.

An alternate, but very common, form of carrier amplifier is the ac-excited circuit shown in Fig. 17-6. In this circuit the transducer is ac-excited by the carrier signal, eliminating the need for the amplitude modulator. The small ac signal from the transducer is amplified and filtered before being

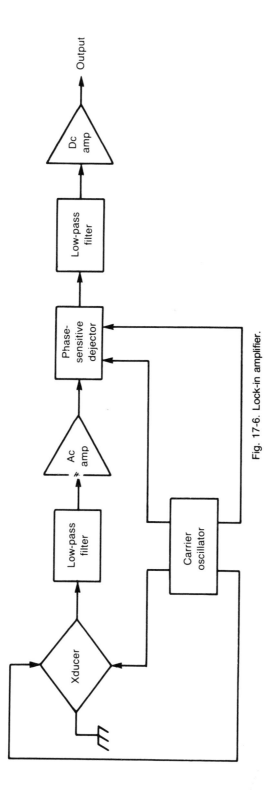

Fig. 17-6. Lock-in amplifier.

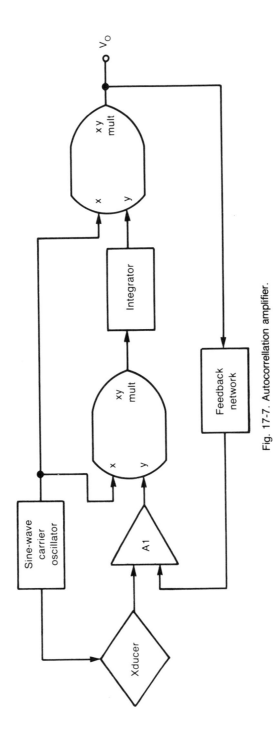

Fig. 17-7. Autocorrelation amplifier.

applied to the PSD circuit. Again, some designs use a bandpass ac amplifier to eliminate odd harmonic response. This circuit allows adjustment of transducer offset errors in the PSD circuit instead of in the transducer, by varying the phase of the reference signal.

LOCK-IN AMPLIFIERS

The amplifiers discussed so far in this chapter produce relatively large amounts of noise, and will respond to noise present in the input signal. They suffer from shot noise, thermal noise, H-field noise, E-field noise, ground loop noise, and so forth. The noise voltage or power at the output is directly proportional to the square root of the circuit bandwidth. The lock-in amplifier is a special case of the carrier-amplifier idea in which the bandwidth is very narrow. Some lock-in amplifiers use the carrier amplifier circuit of Fig. 17-7, but use an input amplifier with a very high Q bandpass. The carrier frequency may be anything between 1 Hz and 200 kHz. The lock-in principle works because the information signal is made to contain the carrier frequency in a way that is easy to demodulate and interpret. The ac amplifier accepts only a narrow band of frequencies centered about the carrier frequency. The narrowness of the bandwidth, which makes possible the improved signal-to-noise ratio, also limits the lock-in amplifier to very low frequency input signals. Even then it is sometimes necessary to time-average (i.e., integrate) the signal for several seconds to obtain the needed data.

Lock-in amplifiers are capable of thinning out the noise, and retrieving signals that are otherwise "buried" in the noise level. Improvements of up to 85 dB are relatively easily obtained, and up to 100 dB is possible if cost is no factor.

There are actually several different forms of lock-in amplifier. The type discussed above is perhaps the simplest type. It is merely a narrow band version of the ac-excited carrier amplifier. The lock-in amplifier of Fig. 17-7, however, uses a slightly different technique. It is called an *autocorrellation amplifier*. The carrier is modulated by the input signal, and then integrated (i.e., timeaveraged). The output of the integrator is demodulated in a product-detector circuit. The circuit of Fig. 17-7 produces very low output voltages for input signals that are not in phase with the reference signal, but produces relatively high output at the proper frequency.

Chapter 18

Electronic Integrator and Differentiator Circuits

T HE MATHEMATICAL PROCESSES OF INTEGRATION and differentiation are part of the signal processing necessary in data-acquisition systems. In this chapter I discuss the operational-amplifier circuits that will output a voltage analog that is proportional to either the time average or instantaneous rate of change of the input signal.

The names for these circuits come from the time when operational amplifiers were intended almost solely as building blocks in analog computers, and refer to the mathematical operations of integration and differentiation. Differentiation is the art of finding the derivative of a curve, that is, its instantaneous rate of change. For the simplest case, a straight line as shown in Fig. 18-1A, the derivative is simply the slope of the line, or $(Y_2-Y_1)/(X_2-X_1)$. In this case we usually write the expression for the slope with the Greek letter "delta" (i.e., delta-y over delta-x) to indicate a small change in x and a small change in y.

For the case of a straight line, the derivative is simple to calculate. But situations are frequently encountered where the line is not straight. Figure 18-1B gives an example of a voltage that varies with time. Here the curve really is a curve and is not the simple straight-line case. If you want to

know the instantaneous rate of change, i.e., the rate of change at a point, then you can take the derivative of a line tangent to that point.

Integration is the inverse of differentiation, and is used to find the area under a curve. In electronics, you might want to take the area under a time-varying voltage curve. Figure 18-2 shows a voltage that represents a pressure transducer output, in this particular case the output of a human blood-pressure transducer. Note the pressure/voltage varies with time from a low ("diastolic") to a high ("systolic") between T_1 and T_2 (which represents one cardiac cycle). If you want to know the mean blood pressure, then you would want to find the area under the curve, as shown by the formula in Fig. 18-2. From this little illustration you can see that the integrator serves to find the time-average of an analog waveform.

Integrators and differentiators affect signals in different ways. Figure 18-3 shows the example of a squarewave (A) applied to the inputs of an integrator (B) and differentiator (C). At time T_1 the squarewave makes a positive-going transition to maximum amplitude. At this time it has a very high rate of change, so the output of the differentiator is very high (see waveform C at T_1). But then the amplitude of the input signal reaches maximum and remains constant until T_2, when it drops back to its previous value. Thus, the differentiator will produce a sharp, positive-going spike at T_1 and a sharp negative-going spike at T_2. These spikes are frequently used in circuits such as timers and zero-crossing detectors.

The integrator waveform in Fig. 18-3 shows a constant positive-going slope between T_1 and T_2. The steepness of the slope is dependent upon the amplitude of the input squarewave, but the line is linear. You can see from curve B in Fig. 18-3 that the squarewave into the integrator produces a triangle waveform. Consider the case of a sinewave

applied to the input of the differentiator. The instantaneous voltage of a sinewave is given by:

$$V_i = V \sin (2 \pi f t) \qquad \textbf{Eq. 18-1}$$

where

V_i is the instantaneous voltage
V is the peak voltage
f is the frequency in hertz (Hz)
t is the time in seconds (s)

The differentiator output voltage is found by

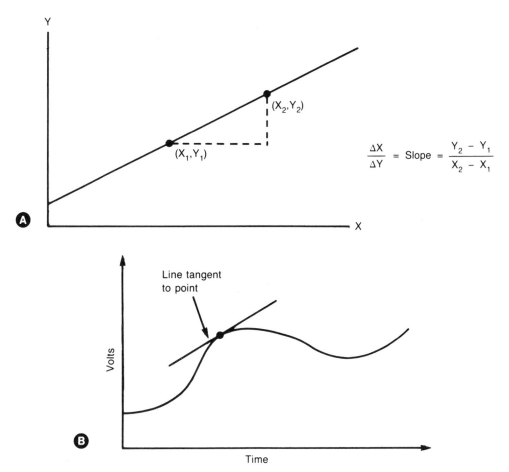

$$\frac{\Delta X}{\Delta Y} = \text{Slope} = \frac{Y_2 - Y_1}{X_2 - X_1}$$

Fig. 18-1. Differentiation.

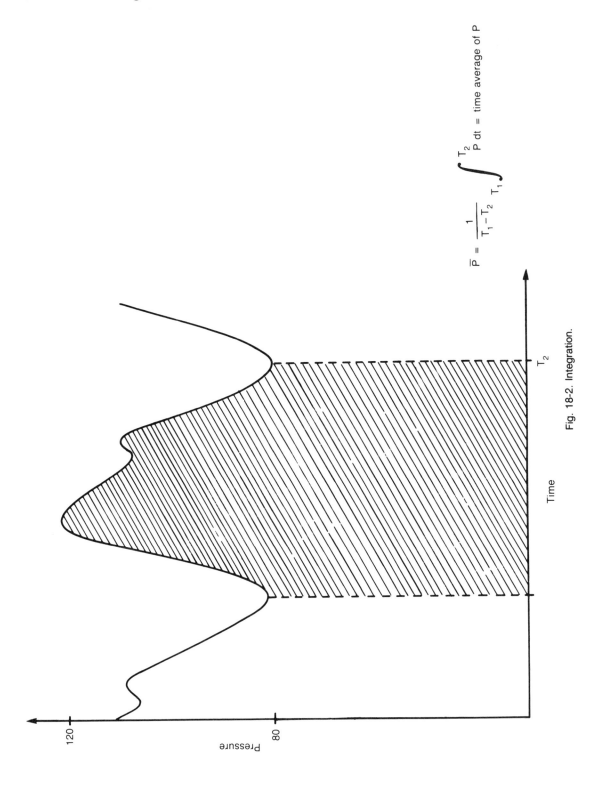

$$\overline{P} = \frac{1}{T_1 - T_2} \int_{T_1}^{T_2} P \, dt = \text{time average of P}$$

Fig. 18-2. Integration.

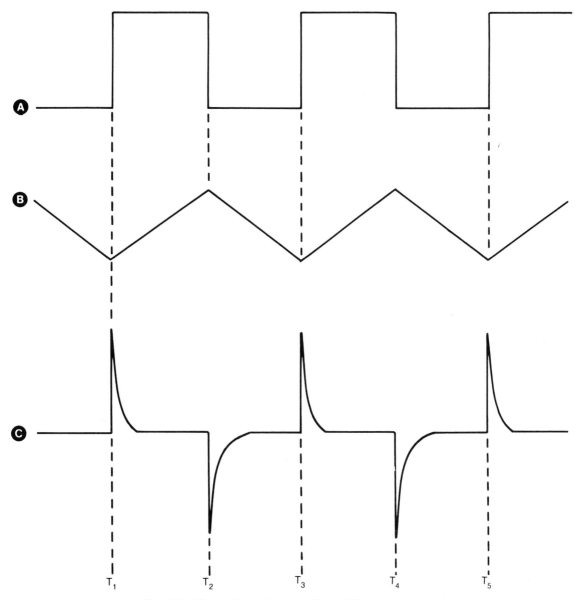

Fig. 18-3. Effect of integration and differentiation on squarewaves.

substituting the sine expression above into the transfer function for a differentiator:

$$V_0 = RC \, dV/dt \qquad \textbf{Eq. 18-2}$$

$$V_0 = RC \, \frac{d(V \sin (2 \pi f t))}{dt} \qquad \textbf{Eq. 18-3}$$

$$V_0 = RC(2 \pi f)(V \cos (2 \pi f T)) \qquad \textbf{Eq. 18-4}$$

Thus, if a sinewave is applied to the inputs of integrators and differentiators, then the result is a sine-wave output that is shifted in phase 90 degrees. The principal difference between the two forms of circuit is in the direction of the phase shift.

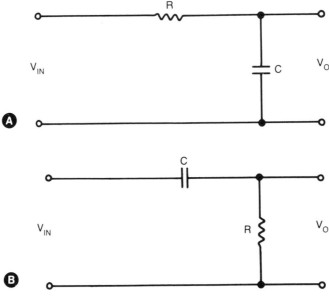

Fig. 18-4. A) RC integrator B) RC differentiator.

Integrators and differentiators can also be viewed in terms of their effects as filters. The integrator is basically a low-pass filter for analog signals, while the differentiator is a high-pass filter.

PASSIVE RC CIRCUITS

Perhaps the simplest form of integrator and differentiator circuits are made from simple resistor and capacitor elements, as shown in Fig. 18-4. The integrator is shown in Fig. 18-4A, while the differentiator is in Fig. 18-4B. The integrator consists of a resistor element in series with the signal line, and a capacitor across the signal line. The differentiator is just the opposite, the capacitor is in series with the signal line, while the resistor is in parallel with the line. You may recognize these circuits as low-pass and high-pass filters, respectively. The low-pass case (integrator) has a −6 dB/octave falling frequency response, while the high-pass case (differentiator) has a +6 dB/octave rising frequency response.

The operation of the integrator and differentiator is dependent upon the time constant of the RC network (i.e., R × C). In most cases you want the integrator time constant to be long (i.e., 1X) compared with the period of the signal being integrated, while in the differentiator you want the RC time constant short (i.e., 1/T) the period of the signal. You can cascade several integrators in order to enhance the effect and also increase the slope of the frequency-response fall-off.

OP-AMP CIRCUITS

The operational amplifier makes it a lot easier to build high-quality active integrator and differentiator circuits. Previously, it was necessary to construct a high-gain transistor amplifier for this purpose. Figure 18-5 shows the basic circuit of the operational-amplifier differentiator. Again there are RC elements, but in a slightly different context. The capacitor is in series with the op-amp's inverting input, while the resistor is the op-amp feedback resistor.

By Kirchoff's Current Law (KCL) we know that the currents (I_1 and I_2) into and out of the summing junction evaluate to zero. From this we know that:

$$I_2 = -I_1 \qquad \text{Eq. 18-5}$$

We also know from elementary circuit theory that

$$I_1 = \frac{D \, dV_{IN}}{dt} \qquad \textbf{Eq. 18-6}$$

and,

$$I_2 = V_O/R \qquad \textbf{Eq. 18-7}$$

Next, by substituting Eqs. 18-6 and 18-7 into Eq. 18-5 it is found that

$$\frac{V_O}{R} = \frac{C \, dV_{IN}}{dt} \qquad \textbf{Eq. 18-8}$$

or, with the terms of Eq. 18-8 rearranged we get

$$V_O = - RC \frac{dV_{IN}}{dt} \qquad \textbf{Eq. 18-9}$$

where V_O and V_{IN} are in the same units (volts, millivolts, etc.), R is in ohms, C is in farads, and t is in seconds.

This is a mathematical way of saying that output voltage V_O is equal to the RC time constant times the derivative of input voltage V_{IN} with respect to time (the "dV_{IN}/dt") part. Because the circuit is essentially a special case of the familiar inverting-follower circuit, the output is inverted, and thus the negative sign.

Figure 18-6 shows the classical operational amplifier version of the Miller integrator circuit.

Again, an operational amplifier is the active element, while a resistor is in series with the inverting input and a capacitor is in the feedback loop. Note that the placement of the capacitor and resistor elements are exactly opposite in both the RC and operational-amplifier versions of integrator and differentiator circuits. In other words, the RC elements reverse roles between Figs. 18-5 and 18-6. That fact will tell the astute observer a little bit regarding the nature of integration and differentiation. The output of the integrator is dependent upon the input signal amplitude and the RC time constant.

You may derive the transfer function for the operational integrator using a procedure similar to the differentiator case:

$$I_2 = -I_1 \qquad \textbf{Eq. 18-10}$$
$$I_1 = V_{IN}/R \qquad \textbf{Eq. 18-11}$$
$$I_2 = C \, dV_O/dt \qquad \textbf{Eq. 18-12}$$

Substituting Eqs. 18-11 and 18-12 into Eq. 18-10 yields

$$\frac{C \, dV_O}{dt} = \frac{-V_{IN}}{R} \qquad \textbf{Eq. 18-13}$$

Integrating both sides of Eq. 18-13 yields

$$\frac{C \, V_O}{dt} \, dt = \frac{-V_{IN}}{R} \, dt \qquad \textbf{Eq. 18-14}$$

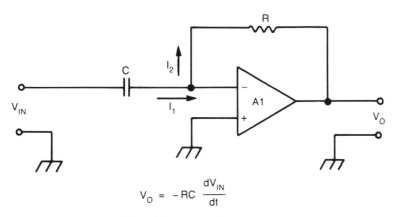

$$V_O = - RC \frac{dV_{IN}}{dt}$$

Fig. 18-5. Active differentiator.

$$C V_O = \frac{-V_{IN}}{R} dt \qquad \textbf{Eq. 18-15}$$

Rearranging Eq. 18-15 and collecting terms gives us

$$V_O = \frac{-1}{RC} \; V_{IN} \, dt + K \quad \textbf{Eq. 18-16}$$

where V_O and V_{IN} are in the same units (volts, millivolts, etc); R is in ohms; C is in farads, and t is in seconds.

This expression is a way of saying that the output voltage is equal to the time-average of the input signal, plus some constant K which is the voltage that may have been stored in the capacitor from some previous operation (often zero in electronic applications).

PRACTICAL CIRCUITS

The circuits shown in Figs. 18-5 and 18-6 are classic ones and appear in all manner of textbooks and magazine articles. Unfortunately, they don't work very well in actual practice (or at all in some cases). The big problem is that these circuits are simplistic because they depend upon ideal operational amplifiers. Unfortunately, the real kind-you-can-go-buy op-amps fall far short of the ideal in the mind of the textbook writer. In real circuits you

find that differentiators ring and oscillate, and integrators saturate very shortly after turn-on.

The problem with op-amp integrators was driven home to me when I worked in a medical school/hospital bioelectronics lab and had to build an electronic integrator for one of the customers of our electronics laboratory. When I used a 741 operational amplifier, the output voltage saturated within milliseconds after turn-on. In fact, saturation came so fast that I initially thought the op-amps were bad. The problem was that the input bias currents of the op amp (which are zero in ideal devices) create a high enough output voltage to fully charge the capacitor in the feedback loop very rapidly.

There is another problem with this kind of circuit, and it magnifies the problem of saturation. This circuit has a very high gain with certain values of R and C. Let's pick an example and see what this gain can mean. The gain of this circuit is given by the term $-1/RC$. What is the gain with a 0.01 μF capacitor (certainly not a "large" capacitor in conventional wisdom) and a 10,000 ohm resistor:

(note: 0.01 μF is 10^{-8}, or 0.00000001 farads)

$$A_V = -1/RC$$
$$A_V = -1/(10,000 \text{ ohms})(0.00000001 \text{ farads})$$
$$A_V = -1/0.0001$$
$$A_V = -10,000$$

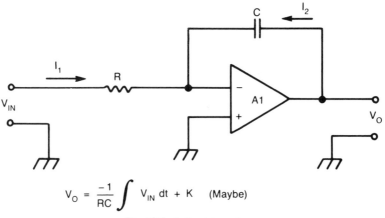

$$V_O = \frac{-1}{RC} \int V_{IN} \, dt + K \quad \text{(Maybe)}$$

Fig. 18-6. Active integrator.

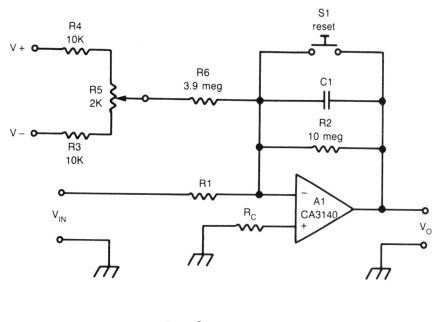

$$V_O = \frac{-1}{R1\ C1} \int V_{IN}\ dt\ +\ K$$

Fig. 18-7. Practical integrator.

In other words, with a gain of −10,000, a +1 volt applied to the input will want to produce a −10,000 volt output. Unfortunately, the operational amplifier output is limited to −10 to −20 volts, depending upon the device and the applied V− dc power supply voltage. For this case the operational amplifier will saturate very rapidly! If you want to keep the output voltage from saturating, then you must either keep the RC time constant under control or prevent the input signal from rising too high (not good!). If the maximum output voltage allowable is 10 volts, then the maximum input signal is 10 volts/10,000 or 1 millivolt. Obviously, the best solution is to keep the RC time constant within bounds.

When I built my first integrator, and found that 741 devices were not suitable, I turned to the high-cost premium-grade devices. At that time, a premium 725 device cost $15, and it suffered the same problems as the 741. The only difference between the $15 premium op-amp and the $0.50 741 device is that on the $15 op-amp the output saturated slowly enough for me to watch it on an oscilloscope or voltmeter—about 4 seconds—instead of nearly instantaneously. Unfortunately, this was still not acceptable. Applying a waveform to the input of even the premium op-amp integrator allowed me to see the output waveform rise up the screen of the oscilloscope and disappear off the top of the screen!

How to Solve the Saturation Problem

Fortunately, there are some design tactics that will allow us to keep the integration aspects while getting rid of the problems. A practical integrator is shown in Fig. 18-7. The heart of this circuit is an RCA BiMOS operational amplifier, type CA-3140, or its equivalent BiFET type. The reason why this works so well is that it has a low input bias current (being MOSFET input). When I tested close to a dozen different op amps for the project, the CA-3140, which cost about $2, outperformed devices costing ten times as much.

Capacitor C1 and resistor R1 in Fig. 18-7 form the integration elements, and are used in the equation. Resistor R2 is used both to discharge C1 to prevent offsets from input signal and the op amp itself from saturating the amplifier and to limit the gain at low frequencies. circuit. The RESET switch is used to set the capacitor voltage back to zero (to prevent a "K" factor offset) before the circuit is used. In some measurement applications the circuit initializes by closing S1 (or a relay equivalent) momentarily.

Because of R2 in the circuit we must place a constraint on Eq. 18-16. The equation is valid only for frequencies greater than or equal to F in Eq. 18-17:

$$F = \frac{10^6}{2 \pi R2 \, C} \qquad \text{Eq. 18-17}$$

where

f is in hertz (Hz)
R2 is in ohms
C is in microfarads (μF)

In both integrators (Figs. 18-6 and 18-7) there is a compensation resistor, R_C, between the noninverting input of the operational amplifier and ground. This resistor cancels the effects of input bias current (see Chapter 15), and has a value equal to the parallel combination of R1 and R2:

$$R_C = \frac{R1 \times R2}{R1 + R2} \qquad \text{Eq. 18-18}$$

The RC time constant of an operational amplifier integrator should be approximately equal to (or greater than) the period of the signal being integrated. The frequency of a signal is the reciprocal of its period:

$$f = 1/t \qquad \text{Eq. 18-19}$$

or, when t = RC:

$$f = \frac{1}{R1 \, C} \qquad \text{Eq. 18-20}$$

and,

$$R1 \, C = 1/f \qquad \text{Eq. 18-21}$$

Example. Assume a 200-Hz, 1-volt (peak) square-wave input voltage, V_{IN}, to the circuit of Fig. 18-7. Find appropriate values for R1 and C. Also find the peak voltage output.

Solution. We know from Eq. 18-21 that R1 $\times$ C is 5 milliseconds (ms):

$$R1 \, C = 1/f \qquad \text{Eq. 18-22}$$
$$R1 \, C = 1/200 \text{ Hz} \qquad \text{Eq. 18-23}$$
$$R1 \, C = 0.005 \text{ sec} = 5 \text{ ms} \qquad \text{Eq. 18-24}$$

Assuming a trial value for C of 0.01 uF (10_7 farads), we find:

$$R1 = 0.005/C \qquad \text{Eq. 18-25}$$
$$R1 = 0.005 \text{ sec}/10^{-7} \text{ F} \qquad \text{Eq. 18-26}$$
$$R1 = 50,000 \text{ ohms} \qquad \text{Eq. 18-27}$$

Thus, for an integrator such as Fig. 18-7 you would want R1 = 50 kohms and C1 = 0.1 μF. The peak voltage is found from these values and the input signal applied to the transfer equation:

$$V_O = \frac{-1}{R1 \, C} V_{IN} \, dt \qquad \text{Eq. 18-28}$$

where

V_O is the output voltage
V_{IN} is the input voltage
R1 is in ohms
C is in farads
t is the period of the input voltage

Plugging in the values from above:

$$V_O = \frac{-1}{(5 \times 10^4 \text{ ohms})(10^{-7} \text{ farads})} (1 \text{ V}) dt$$
$$\text{Eq. 18-29}$$

$$V_O = \frac{-1}{0.005} (1 \text{ V}) dt \qquad \text{Eq. 18-30}$$

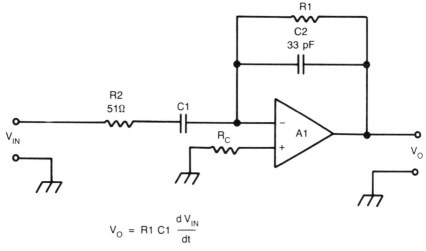

$$V_O = R1 \; C1 \; \frac{d \, V_{IN}}{dt}$$

Fig. 18-8. Practical differentiator.

$t = 0.005$ sec

$V_O = (-200)(1 \text{ V})\big|_0$ **Eq. 18-31**

$V_O = (-200)(1 \text{ V})(0.005 \text{ sec}) - [(-200)(1 \text{ V})(0)]$ **Eq. 18-32**

$V_O = (1 \text{ V}) - (0) = 1$ volt   **Eq. 18-33**

There is still a minor drift problem, so potentiometer R5 is added to cancel it. This component adds a slight countercurrent to the inverting input

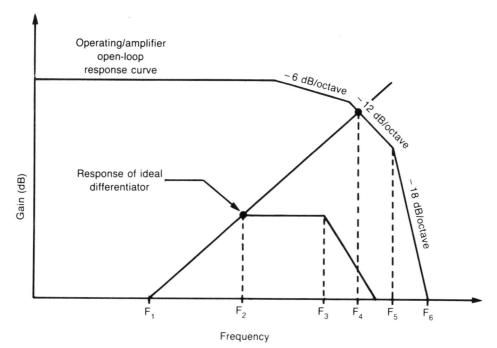

Fig. 18-9. Differentiator frequency-response plot.

through resistor R6. To adjust this circuit, set R5 initially to mid-range. The potentiometer is adjusted by shorting the V_{IN} input to ground (setting $V_{IN} = 0$), and then measuring the output voltage. Press S1 to discharge C1, and note the output voltage should go to zero. If it does not, turn R5 in the direction that counters the change of V_O after each time S1 is pressed. Keep pressing S1 and then making small changes in R5 until you find that the output voltage stays very nearly zero, and constant, after S1 is pressed (there will be some long-term drift normally).

Figure 18-8 shows the practical version of the differentiator circuit. The differentiation elements are R1 and C1, and the previous equation for the output is used. Capacitor C2 is a small value unit (1 to 100 pF), and is used to alter the frequency response of the circuit to prevent oscillation on fast rise-time inputs and to reduce the high-frequency noise in the output signal. Similarly, a snubber resistor in the input (R2) also limits this problem. The operational amplifier can be almost any type with a sufficient slew rate.

The snubber resistor value must be set in consideration of the C1 capacitance value and the maximum frequency at which the device operates as a differentiator. This frequency is defined as:

$$f = \frac{1}{2 \pi R2 \, C} \qquad \textbf{Eq. 18-34}$$

Or, you can rearrange Eq. 18-34 to find R2:

$$R2 = \frac{1}{2 \pi f \, C} \qquad \textbf{Eq. 18-35}$$

Chapter 19

Signal-Acquisitions Problems and How to Solve Them

WHEN THE DATA-ACQUISITION SYSTEM IS DEAL-ing with large signals, perhaps those in the over-100-millivolts range, and in an electrically clean environment, there are few problems to be solved. But when weak signals and/or a noisy electrical environment exist, there might be quite a few problems in either the accuracy or very existence of collected data. In this chapter I discuss a few of the more common forms of problems, and their usual solutions.

NOISE

Noise can be defined as any unwanted signal, even though a somewhat narrower definition is sometimes sought. Noise signals can mix with valid signals to form erroneous artifactual signals that negate the value of the data collected—unless steps are taken to either limit or eliminate the noise. With some of the methods shown below you can at least reduce the effects of noise to a point where it can safely be ignored. By noise I mean any signal that is not part of the desired signal. Several different forms of noise signals can be recognized: white noise, impulse noise, and interference noise.

White noise contains all possible frequencies, and so gets its name from analogy to white light,

which contains all colors. Such noise is also called *Gaussian noise,* although in reality it is neither "white" nor "gaussian" unless there are no bandwidth limits placed on the system. True white noise has a bandwidth from dc to daylight and beyond. In most data acquisition systems, however, there are bandwidth limitations so the noise is actually pseudo-gaussian "pink" noise. True gaussian noise can be eliminated completely by low-pass filtering because it by nature integrates to zero, given sufficient time. Bandwidth-limited noise, however, does not integrate to zero, but to a low value. The effect of low-pass filtering on pink noise is therefore not total.

An analogy to pseudo-gaussian or pink noise is the hiss heard between stations on an FM broadcast band receiver. Much of the noise in instrumentation systems is due to thermal sources and has an rms value of

$$V_N = (4KTBR)^{1/2}$$

where

V_N is the noise signal, in volts
K is Boltzmann's constant $(1.38 \times 10^{-23}$ joules per degree Kelvin).

T is the temperature in degrees Kelvin

B is the bandwidth in hertz

R is the circuit resistance in ohms

Example. Find the rms amplitude of the noise signal at room temperature (i.e., 300 degrees Kelvin) if the bandwidth of a circuit is 1000 Hz and the circuit resistance is 100,000 ohms.

$$V_N = (4KTBR)^{1/2}$$
$$V_N = [(4)(300)(1000)(1.38 \times 10^{-23})(100,000)]$$
$$V_N = [1.66 \times 10^{-23}]^{1/2}$$
$$V_n = 1.3 \times 10^{-6} \text{ volts}$$

In the example above, 1.6 microvolts of noise was created by no more than molecules running around a resistor in response to environmental temperature. Although this signal may appear to have a very low amplitude, keep in mind that many signals found in practical systems have the same order of magnitude. For example, the human electroencephalograph (EEG) machine records minute scalp potentials derived from brain electrical activity that may have components as low as 1 to 2 microvolts and peak amplitudes in the 10 to 100 microvolt range. In that application, 1.6 μV represents a significant artifact.

The answer to this type of problem is to keep circuit impedances in the early stages — i.e., those stages that most of the gain follows — very low so that the R term is reduced to minimum. Additionally, low-pass filtering, bandpass filtering, or other methods might be employed to keep the bandwidth term low. In some cases the amplifier or other circuit is cooled to reduce the temperature term in the equation. Although this latter is unusual, there are cases where it has to be done.

There are also several sources of noise that are peculiar to solid-state amplifiers, such as shot noise, Johnson noise, and flicker noise. In low-cost amplifiers these noise sources can add up to a significant effect. Although low-pass filtering offers relief, it is better to specify a low-noise amplifier for the earliest stages in the system. Friis' equation tells us that low-noise amplifiers in the input stages provide most of the noise relief for the entire sys-

tem. It is for this reason that satellite communications or TV earth stations use low-noise amplifiers (LNA) as preamplifiers for the dish antenna.

OTHER NOISE PROBLEMS

Impulse noise is due to local electrical disturbances such as arcs, lightning bolts, electrical motors, and so forth. In this same category is general electromagnetic interference (EMI) problems. Such interference is usually caused by nearby radio transmitters or other rf sources. It is not usually possible to force the transmitter off the air, even when it is an amateur, because they are licensed by the government to be there, while you are not. From an engineering point of view, your equipment might be very expensive and quite good, and still be junk from an EMI point of view. The purpose of any electronic equipment is two-fold: 1) it must respond to proper signals, and 2) it must reject improper signals. It is point 2 where most improperly designed equipment fails.

Shielding and filtering of signal lines is the key to EMI problems. Figure 19-1 shows a hypothetical instrument with several of the possible correction modalities used. First, note that the entire instrument is built inside of a shielded metal box, and the box is grounded. Points of entry and exit are passed through feedthrough "EMI filter" capacitors (500 pF to 0.01 μF). Each stage is isolated from other stages by a resistor, and has its own decoupling capacitor (C5 and C7). The main power bus is decoupled (C6) and has a series radio-frequency choke (RFC2) to prevent rf that gets past C2 from interfering with the operation of the circuit. The input leads are similarly filtered with RFC1 and C4. The input resistance (R_{IN}) of the amplifier and capacitor C4 also form a low-pass filter with a frequency response that rolls off at a -3 dB/octave rate from the -3 dB point defined by:

$$f(Hz) = \frac{1,000,000}{6.28 \, C4 \, R_{IN}}$$

where

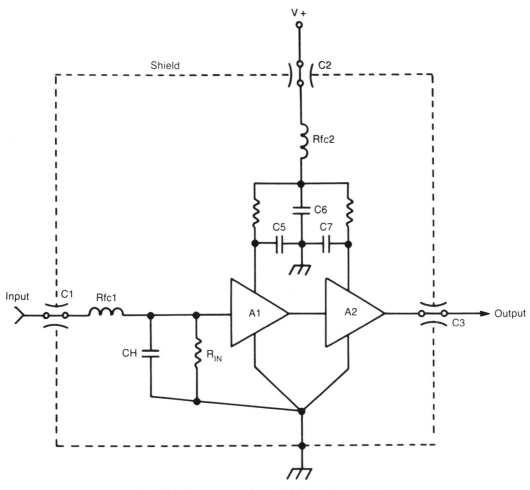

Fig. 19-1. Proper grounding, shielding, and decoupling.

f is the frequency in hertz
C4 is in microfarads
R_{IN} is in ohms

Potential sources of interference are noise and EMI signals on the power lines. I can recall digital instrumentation and computers in a medical school building that acted in a schizophrenic manner until it was identified that the ac power lines were the source of the problem. Where sensitive scientific instruments are used, you might want to consider designing the electrical system so that it is either isolated from the building system, or has a separate system that keeps a separate neutral and ground

conductor all the way back to the service entrance of the building.

Figures 19-2 and 19-3 show methods for dealing with power-line noise. Figure 19-2 shows an LC power-line filter. These devices are shielded low-pass filters and are mounted inside of equipment as close as possible to the point where ac enters the cabinet. Some filters are available molded into the ac chassis connector. External to the filter is a metal oxide varistor (MOV) device used to suppress ac line transients above about 155 volts peak (some can reach 2000 volts for 30 microseconds).

The Topaz transformer in Fig. 19-3 performs

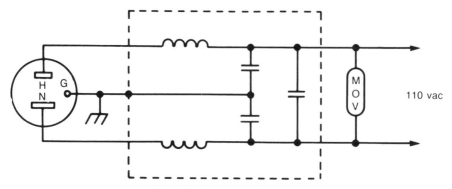

Fig. 19-2. Ac-line EMI filtering.

two functions. First, it isolates the equipment electrical system from the main electrical system. Second, it frequency limits the system to prevent high-frequency transients and pulses from passing into the equipment. It is my opinion, shared by many other engineers, that no computerized or other digital equipment — and many types of analog equipment — should be operated in a noisy environment without one of these transformers. If the equipment is life-support, or life-saving, as it often is in medical applications, then it is probably engineering malpractice to design a piece of equipment without the transformer.

OTHER PROBLEMS

Other electrical devices nearby, as well as the ubiquitous 60-Hz field from building wiring, can induce signals in amplifier inputs. It is wise to use

Fig. 19-3. Ac isolation amplifier.

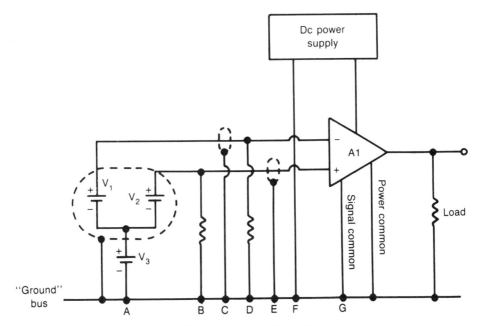

Fig. 19-4. The ground-loop problem.

only differential amplifiers for these applications because of their high common-mode rejection ratio. Signals from desired sources are connected as differential signals (V_1 and V_2 in Fig. 19-4) across the two inputs, while interference from the 60-Hz lines tends to be common mode because it affects both inputs equally.

It is sometimes possible, however, to manufacture a differential signal from a common-mode signal. There are two methods to do this, and both involve the improper use of shields. One source of the problem is the ground loop, as shown in Fig. 19-4. This problem arises from the use of two many grounds. In this example the shielded source, shielded input lines, the amplifier, and the power supply are all grounded to different points on the ground plane. Power-supply dc currents flow from the power supply at point "D" to the amplifier power-supply common at point "F", and produces a voltage drop along the way. Similarly, other sources also cause ground plane voltage drops. These voltages are called ground-loop signals, and form valid differential signals from the amplifier's point of view.

The cure for ground loop signals is shown in Fig. 19-5. In this example we see that all of the ground connections within the equipment are routed to a single, common grounding point. This effectively eliminates the ground loop voltage drops. Also, note that a single shield around both lines is used, rather than allowing each line to have its own shield. In a moment we will deal with additional shielding schemes.

When designing a new system, and you have the opportunity to design or specify the design of the printed wiring boards, then make sure that single-point grounding is enforced. Figure 19-6A shows a method for minimizing the noise problems in a circuit board. Note that there are three different grounds: power supply, digital signal, and analog signal. All three grounds are joined together at the card-edge connector and then spread out to their respective circuits.

Grounding problems also affect collections of equipment, as well as single circuits. In Fig. 19-6B you see a prescription for disaster. The grounds are daisy-chained together and grounded to earth ground at a point close to only one piece of equipment (D). The correction for this problem is shown in Fig. 19-6C: the star ground. Ground conductors

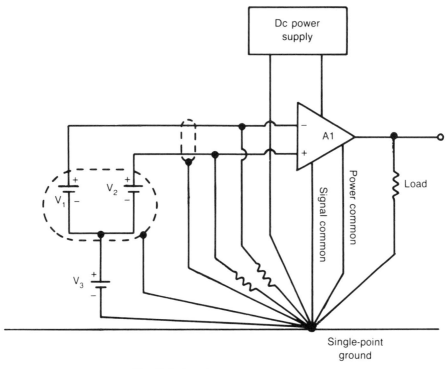

Fig. 19-5. Cure for ground-loop problem.

from all four pieces of equipment are brought together at a common point, which is then grounded.

Figure 19-7 shows the causes and cures for another form of artifact: common-mode signals manufacturing differential signals. The circuit in Fig. 19-7A uses standard single shielding, but the equivalent circuit shown in Fig. 19-7B reveals the problem. The shield produces a capacitance to ground with the input wires (C1 and C2), and there are always cable ohmic and source resistances, represented in Figure 19-7B as R1 and R2. The system works well if R1 = R2, and C1 = C2, but even small imbalances in the RC networks will allow V_{CM} to manufacture a differential signal. In this case it is found that V_{C1} does not equal V_{C2}, so the amplifier sees what it believes to be a valid input signal.

A guard shield (Fig. 19-7C) circuit can be used to overcome this problem. The guard shield is driven by signals from the two input lines through high-value resistances, Ra and Rb, and in many cases a common-mode amplifier. This tactic has the effect of placing both sides of the cable capacitances at the same potential, so $V_{C1} - V_{C2} = 0$. The outer shield is not strictly necessary but is highly recommended.

ISOLATION AMPLIFIERS

There are certain applications where it is advisable to use a high degree of isolation between the main system and the signals acquisition devices. In some cases safety is the issue. In medical applications, for example, it is possible to kill a patient with as little as 20 μA (or thereabouts, the exact number is controversial) of 60-Hz current. So it becomes necessary to prevent such currents from the instrument mainframe from hurting the patient. In certain industrial applications, there might be either high voltages present, or high noise levels. In all of these cases, you might want to use an isolation amplifier.

Figure 19-8 shows the basic circuit symbol for

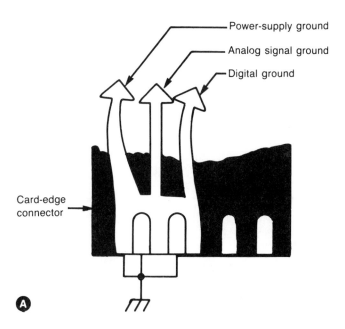

Power-supply ground

Analog signal ground

Digital ground

Card-edge connector

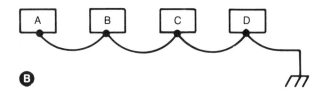

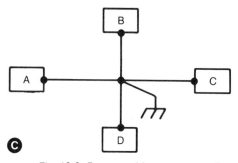

Fig. 19-6. Proper and improper grounding.

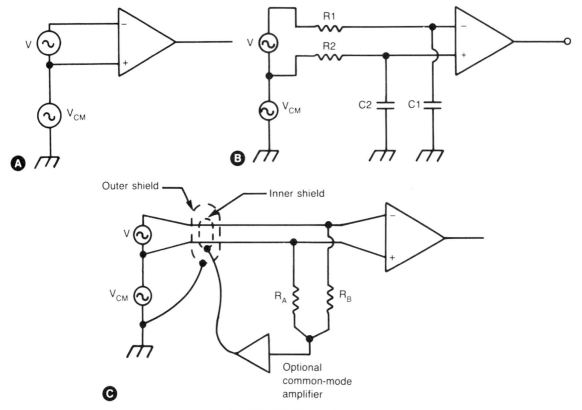

Fig. 19-7. Shielding problems.

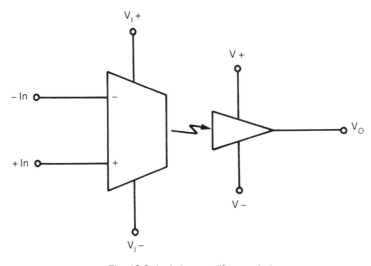

Fig. 19-8. Isolation amplifier symbol.

an isolation amplifier. The input side and output sides are totally isolated from each other electrically (the connection is made by either an ac carrier or a optical device). The isolation is maintained by keeping the two power-supply systems totally separate. Typically, V+ and V− are connected to a power supply derived from the ac power lines. V_{I+} and V_{I-} are derived from an isolated supply, or batteries.

Chapter 20

Introduction to Microcomputers

MICROPROCESSORS (AND THE MICROCOMPUTERS that they have spawned) have revolutionized the electronics industry. Once only a few electronics engineers and technicians knew anything at all about computers: it was an arcane area in the backwaters of the profession. But today it is almost a truism that all forms of engineers and subengineering technical personnel must be aware of the microcomputer and its applications. Nowhere is this more evident than in electronics engineering, and especially in the field of data acquisition for research, measurement, instrumentation, and control.

The advantage of the microprocessor or microcomputer is that it allows us to "computerize" our designs at low cost. So what, you say? Well, consider the advantages of computerization. You gain a much greater degree of control over factors such as drift when you use the digital implementation of a design. In addition, if you "universalize" the design sufficiently, it is possible to gain immense reconfigurability in the digital implementation. Sometimes one finds that relatively major "redesigns" can be accomplished with only a little hardware and some software patches. On older designs, i.e., using the analog method, even a

minor change in the design configuration could have profound impact on the physical hardware: new printed circuit boards had to be laid out, new circuitry tested, and so forth. With the microcomputer implementation, however, it is sometimes possible to make only minor hardware changes to accomplish big changes in device application.

The microprocessor is a large-scale integrated circuit (LSI), typically in a 40-pin DIP package. The microprocessor IC contains all of the circuitry needed to make the central processing unit (CPU) of a programmable digital computer (in addition, some "single-chip computer" chips also provide limited internal memory and I/O capability, e.g., the Intel 8048-family). Most microprocessors, especially the early designs, required at least a few other external chips in order to make a complete digital computer.

The 8048 by Intel is an example of a "single-chip computer" that contains all of the microprocessor CPU functions found in any μP chip, plus a limited amount of internal random access read/write memory (RAM), 2K of read only memory, and a pair of input/output (I/O) ports. This chip is not merely a microprocessor, but also a complete single-chip computer. If your application requires no capability other than the internal memory and I/O, then the 8048 is capable of providing your entire computer needs. Another single-chip computer in the Intel 8048-family is the 8051 device. This chip is of particular interest to data acquisitions people because it contains an on-board A/D converter.

The purpose of this book is to introduce you to computer-oriented data-acquisition systems. Some of the computers that are covered in these pages are sold complete with cabinet and keyboard, while others are bare-board "O.E.M." (original equipment manufacturer) devices that are stripped down and intended for use as a component in some other system.

The computers in this book are primarily small computers that are frequently used as a subsystem in a larger context. One example is the Apple IIe microcomputer. This instrument has found its way into a lot of data-acquisition systems because of the flexibility in applications allowed by the Apple IIe (and also II and II+, but not IIc) plug-in expansion slots. Also arguing for its inclusion is the fact that the 6502 microprocessor used in the Apple II has become one of the most popular microprocessors in control and instrumentation applications. While this situation is more than likely an outgrowth of the phenomenal popularity of the MOS Technology, Inc., *KIM-1* and the Rockwell AIM-65 (which used the KIM-BUS), it is also true that many engineers use the Apple II as the "development system" on which the software is developed and tested. The Apple II can also be equipped with a ROM programmer, and so will allow the engineer or other worker to program the "firmware" that will drive a smaller 6502-based O.E.M. single-board computer that is actually placed in the final product.

A LITTLE MICROPROCESSOR HISTORY

Intel Corporation is usually credited with the invention of the microcomputer in 1972. The story is usually told that Intel was contracted to design and produce a four-bit programmable "controller" that would allow the user to control traffic signal lights. That unit supposedly became the now obsolete (but still seen occasionally) 4004 four-bit microprocessor device. Oddly enough, the original contractor (so the industry legend goes) did not want the device so Intel was free to go on the open market with it . . . to the betterment of the rest of us. The next unit was the 8008 device, which was basically an eight-bit 4004. Later, the now-famous 8080A device was designed and offered to the market. It was the 8080A that became THE microprocessor for many thousands of engineers, technicians, and others who were apparently eager to break into the field. Although initially very expensive, the 8080A opened the way for the phenomenal explosion of hobbyist and other "personal computers." It is amazing how the once extremely costly 8080A device has plummeted in price to the point where they are now all but giving it away.

After the 8080A became a huge success, another company got onto the bandwagon. Zilog, Inc. began to produce the all but legendary Z80 microprocessor. The Z80 was designed using the same general philosophy as the 8080A, and in fact, was able to use the entire 8080A repertoire of machine code instructions — plus a few dozen additional instructions that the 8080A did not recognize.

The Motorola 6800 was probably the next device on the market, and it used a different architectural philosophy than either the 8080A or the Z80 devices. The MOS Technology family of 65xx devices (of which, the 6502 is the most famous but not only example) was next on the scene, and became immensely popular. It is similar in architecture to the 6800 series. The Apple II is based on the 6502 device, and the Commodore 64 (and more recently the Commodore 128) is based on the 6510 microprocessor chip — a 6502 derivative.

Currently, almost every semiconductor company makes at least one microprocessor either of their own or under a "second source" license from a primary vendor.

SOME DEFINITIONS

The terminology of microcomputer and microprocessors is used by almost everyone today, but even so it bears a little repeating since there are so many who tend to use the words incorrectly.

The list below is somewhat random, and is based on my own experiences, i.e., terms that I have misused or heard others misuse.

Programmable Digital Computer

A programmable digital computer contains a "memory" section that is used to store both data and instructions. The central-processing unit (CPU) in a programmable digital computer responds to a certain set of binary numbers that it recognizes as "commands" or "instructions." These binary numbers are stored in memory. The job that the computer will perform is determined by these instructions and both the manner and order of their use in the "program." Because the programmer can insert these instructions in memory, the computer is said to be *programmable*. It is "digital" because it works with binary devices (such as electronic switches) that can implement binary (i.e., base 2) arithmetic. The alternative to programmable computers is non-programmable or fixed-programmed computers, while the alternative to digital computers is analog computers (an all but obsolete form). These latter forms were once used extensively, and consisted of amplifiers, integrators, differentiators (based on operational amplifiers) and a lot of other analog electronic circuitry.

Bit

The word "bit" is an acronym made up from letters of the words 'BInary digiTs." A "BIT" is, then a digit of the binary, or base-2, numbers system: either 0 or 1. These digits are represented by discrete voltages or currents in the electronic circuitry. For example, in the popular TTL logic system of digital devices (74xx/74xxx devices), the zero is represented by a voltage from 0 to 0.8 volts, while the one is represented by a voltage of +2.4 to +5 volts.

Byte

A "word" consisting of eight bits; e.g., 11010011 is an example of an eight-bit, or "one byte," binary number.

Microprocessor

A *microprocessor* is an integrated circuit that contains all of the necessary components to form the central processing unit of a programmable digital computer. The microprocessor is not to be confused (as it often is) with the microcomputer. The latter term is used to denote a fully programmable digital computer that uses a microprocessor as the CPU. Thus, a microcomputer will contain a microprocessor plus whatever additional circuitry is needed to make it capable of operating as a full computer.

Microcomputer

A *microcomputer* is the smallest element that is truly identifiable as a full-fledged programmable digital computer. It is, by definition, built around a microprocessor CPU rather than other chips. It will contain, in addition to the microprocessor chip, other chips to fill out the functions of the computer. For small computers, these extra chips might be only two or three LSI special-purpose devices that contain a small amount of RAM, perhaps some ROM and, of course, the I/O functions. On larger machines the additional chips may number 50 to 100, or even more. Just what comprises a microcomputer varies somewhat. For example, the small 4.5×6 inch O.E.M. printed circuit board computer is a microcomputer, as are the Apple II and IBM-PC. But so is the Timex TS-1000 and others of similar ilk. Yet at the same time, we see huge rack-mounted computers that look more like minicomputers also designated (correctly!) as "microcomputers." The reason? They're based on a microprocessor CPU chip.

Minicomputer

The minicomputer has traditionally been regarded as a scaled-down mainframe data processing computer. Examples of true minicomputers include the non-obsolete (but once ubiquitous) Digital Equipment Corporation PDP-8 and their nearly legendary PDP-11 family of computers. These machines did not use microprocessor chips in the CPU, but rather discrete logic elements, mostly

TTL. It is not uncommon for a minicomputer to require 30 to 40 amperes of current from the +5 volt, regulated-dc power supply.

In the past, minicomputers have been more powerful than microcomputers. They would, for example, sport bit lengths of 12 to 32 bits rather than the 4 or 8 bits found in microcomputers. In addition, they would operate at clock speeds of 8 to 12 MHz rather than 1 to 3 MHz used in the microcomputer. Today, however, the distinctions have faded and are now almost indistinct. We now see microcomputers that are housed in rack-mounted cabinets six-feet high, and minicomputers that are desktop units. Microcomputer clock speeds now approach those previously boasted of in minicomputers only. The minicomputer may also use an LSI CPU chip that looks suspiciously like a microprocessor, albeit of special manufacture. The DEC PDP-11 format was implemented in the LSI-11 microprocessor chip and wound up in such computers as the Heath H11. It is now quite difficult to draw the line between the minicomputer and the microcomputer.

Mainframe Computers

The big horse used in the data-processing department of large companies and most banks is called a *mainframe computer*. The IBM 370 is an example of a large mainframe computer. Mainframe computers are derisively called "dinosaurs" by ignorant elitists among the microcomputer fraternity. But be well aware, however, that this dinosaur is not about to become extinct like its reptillian namesake. If the mainframe computer is a "dinosaur," then the microcomputer is a mere lizard!

Interfacing

Interfacing is probably among the most abused words in the English language. Salesmen have been known to write to me stating that they would like to come down and ". . . interface with me." What's that? They're going to insert an RS-232 connector in my mouth and plug in? I usually discard such letters immediately. Any docu-

ment that comes across my desk with "interface" or its derivatives used in that manner will get a C− for starters.

Interfacing (when the word is correctly used in a computer equipment context), is the art of taking a microcomputer and wedding it with other circuits and equipment in order to solve some particular problem. At least one author has defined interfacing as the art of replacing hardware logic with software, and that is a reasonable explanation in my opinion. It is interesting to note that interfacing is a relatively simple process involving understandable and reasonable rules and procedures; it doesn't ordinarily require loads of "smarts." Yet, it is often in the art of interfacing that engineers, technicians, and other luminaries make their reputation among the non-technical personnel. In a university research setting, for example, one can make large strides up the respect ladder by being able to work the black magic that makes a computer control an experiment or inprove an instrument.

Selecting the Right Microprocessor

The gut-feeling selection of a microprocessor will usually lead us to the chip with the greatest capability, the longest data bus word length, and the most superior instruction set. But that might also be the least economical choice! The selection of a microprocessor chip should be made with regard to the application at hand (seems reasonable, but for many it is a revolutionary idea!), and not on an emotional level at which the desire for the "best" supersedes everything else. If a brand new 16-bit microprocessor chip seems exciting, then take a second look at the situation before becoming committed. If, for example, an 8-bit microprocessor will do the same job and still cover all planned or anticipated expansion, then why go to the trouble and expense of a 16-bit machine? But on the other hand, if the application and its planned growth strains the 8-bit machine, then it might be cheaper in the long run to go ahead and spend the money up front on the 16-bit device.

Proper regard must be given to the potential for expansion of the system. If you lock yourself into a machine that does not operate fast enough, or have a good enough instruction set, or cannot support sufficient memory, then you will be hard pressed to make the machine deliver additional capability when additional features must be called for. Also, consider the matter of memory size. It turns out that one never has enough memory. Some of this problem is due to the natural tendency for programmers to use up the entire memory of the computer. In fact, one computer designer claims that it is wise to tell the programmers about only half of the memory that is really available!

Otherwise, they will use up all of what you allocate despite your protestations that they should leave room for expansion. If it turns out that the programmer genuinely requires additional memory, then it is possible to "add" memory without too much bloodletting.

Although the material in this book is little more than a matter of definitions, it is important to keep them in perspective in order to improve communications . . . something that is always of benefit.

Chapter 21

Operation of the Microcomputer

MOST READERS OF THIS BOOK ARE, BY THIS time, relatively familiar with the operation of the programmable digital computer. Even so, there are many for whom the computer is still little more than a complicated black box. All too often microcomputer books assume a familiarity with computers that is simply not there in all readers, even though most engineers and technicians are "computer literate" (to use a cliche!). Thus, we have a need for a section that discusses the programmable digital computer as a generic device. The "machine" of this section is a universal computer, representative of the entire set of programmable digital computers, and not any one manufacturer's offerings. It is a hypothetical machine, not available anywhere. It, like an androgynous being, has elements of all kinds within its body. We call this hypothetical computer the TAB Automatic Computer, or "TABAC" for short.

THE TAB AUTOMATIC COMPUTER (TABAC)

Like any programmable digital computer, TABAC has three main parts: the central processing unit (CPU), memory, and *input/output* (I/O). There are other functions found in certain specific machines, but many of these are either special applications of these main sections, or are too unique to be considered in a discussion of a general "universal" machine.

The central processing unit (CPU) controls the operation of the entire machine. It consists of several necessary subsections which are described in greater detail below.

Memory can be viewed as an array of "pidgeon holes," or "cubbyholes" such as those used by postal workers to sort mail. Each pidgeon hole represents a specific address or the letter carrier's route. An address in the array can be uniquely specified (identifying only *one* location by designating its *row* and *column*. If you want to specify the memory location (i.e., pidgeon hole) at row 3 and column 2, then you would create a row-3, column-2 address number, which in this case is "32."

Each pidgeon hole represents a unique location in which to store mail. In the computer the memory location stores not pieces of paper, but a single binary "word" of information. In an eight-bit microcomputer, for example, each memory location will store a single eight-bit binary word (e.g., 11010101). The different types of memory device are not too terribly important to us here, except in that most general terms. To the CPU, and our

211

present discription of how it operates, it doesn't matter much whether memory is random access read/write memory (RAM), read only memory (ROM), dynamic or static.

There are three main lines of communications between the memory and the CPU: address bus, data bus, and control logic signals. These avenues of communications control the interaction between memory and I/O, on the one hand, and the CPU on the other, regardless of whether the operation is a read or write function. The address bus consists of parallel data lines, one for each bit of the binary word that is used to specify the address location. In most eight-bit microcomputers, for example, the address bus contains sixteen bits. A 16-bit address bus can uniquely address up to 2^{16}, or 65,536 different eight-bit memory locations. This size is specified as "64K" in computerese, not "65K" as one might expect. It seems that lower case "k" is used almost universally in science and engineering to represent the metric prefix *kilo,* which represents the multiplier 1000. Thus, when someone tells you that their computer has "64K" of memory, you might expect it to contain 64 × 1000, or 64,000 electronic pidgeon holes. But you would be wrong in that assumption. For long ago computerists noted that 2^{10} was 1,024, so determined that their "k" would be 1,024, not 1,000! This means that a 64K computer will contain a total of 64 × 1024, or 65,535 electronic pidgeon holes in which to stuff data. To differentiate "big K" (1024) from "little k" (1000), standard computerist shorthand uses upper case "K" rather than lower case "k."

Be aware that the size of memory which can be addressed doubles for every bit added to the address bus. Hence, adding one bit to our sixteen-bit address bus creates a seventeen-bit address bus that is capable of addressing up to 128K of memory. Some "8-bit" machines which have 16-bit address buses are made to look larger by tactics which create a pseudo-address bus. In these machines several 64K memory banks are used to simulate continously addressable 128K, 256K, or more machines.

The data bus is the communications channel over which data travels between the main register

(called the *accumulator* or *A-register* in the CPU and the memory. The data bus also carries data to and from the various input and/or output ports. If the CPU wants to read the data stored in a particular memory location, then that data is passed from the memory location to the accumulator (in the CPU) by way of the data bus. Memory write operations are exactly the opposite: data from the accumulator is passed over the data bus to a particular memory location.

The last memory signal is the *control logic* or *timing* signal. These signals tell memory if it is being addressed, and whether the CPU is requesting a read or write operation. The details of the control logic signals vary considerable from one microprocessor chip to another, so there is little that I can say at this point. Later in this book you will be introduced to both the 6502 and Z80 standard signals (which are representative of two different architectures), but for others I recommend either the manufacturer's literature, or, my own book entitled *8-bit* and *16-bit Microprocessor Cookbook,* TAB Books catalog #1643.

The input/output (I/O) section of the TABAC computer is the means by which the CPU communicates with the outside world. An input port will bring data in from the outside world and then pass it over the internal data bus to the CPU, where it is stored in the accumulator. An output port is exactly the opposite: it passes accumulator data from the data bus to the outside world. In most cases the "outside" world consists of either peripherals (e.g., printers, video monitors) or communications devices (e.g., MODEMs).

In some machines there are separate I/O instructions that are distinct from the memory-oriented instructions. For example, in the Z80 machine there are separate instructions for "write to memory" and "write to output port" operations. In Z80-based computers, the lower eight bits (AO–A7) of the address bus are used during I/O operations to carry the unique address of the I/O port being called (the accumulator data will pass over both the data bus and the upper eight bits — A8 to A15 — of the address bus). Since there are eight bits in the unique I/O address, you can use up to 2^8, or 256 different I/O ports that are numbered from 000 to 255. In the 6502-based computer, there are no distinct I/O instructions. In these machines (e.g., the Apple II), the I/O components are treated as memory locations. This technique is called *memory-mapped I/O*. Input and output operations then become memory-read or memory-write operations, respectively.

Central Processing Unit (CPU). The CPU is literally the heart and soul of any computer. Although there are some differences between specific machines, all will have at least the features shown in our TABAC computer diagram of Fig. 21-1. The principal subsections of the TABAC computer are: accumulator or A-register, arithmetic logic unit (ALU), program counter (PC), instruction register (IR), status register (SR) or processor status register (PSR), and control logic section.

The accumulator is the main register in the CPU. With a few exceptions, all of the instructions use the accumulator as either the source/destination of data or the object of some action (e.g., unless otherwise indicated, and ADD instruction always performs a binary addition operation between some specified data and the data in the accumulator).

Although there are other registers in the CPU (the Z80 is loaded with them!), the accumulator is the main register. The main purpose of the accumulator is the temporary storage of data being operated on or transferred within the machine. Note that data transfers to and from the accumulator are non-destructive. In other words, it is a misnomer to tag such operations as "transfers," because in actuality they are "copying" operations. Suppose, for example, the program calls for us to "transfer" the hexadecimal number $8F from the accumulator to memory location $A008. After the proper instruction ("STA $A008" in 6592 assembly language) is executed, the hex number $8F will be found in both the accumulator and memory location $A008.

It is important to remember that accumulator data changes with every new instruction! If you have some critical datum stored in the accumulator, it is critical that you write it to some location in memory for permanent storage. This function is sometimes performed on the "external stack" or in some portion of memory set aside by the programmer as a "pseudo-stack."

The arithmetic logic unit (ALU) contains the circuitry needed to perform the arithmetic and logical operations. In most computers the arithmetic operations consist of addition and possibly subtraction, while the logical operations consist of AND, OR, and exclusive OR (XOR). Note that even subtraction is not always found! In some computers there is no hardware arithmetic function other than addition. The subtraction function is performed in software using two's complement arithmetic (a method of making the computer think its actually adding!). Multiplication and division are treated as multiple additions or subtractions unless the designer has thought to provide a hardware multiply/divide capability.

The program counter (PC) contains the address of the next instruction to be executed. The secret to the success of any programmable digital computer such as TABAC is its ability to fetch and execute instructions sequentially. Normally, the PC will increment appropriately (1, 2, 3, or 4) while executing each instruction (i.e., 1 for a one-byte instruction, 2 for a two-byte instruction, etc.). For example, the LDA,N instruction on the 6502 microprocessor loads the accumulator with the number "N." In a program listing, you will find the number "N" stored in the next sequential memory location from the code for the LDA portion (called the operations code, or "op-code" for short):

$$0205 \quad \text{LDA}$$
$$0206 \quad \text{N}$$

At the beginning of this operation, the PC contains "0205," but after execution the PC will contain the number "0206" because LDA,N is a two-byte instruction.

There are several ways to modify the contents of the PC. One way is to let the program execute sequentially: the PC contents will increment for each instruction. You can also activate the *reset* line, which forces the PC to either location $0000, or some other specific location (often at the other end of the memory, for example $FFFA in the 6502). Another method is to execute either a JUMP or a conditional-JUMP instruction. In this latter case the PC will contain the address of the "jumped-to" location after the instruction is executed. Finally, some computers have a special instruction that will load the PC with a programmer-selected number. This "direct-entry" method is not available on all microprocessors, however.

The instruction register is the temporary storage location for the instruction codes that were stored in memory. When the instruction is fetched from memory by the CPU, it will reside in the instruction register until the next instruction is fetched.

The instruction decoder is a logic circuit that reads the instruction register contents, and then carries out the intended operation.

The control logic section is responsible for the "housekeeping" chores in the CPU. It issues and/or responds to control signals between the CPU and the rest of the universe. Examples of typical controls signals are: memory requests, I/O requests (in non-memory-mapped machines), read/write signalling, interrupts and so forth.

The *status register,* also sometimes referred to individually as the "status flags," is used to indicate to the program and sometimes the outside world the exact status of the CPU at any given instant. Each bit of the status register represents a different function. Different microprocessor chips use different sets of status flags, but all will have at least a *carry flag* (C-flag), and a *zero flag* (Z-flag). The C-flag indicates when an arithmetic or logical operation results in a "carry" from the most significant bit of the accumulator (B7), while the Z-flag indicates when the result of the present operation is zero (00000000); typically, Z = 1 when result is zero, and C = 1 when a carry occurs.

We have now finished our tour of the CPU of the TABAC computer. This discussion in general terms also applies to most microprocessors. Although various manufacturers use different names for the different sections, and some will add sec-

tions, almost all microprocessors are essentially the same inside of the CPU — at least in respect to basics.

Operation of TABAC

A programmable digital computer such as TABAC operates by sequentially fetching, decoding, and then executing instructions stored in memory. These instructions are stored in the form of binary numbers. In some early machines there were two memory banks, one each for program instructions and data. The modern computer, however, uses a single memory bank for both instructions and data.

How, one might legitimately ask, does TABAC know whether any particular binary word fetched from memory is data, an instruction or a binary representation of an alphanumeric character (e.g., ASCII)? The answer to this instruction is key to the operation of TABAC: cycles!

TABAC operates in *cycles.* A computer will have at least two discrete cycles: instruction fetch and execution (in some machines the process is more sophisticated, and cycles are added). While the details differ from one machine to another, the general operation is similar for all of them.

Instructions are stored in memory as binary numbers ("op-codes"). During the instruction fetch cycle, an op-code is retrieved from the memory location specified by the program counter, and stuffed into the instruction register inside the CPU. The CPU assumes that the programmer was smart enough to arrange things according to the rules, so that the datum fetched from location ABCD during some instruction cycle in the future is, in fact, an instruction op-code. It is the responsibility of the programmer to arrange things in a manner so as to not confuse the computer.

During the first cycle, an instruction is fetched and stored in the instruction register (IR). During the second (i.e., next) cycle, the instruction decoder circuit inside of the CPU will read the IR, and than carry out the indicated operation: this is called the *execution cycle,* while the first cycle is the *instruction fetch/decode cycle.* The CPU then enters the next instruction fetch cycle and the process is repeated. This process is repeated over and over as long as TABAC is operating. Each step is synchronized by an internal "clock" that is designed to insure things remain rational.

From the above description, you might be able to glimpse a truth concerning what a computer can or cannot do. The CPU can shift data around, perform logical operations (AND, OR, XOR), add two n-bit numbers (sometimes—but not always—subtract as well), all in accordance with a limited repertoire of instructions encoded in the form of binary words. Operations in TABAC (and all other microcomputers) are performed sequentially through a series of discrete steps. The secret to whether or not any particular problem is suitable for computer solution depends entirely on whether or not a plan of action (called an "algorithm") can be written that will lead to solution through sequentially executed steps. Most practical instrumentation, control, measurement or data-processing problems can be so structured—a fact which accounts for the meteoric rise of the microcomputer in those fields.

In the chapters to follow I discuss the practical use of the microcomputer in data acquisition systems. You will learn the fundamentals of interfacing, both in general and in particular for the Apple IIe and IBM-PC (the two most popular computers in this field). I will also consider the basics of data-conversion circuits, because digital-to-analog converters (DACs) and analog-to-digital converters (ADC or "A/D") are at the very heart of these systems.

Chapter 22

Selecting a Computer for Data Acquisition

ALTHOUGH THERE ARE A VERY LARGE NUMBER OF computers on the market, the machines suitable for data acquisition are fewer in number. There are several choices: special purpose machine, single-board computer, or a microcomputer (also called personal computers).

For certain industrial applications, especially in environments where unaltered microcomputers are not suitable, special-purpose computers are required. These instruments are designed specifically for factory or other harsh-environment applications. An example of such a machine is the rack-mounted industrial computer sold by ICS Computer Products P.O. Box 23058, San Diego, CA 92123).

Single-board computers are used as plug-in modules in other instruments. You can buy a complete computer on a small printed-circuit card from any of several sources including John Bell Engineering, Pro-Log, and others.

Regardless of the model of computer selected, you must consider several factors: operating speed, mass storage capacity, memory requirements, number and type of I/O ports, and the availability of plug-in modules with compatible software.

In the rest of this chapter, I discuss the various personal computers. The main selection criterion was a plug-in capability. With the exception of the AIM-65, this criterion is met by the IBM-PC and the Apple IIe.

IBM-PC

The IBM Personal Computer (IBM-PC) has become the de-facto standard for small computer data processing. This computer is based on the powerful Intel 8088 microprocessor chip. The 8088 uses a 16-bit internal architecture identical to the 8086 device, but has an 8-bit I/O structure similar to the 8085 and other Intell eight-bit machines.

One of the principal advantages of the IBM-PC is the large amount of hardware and accessories that are available, both from IBM and after-market vendors. Many accessories are in the form of plug-in cards that fit into one or more of the five slots on the mother board. The original IBM-PC used four 16K banks of memory on the motherboard, for a total of 64K. Add-on plug-in boards can add memory to a total of 640K. The later IBM-PC (identified by a "circle-B" on the rear panel) uses four 64K banks of memory chips for a total of 256K on board, and still allows external memory (384K) for a total of 640K.

The standard IBM-PC comes with one 5.25-inch full size, double-sided, double-density (DS/DD), 320 kilobyte floppy-disk drive, and provision for a second 5.25-inch drive in the same chassis. Some vendors offer half-height 5.25-inch drives and 5 and 10 megabyte hard disk drives in the same size as the regular floppy-disk drives. As a result, some users have configured machines to contain two half-height 5.25-inch floppy disk drives in one drive position, and a 5 to 10 megabyte hard disk in the other slot. These drives are labelled A:, B: and C: for programming purposes. Users of the AST Six-Pack, and certain other plug-in cards that provide additional memory, may use software such as "Super Drive," which forces an area of RAM to act as a disk drive — at a lot faster speed than mechanical drives. The IBM-PC operating system sees the RAM disk as a regular disk drive. A caution, however, is that power failures will destroy RAM disk data, but not real disk data.

A typical IBM-PC for control or data acquisition purposes will have a collection of cards that provides for a parallel printer port, a monochrome video monitor, a serial communications port, and an A/D converter board. One system, for example, uses the IBM monochrome monitor/parallel printer card, a disk drive card (handles two floppy disks), an AST Six-Pack, and a TecMar A/D converter board (or the equivalent). The AST Six-Pack is a multifunction board that permits up to 384 kilobytes of additional RAM memory. It also has a time of day and calender function.

There are several connectors on the IBM-PC that are useful for interfacing. The standard connectors, such as RS-232C serial communications or printer port, or the Centronics parallel printer port, are covered elsewhere in this book and so will not be discussed further here (see relevant chapter).

THE APPLE II FAMILY MACHINES

Since the mid-1970s when microcomputers "took off," there have been literally hundreds of models offered on the market. Some machines were mere copies of others (how many "S-100" copies of the Altair were there?). Others were poorly conceived and filled no market niche (or only a small, unprofitable niche). Still others failed for lack of proper marketing. But there are several industry standards that caught fire and are now found almost everywhere; the Apple II is one of them.

The legendary Apple II microcomputer is one of the primary industry standards, and has been around in one model or another for as many years as any microcomputer. This machine was one of the first really small, desktop microcomputers and has been extremely successful. The original machine came with 48K of memory, with the memory space between 48K and 64K taken up by internal programming. Currently, 128K models are available. The Apple II is well known for simple peripheral drive devices because it replaces hardware in these circuits with software. The Apple is based on the 6502 microprocessor chip.

During the period when Apple II popularity skyrocketed I was employed in a university medical center performing instrumentation chores — first as an electronics technician and later as an engineer. Prior to the Apple II only a few of the wealthiest research laboratories could afford a dedicated computer. There were some exciting minicomputers, but they were relatively expensive. In our institution there was a couple of Digital Equipment Corporation (DEC) PDP-11's, a handful of older PDP-8 machines, and an IBM 360 behemoth available if you converted your data to digital format on tapes or — shudder — punched cards. The Physiology Department had an expensive toy that allowed a scientist to "digitize" analog data presented on graph paper. Each researcher would use an x-y or y time recorder to acquire data, and then hand enter it using the digitizer table. When the Apple II came along and plug-in analog/digital converter cards were available, this situation changed: direct analog data entry became possible.

The Apple II was cheaper than an equivalent S-100 machine. Although it was arguable that S-100 machines could be configured to be more powerful, much of that capability was wasted: "Better is the enemy of good enough." As the prices of multichannel A/D converters dropped, more and more experiments could be computerized. The Apple II, it was found, could (at low cost) acquire the data, process it where needed, do the statistical analysis, and then output the results in graphic, tabular, or analog form. Because digitized data was permanently stored, curves taken on the same timebase could be overlayed on the same piece of graph paper (using different color inks as needed), rather than on parallel (hard-to-read) strips of multichannel graph paper. Although there are now "better" computers, the Apple II is still "good enough" for most applications.

There have been many "Apple look-alikes" on the market, all using the 6502 chip and claiming Apple compatability. Be a little careful when selecting a non-Apple look-alike that claims compatibility, because some are "compatible — *except . . .*" and that "except" can be profound. There are also Apple counterfeits on the market. Most of these are manufactured in southeast Asia and are offered for sale as if they were Apples — not merely *Apple* compatible. They look like Apples, carry the Apple logo, and are offered as Apples. But most of them are second-quality machines. Even where they are made well, with good quality components, there is always the problem of obtaining service for the machine: I doubt that the Apple service shops would be helpful when presented with a counterfeit of their product!

The Apple II machine uses outboard 5.25-inch disk drives, while the Apple III contains a single built-in 5.25-inch drive. Most Apple dealers offer a package that includes one disk drive, and offer the second drive as an option.

The main reason for including the Apple computers in this book is that it is one of the most popular machines in use. Despite its age, it remains popular. This machine is a low-cost, viable alternative for laboratories, industrial process controls, and other similar applications that are often performed by more expensive but largely wasted computers. The Apple II is also often recommended as the development system for 6502-based O.E.M.

single-board computers and controllers. The idea is that you will develop the assembly language program for the O.E.M. computer on the Apple II, and then transfer it to a read only memory (ROM) for insertion into the ROM socket of the small computer. This system allows a flexible system that can easily be changed to reflect new situations.

One reason that the Apple is so popular is that it contains eight printed circuit card-edge connectors on the motherboard that accept plug-in printed circuit cards. These sockets are used by the interfacer, the customizer, and so forth. The sockets have 50 pins, 25 on each side of the board. Pins 1 through 25 are on the component side, while 26 through 50 are on the "wiring side." There are a number of non-Apple accessory cards that can be plugged into Apple II's. These include extra ROM cards, I/O cards of one sort or another, A/D or D/A converters, and so forth. The Apple II can be custom configured with plug-in cards to meet your particular needs.

KIM-1/AIM-65 FAMILY MACHINES

Although no longer well known in computer stores, one of the most popular microcomputers, especially in control and data aquisition applications, was that of the MOS Technology, Inc., KIM-1 machine. This computer, plus Rockwell's AIM-65 machine (which evolved from the KIM-1) are the basis for many current control systems, data-aquisition systems, and instruments. Many "microcomputer-controlled" instruments use a small OEM single card computer that is based on the KIM-1 bus. Such systems often use the relatively sophisticated AIM-65 as the software development and hardware proofout machine, and then transfer machine code to a Read Only Memory (ROM) on the small computer board. The KIM-1 was the first microcomputer used by many now-experienced designers, technical people of assorted descriptions, and just plain old "hobby hackers." Although probably intended as a trainer to introduce engineers to the 6502 microprocessor produced by the same company, the KIM-1 soon became a run-away best seller. At a price of $149.95

it was the best buy on the market. Many engineers cut their microcomputer teeth on the little KIM-1.

The KIM-1 begat a cult of devotees that published numerous articles, designed an endless array of 6502-based software and produced a lot of add-on extras for the little KIM-1 single-board computer. It seems that the KIMmy-cult people were exceptionally given to humor in their presentations, and attached a somewhat feminine gender to the neuter computer. For example, when one chap designed a cabinet for the KIM-1 microcomputer, he entitled the article about it "A Dress for KIM."

The KIM-1 is hard to find anymore, but its legacy lives on in other computers. The SYM-1 by Synertek Systems Corporation was a directly compatible KIM clone, while the Rockwell AIM-65 is an advanced KIMmy. Even the Ohio Scientific Superboard II was something of a derivative of the KIM-1, even though considerably more advanced than the early KIM-1.

The KIM-1 concept is a single-board computer of simple design, based on the popular 6502 microprocessor. The input was via a multipurpose keypad mounted directly to the printed circuit board. The computer had either 2K or 4K of memory on board, so encouraged programmers to use efficient programming techniques that are conservative of memory. Additional memory could be interfaced via one of the connectors on the board.

There were also two I/O ports available on the KIM-series of computers. The I/O channel was controlled by a 6522 integrated circuit called a *versatile interface adapter* (VIA). This device offers two ports, which can be configured under

Table 22-1. Memory Map for the 6522 Versatile Interface Adapter (VIA)

Location	Function
A000	Port B output data register (i.e., port-B I/O port)
A001	Port-A output data register (i.e., port-A I/O port)
A002	Port-B data direction register (DDRB)
A003	Port-A data direction register (DDRA)
A004	Timer T1 write T1L-L
A005	Timer T1 write T1L-H & T1C-H, clear interrupt flag
A006	Timer T1 write T1L-L
A007	Timer T1 write T1L-H
A008	Timer T2 write T2L-L
A009	Timer T2 write T2C-H, clear T2 interrupt flag
A00A	Shift register (SR)
A00B	Auxilliary control register (ACR)
A00C	Peripheral control register (PCR)
A00D	Interrupt flag register (IFR)
A00E	Interrupt enable register (IER)
A00F	Port-A output data register (ORA)—no handshake

program control as either input or output on a bit-for-bit basis. It is not necessary to make either port A or port B all input or all output. You can, for example, make bit 0 of port A an input line, while bit 1 (or any other bit) of port A is configured as an input. The difference is programmable, and can be reconfigured at will. The 6522 contains two *data direction registers,* one each for port A and port B (these are labeled DDRA and DDRB). Writing a 1 to a bit in the DDR makes the corresponding I/O port bit an output, while a 0 renders it an input.

The 6522 registers are all memory-mapped in a 6502-based machine. In the Rockwell AIM-65, for example, the following 6522 functions are mapped at the indicated locations in page-A0 of memory as shown in Table 22-1.

Chapter 23

Computer Interfacing Basics

PROGRAMMABLE DIGITAL COMPUTERS ARE AT THE heart of data acquisitions-systems today. Although once used only in the largest systems, the low cost, small size, and flexibility of modern "personal computers" have made them indispensible for data acquisitions chores. Consider some examples. A well-known Apple IIc television commercial showed the tiny machine controlling a whole factory. That same Apple IIc (or other true portables) can go afield with a scientist to not merely acquire data, but give it a preliminary statistical scrubbing as well—a trick that might save a trip back because of bad data. Several years ago I saw the massive studios at the Christian Broadcasting Network controlling video switching and lighting chores at their satellite facility in Virginia using an old S-100 microcomputer. The examples could go on almost endlessly.

The key to using a digital computer in automatic data acquisitions systems is proper interfacing between it and the outside world—instruments, A/D converters, printers, etc. Although the topic of computer interfacing could easily occupy several books, I will give a brief synopsis here.

The world of digital electronics is not as simple as we might believe. Although the concept of logic families dictates that there must be standards for interfacing, there are still certain difficulties when transforming from one standard to another. In addition, the various computer busses are often incompatible. For example, one favorite personal computer advertises an "RS-232 input/output port." But when the user's manual is read, however, we find that it is not quite an RS-232 standard port—it is a "TTL compatible RS-232" port. In fact, if a real RS-232 device is plugged in, damage will result. That computer uses the RS-232 DB-25 connector and pinout definitions, but the voltage levels that represent HIGH and LOW are non-standard. The RS-232 uses +5 to +15 volts for LOW and −5 to −15 volts for HIGH, while TTL uses 0 to +0.8 volts for LOW and +2.4 to +5 volts for HIGH—hardly compatible with RS-232. In this section I discuss the various methods for translating between digital integrated circuit logic families and computer busses.

THE INTERCONNECTION PROBLEM

Figure 23-1 shows the interconnection problem in simple terms. We see a voltage scale running from −15 to +15 volts dc, and the ranges for

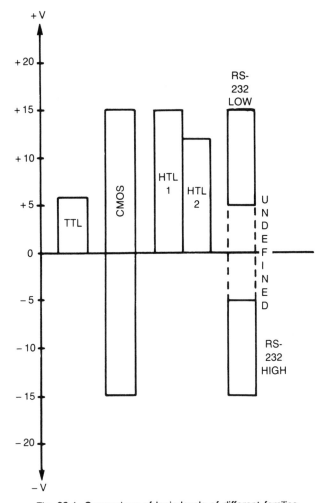

Fig. 23-1. Comparison of logic levels of different families.

each individual form of logic device or standard bus. The TTL device operates on a range of 0 to +5.2 volts dc, with LOW being defined as 0 to +0.8 volts and HIGH being greater than +2.4 volts.

The CMOS family can be operated on any power-supply potential from $-/+4.5$ to $-/+15$ volts dc, and the HIGH and LOW voltages need not be equal. The signal voltages used depend upon the actual dc power supply voltages used in the circuit, with the HIGH-to-LOW and LOW-to-HIGH transition point being one-half the difference. For example, when $V+ = V-$ the output transition occurs when the input signal crosses zero. If the two potentials are unequal, then the transition point is the mid-point voltage. For example, when $V+ = 12$ volts and $V- = -10$ volts are used, then the trip point is +1 volt. In general, however, it is common practice to make $V+ = V-$ unless one of them is zero. In the event one potential is zero, then the trip point is one-half the other voltage (see Table 23-1).

Certain forms of bipolar high threshold logic (HTL), also called high-noise immunity logic (HNIL), use logic levels of 0 volts for LOW and either +12 or +15 volts for HIGH. Neither of

Table 23-1. Voltage Trip Points

V+	V−	Trip Point
+5	0	+2.5
+10	0	+5.0
0	−5	−2.5
0	−10	−5.0
+12	−12	0

these logic levels are compatible with either TTL or RS-232, but may be compatible with certain CMOS circuits if the correct dc power supply-potentials are selected.

The RS-232C standard supposedly spells out the levels and pinouts for a standard interconnection for computers, peripherals, MODEMs and other devices. In this standard (covered earlier in detail) HIGH is defined as a voltage between −5 and −15 volts, while LOW is defined as a voltage between +5 and +15 volts. Typically, personal computer devices use −12 and +12 volts for these levels and may or may not respond to lower potentials depending upon how standard is the "standard interface." The area between −5 and +5 volts is undefined.

DIGITAL LOGIC FAMILIES

Before you can properly appreciate the problems of interfacing elements of digital circuits, you must first understand how those elements work. In this section I discuss the popular IC digital logic families, including a couple of obsolete types that are still in use, so that you will become familiar with their operation.

There are several different families of digital logic devices. The definition of a "family" is somewhat vague, but in general, a logic family is a group of digital IC logic elements that: 1) use similar technology, and 2) can be interfaced with each other without regard for external circuitry. The first criterion, i.e., "similar technology," means that the devices use the same sort of internal circuitry and active devices (npn/pnp bipolar transistors, MOSFET transistors, etc.). The second criterion means that the inputs and outputs of the various devices can be interconnected with each other using only a piece of wire or printed circuit track. Thus, you will find that the input circuits and output circuits are standardized in order to facilitate interfacing.

In this section I consider the most popular logic families of the past ten years or more: RTL, DTL, TTL, CMOS, and HNIL/HTL. Some of these are obsolete but are still found in older equipment, while others are still in current use.

SPEED-VS-POWER

There is always a trade-off between speed and power consumption in digital logic devices. The factors that affect operating speed include the resistance and capacitance elements of the transistors in the device — in other words, an internal RC time constant. In order to increase the operating speed, you have to reduce the RC time constant — which in practical terms means reducing the resistive element of the RC combination (the capacitance is harder to reduce). Unfortunately, reducing the series resistance in a circuit also makes the current higher for any given applied voltage — and that increases power dissipation. Thus, you will find that CMOS devices draw considerably less current than TTL, but the operating speed is also considerably lower.

INVERTER CIRCUITS

Before getting deeper into the subject of digital logic families, let's consider the inverter circuit (also called NOT gate or complementer). In our discussion of the various families, I will use inverter circuits as our example. Figure 23-2 shows an inverter. Figure 23-2A shows the inverter circuit symbol, which consists of a triangle with a circle (indicating inversion) at the output. The rules of operation for an inverter are simple:

1. A HIGH on the input produces a LOW on the output, and
2. A LOW on the input produces a HIGH on the output

Figure 23-2B shows the operation of these rules on a square-wave input: note that the output is always the opposite (i.e., "complement") of the

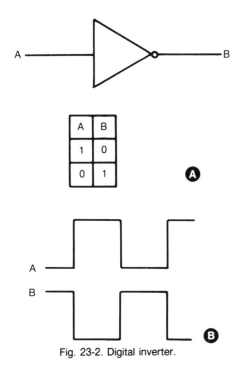

A	B
1	0
0	1

Fig. 23-2. Digital inverter.

input (note: the complement of binary digit "1" is "0," and vice-versa).

RESISTOR-TRANSISTOR LOGIC

Resistor-transistor-logic (RTL) is one of the oldest forms, and is now obsolete. You will, however, still find RTL devices in some older equip-ment. The RTL devices used uA-9xx and MC-7xx type numbers (Fairchild and Motorola, respec-tively).

Figure 23-3 shows a typical RTL inverter cir-cuit. This inverter circuit consists of a single npn bipolar transistor in a grounded inverter configura-tion. The output is taken from the collector of the transistor, and the input is applied to the base circuit. There are series resistors in the base and collector lines. The RTL device operates from a single, monopolar, dc power supply of +3.6 volts (with +4 volts being maximum).

Note that series resistances in the collector and base lines means a certain amount of protec-tion for the RTL device. Provided that there is not more than +4 vdc on the printed-circuit board, no combination of opens or shorts will destroy the RTL device! The RTL device had a relatively low operating speed, on the order of 3 to 6 megahertz. The power requirements were moderate, but not low.

The RTL family is a "saturated" logic family because the transistor is always either fully on or fully off. When a HIGH is applied to the input, transistor Q1 is forward biased hard on, so the collector resistance is very small. This condition places the output terminal only a few tenths of a volt above ground potential, and the entire +3.6 volts from the power supply is dropped across R1.

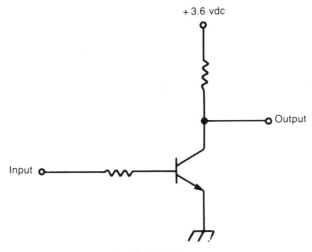

Fig. 23-3. RTL inverter.

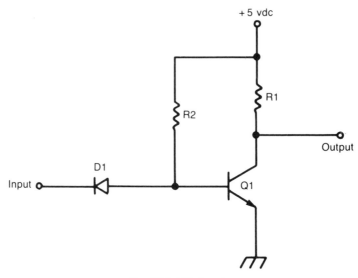

Fig. 23-4. DTL inverter.

The power dissipation is highest in this condition. Alternately, when a LOW is applied to the input, the transistor is turned off (unbiased). There is no current flow from collector to emitter, so no voltage drop exists. The output terminal will present the full +3.6 volts to the outside world.

RTL devices are not recommended for new designs, and indeed some semiconductor manufacturers no longer stock any RTL devices whatsoever.

DIODE-TRANSISTOR LOGIC

Diode-transistor logic (DTL) is another obsolete logic family. These devices carried MC9xx-series type numbers (and certain "house numbers" also). The DTL inverter is shown in Fig. 23-4. The collector-emitter circuit is similar to the RTL family mentioned above. The npn transistor is in the common-emitter configuration, with the output taken from the collector and a series resistor between the collector and V+. Added to the circuit (with respect to its RTL antecedent) is a bias resistor (R2) and a pn diode in series with the input and base line. When the input is HIGH, the diode (D1) is reverse biased and so will not conduct. In this case the current from resistor R2 forward biases transistor Q1 and drives it into saturation. The output line is placed at a potential of only a few

tenths of a volt above ground potential and so is LOW. The entire V+ voltage is dropped across R1 in this case, and the power dissipation is maximum. When the input is LOW, on the other hand, the diode (D1) is forward biased and so conducts. Under this condition the transistor is unbiased, and so no current flows from collector-to-emitter. The output is therefore HIGH and will be at a potential equal to V+.

TRANSISTOR-TRANSISTOR LOGIC

Perhaps the most popular logic family is the transistor-transistor-logic family (also known as TTL, or T²L). TTL devices operate from a single monopolar DC power supply of +5 volts. The normal operating range is +4.75 vdc to +5.2 vdc, but tighter tolerances are sometimes recommended. Although the specification sheets will not generally offer this information, some complex function TTL IC devices will not work properly at potentials close to the minimum. For these devices it is recommended that the dc power supply be greater than 4.85 volts dc. The maximum voltage should be less than +5.05 volts dc in order to keep power dissipation (and thus heat-caused reliability problems) to a minimum. The dc power supply used for TTL devices must be regulated.

TTL "standard" devices carry type numbers

in the 74xx and 74xxx series (e.g., 7402). Military-grade TTL devices are designated with 54xx and 54xxx type numbers. In other words, a "5402" is nothing more than a "7402" in uniform. Generally, 54xx/54xxx devices are in ceramic packages, while 74xx/74xxx devices are in plastic DIP packages. In addition, the 54xx/54xxx devices are often found in flatpack, rather than DIP, packages. There are also a number of "house type number" TTL devices on the market, but these generally obtain 74xx/74xxx numbers if they prove popular.

Figure 23-5 shows the normal values of input and output signal levels used in TTL devices. The LOW condition (logical 0) is defined as a potential of zero to +0.8 volts dc, while the HIGH condition (logical 1) is defined as a potential greater than +2.4 volts, and less than +5 volts (the supply voltage value). Normally, TTL outputs are at a potential around +3 volts when in the HIGH con-

dition. The range of potentials between +0.8 volts and +2.4 volts (the upper end of the LOW range to the lower end of the HIGH range) is undefined. When signals are in this range, the device (or devices it drives) will behave unpredictably.

The voltage of signals should not go negative, or the device may be destroyed. That's why it is labelled the "zap zone" in the illustration. Similarly, the voltage of the signal should not range above +5.2 volts, or the device could suffer permanent damage. These conditions do not occur when the TTL device is interfaced only with other TTL devices, but sometimes do occur when interfacing between TTL and other logic families is attempted.

A typical TTL inverter circuit is shown in Fig. 23-6. Note that it is considerably more complex than the previous inverter circuits. For one thing, the output uses a totem-pole circuit with an isolation diode. The two output transistors (Q3 and Q4) are connected in series with each other and isola-

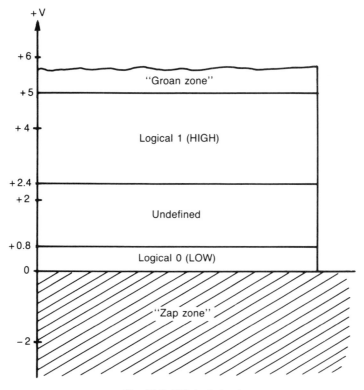

Fig. 23-5. TTL logic levels.

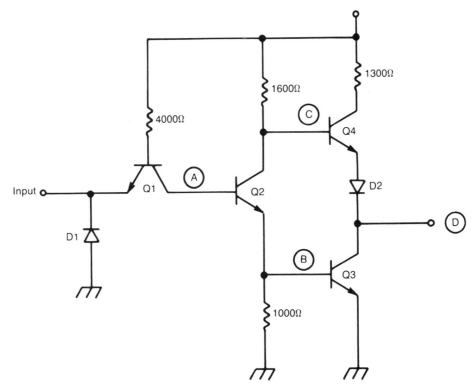

Fig. 23-6. TTL inverter.

tion diode D2. These two transistors are driven by a third transistor, Q2, but out of phase with each other. Note that Q3 is driven from the emitter of Q2, while Q4 is driven from the collector of Q2. Because the collector and emitter of a common-emitter configured transistor produce out-of-phase signals, you will find that Q3 will be turned on hard when Q4 is off, and vice-versa.

Transistor Q1 is the input transistor. The emitter of this transistor forms the input of the inverter, which is configured as a current source. The standard TTL input will source 1.8 milliamperes when LOW, and zero current when HIGH. Let's consider the circuit action. When the input is LOW, the emitter of Q1 is grounded. Since the base of Q1 is permanently forward biased, the collector of Q1 drops LOW. Thus, transistor Q2 is turned off because its base is connected to the collector of Q1 (which is at a potential only a few tenths of a volt higher than ground potential). Under this condition, the collector of Q2 is HIGH,

forcing Q4 into saturation. At the same time, Q3 is turned off, so the output will be HIGH (it is connected to +5 volts through D2, the 1.3k resistor and the collector-emitter path of Q4). Just the opposite situation obtains when the input is HIGH. Transistor Q1 sees its emitter HIGH, and so its collector will also be HIGH. Under this condition, Q2 is saturated, so its collector is at a low potential and its emitter is at a high potential. The base of Q3 is forward biased, so the collector of Q3 finds itself at ground potential — thus producing a LOW output condition.

The typical (or "standard") TTL output will drive ten standard TTL inputs. This spec means that it will sink 18 milliamperes. The term for drive and load requirements are called "fan-out" and "fan-in," respectively. The "fan" specification reduces drive and load conditions in terms of how many standard inputs are concerned. A "fan-in" of one is defined as the load imposed by one standard TTL input, which means a current source of 1.8

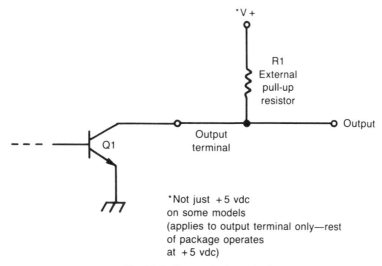

Fig. 23-7. Open-collector output.

mA at TTL HIGH logic levels. When it is stated that most TTL outputs have a fan-out of ten, it is implied that the output will drive ten standard TTL inputs without any trouble.

A slightly different TTL device is shown in Fig. 23-7. This is an example of an open-collector TTL output. The output transistor is in the grounded-emitter configuration, but the collector is left floating at the output terminal. In order for this device to work properly, there must be a load resistor (or other device) connected between the output and V+. This resistor (R4) is usually called a "pull-up resistor."

One of the most common forms of open-col-

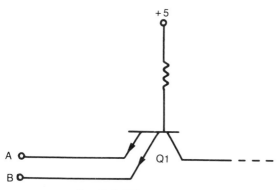

Fig. 23-8. TTL multiple input.

lector TTL devices is the *hex inverter*. This IC contains six inverter circuits that are completely independent of each other, except for the +5 vdc power supply and ground connections. The type numbers are 7405, 7406, 7407, 7416, and 7417. Some of these devices can operate at potentials higher than +5 volts, which means that they have possibilities for interfacing. Because the current level of some of these devices is higher than 18 mA, you can use them as high fan-out drivers.

Additional inputs, such as might be found in the 7400 two-input NAND gate, are obtained by adding emitters to the input transistor, as shown in Fig. 23-8. This is the standard circuit for TTL NAND gates.

TTL SUB-FAMILIES

There are several sub-families in the TTL group. These sub-families are designed to provide certain special characteristics such as high-speed, low power consumption and so forth. Figure 23-9 shows the input circuit for a Schottky TTL device. There is a Schottky zener diode across each input line. These diodes are high-speed types with a low forward voltage drop, and are used to prevent the input transistor from saturating. Thus, the propagation time is shortened because of shortened charge storage time inside of the device. Schottky

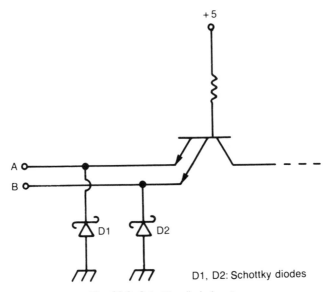

Fig. 23-9. Schottky diode inputs.

TTL devices operate at speeds up to 125 MHz. The Schottky TTL devices are identified by type numbers of the series 74Sxx and 74Sxxx.

Low-powered TTL Schottky devices (Fig. 23-10) bear type numbers in the 74LSxx and 74LSxxx. These devices are not as fast as 74S devices but are somewhat faster than regular TTL devices. The principal advantage is that 74LS chips operate with as little as one-fifth the power required of normal TTL devices bearing the same number (e.g., 7400 versus 74LS00). Like the 74S series, the 74LS series uses Schottky diodes in the input circuit.

High-power TTL devices are indicated by

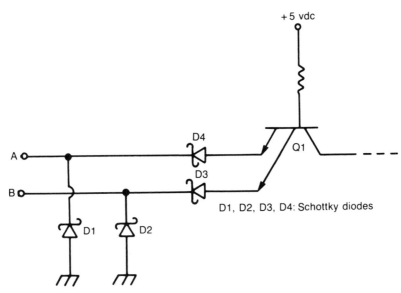

Fig. 23-10. LS-input.

**Table 23-2. Speed and Power
Requirements for TTL Sub-Families**

Series	Power Per Gate (mW)	Operating Frequency (MHz)
74xx	10	30
74Hxx	22	50
74Lxx	1	3
74Sxx	20	125
74LSxx	2	45

type numbers of the 74Hxx and 74Hxxx series. These devices operate to 50 MHz. These devices require higher power for higher speed operation.

Table 23-2 shows the relative speed and power requirements of the various TTL sub-families.

HIGH-THRESHOLD LOGIC

High-threshold logic (HTL), also called high-noise-immunity logic (HNIL), is used when there is a very noisy environment that would upset TTL or some other logic family. The HTL family is based on bipolar transistor technology, like TTL, but uses considerably different operating levels.

Figure 23-11 shows a typical HTL inverter circuit. Transistors Q1 and Q2 form a simple totem-pole output circuit that will provide the inverting action. When the base of Q1 is HIGH, it is in saturation so its collector is at a potential of only a few tenths of a volt. Under this condition, the

output of the HTL inverter is LOW. Similar, when the HTL input (base of Q1) is LOW, the transistor is cut off so the collector will be essentially at V+ (i.e., HIGH). The high threshold is provided by the zener diode in series with the base circuit.

The logic levels used in HTL logic are 0 volts for LOW and +12 or +15 volts for HIGH. Note that the HIGH condition is represented by potentials in the +10 to +12 volts region, thus affording a large noise immunity. There are two versions of HTL available. One uses 0 and +12 vdc as the logic and supply levels, while the other uses 0 and +15 vdc as the logic/supply levels.

TRI-STATE LOGIC

There are a lot of digital logic devices on the market which are called "tri-state." Normal digital logic devices are two-state (i.e., binary). This designation means that an output can only be either HIGH or LOW; there is no in-between. But a *tri-state* device has a third state: disconnected (see Fig. 23-12). Switch S1 represents the normal binary condition of any TTL device (in this case an inverter). When the input is LOW, the switch (S1) is connected to the +5 vdc through a low resistance. Alternatively, when the input is LOW the switch is connected through a low resistance to ground. The tri-state condition is provided by

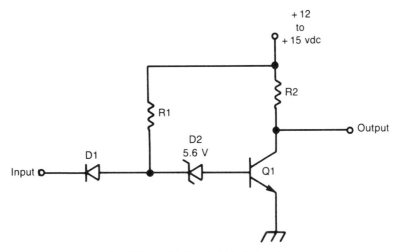

Fig. 23-11. High-threshold logic inverter.

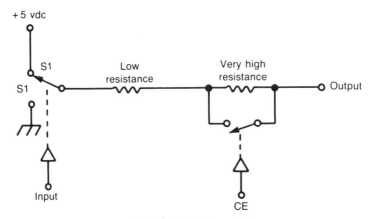

Fig. 23-12. Tri-state output.

switch S2. When the chip select (CS) is LOW, switch S2 is closed, so the output is connected to the inverter output (i.e., pole of S1). But when CS is HIGH, switch S2 is open and there is a very high resistance between the output and the inverter. In effect, the output of the device (the pin on the IC) is disconnected from the internal circuitry.

The main use of the tri-state logic device is in bussing together several outputs on the same line. Thus, you will be able to use several devices that are each turned on only when their respective CS lines are LOW. Microcomputers typically use tri-state logic to connect the INPUT PORT lines to DATA BUS lines in order to prevent loading the lines of the data bus by action on the input port.

COMPLEMENTARY-SYMMETRY METAL OXIDE SEMICONDUCTOR (CMOS)

One of the most popular of the modern families of digital IC logic devices is the complementary symmetry metal oxide semiconductor (CMOS) family. Although considerably slower than TTL devices, CMOS devices are widely used because they use considerably less current than TTL devices. While the TTL device might require milliamperes to operate, the same function in CMOS will operate on microampere — that's 1000 times (three orders of magnitude!) less current. CMOS devices are made using MOSFET transistors instead of bipolar npn/pnp transistors. Before dealing with the details

of CMOS logic elements, let's first review the MOSFET transistor.

MOSFET TRANSISTORS

The transistors used in CMOS devices are metal oxide semiconductor field effect transistors, or "MOSFETs," also sometimes called insulated gate field effect transistors, or "IGFETs." These transistors differ markedly from pnp and npn bipolar transistors used in other digital logic families. There are two basic types of MOSFET transistor: depletion mode and enhancement mode. Each of these are found in two different polarities (n-channel and p-channel), making a total of four different varieties:

1) P-channel depletion mode
2) N-channel depletion mode
3) P-channel enhancement mode
4) N-channel enhancement mode

Figure 23-13 shows the basic depletion mode MOSFET. There are three electrodes on this device: drain, source and gate. The drain and source are merely ohmic (metal) contacts at either end of a semiconductor "channel." In this example, an n-type semiconductor is used, so this device is an n-channel depletion mode MOSFET. The gate is a metallized contact that is separated from the channel by a thin layer of insulating metal oxide material (hence the name "insulated gate").

A depletion mode transistor is a normally on device. That is, when the voltage applied to the gate is zero, current will flow in the channel at a level that is determined only by the Ohm's law relationship of the drain-source voltage and the channel resistance.

Applying a potential to the gate (of the correct polarity, which depends upon the type of material) causes charge carriers to be driven away from the gate region, leaving a depletion zone that is essentially free of carriers. Thus, the material in the depletion zone becomes a high-resistance insulator. The depletion zone is created by the effect of the electrical field generated by the gate voltage. Changing the size of the depletion zone effectively changes the size (and thus the dc resistance) of the charge carrier channel. You will find, therefore, that the resistance between source and drain varies with the voltage applied to the gate.

Figures 23-14A and 23-14B show an enhancement mode MOSFET (n-channel) under two separate conditions. In Fig. 23-14A the gate voltage is zero. Since this type of MOSFET is normally off, the enhancement (i.e., conductive) zone is minimum and the d-s channel resistance is maximum. In Fig. 23-14B, on the other hand, a voltage is applied to the insulated gate, which creates an electrical field within the channel. This potential draws charge carriers toward the gate, thereby decreasing the d-s resistance (this means that the channel is "enhanced").

In both types of MOSFET the channel resistance is varied by the voltage applied to the gate. In an analog circuit, this resistance may vary continously between limits according to the applied signal voltage. Digital circuits, however, have only two signal levels (HIGH and LOW), so the MOSFETs used in CMOS devices tend to be either all the way

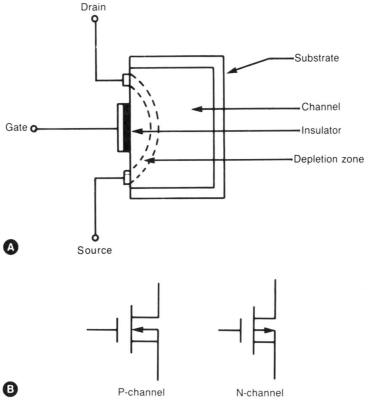

Fig. 23-13. MOSFET transistor (depletion).

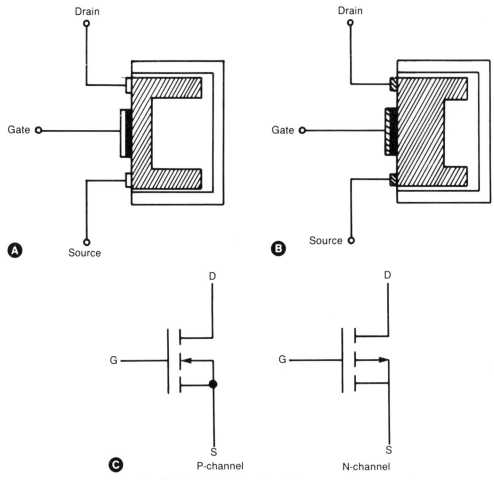

Fig. 23-14. MOSFET transistor (enhancement).

on or all the way off. The usual schematic diagram symbols for the depletion mode and enhancement mode MOSFET transistors are shown in the two figures.

CMOS DIGITAL DEVICES

The word "complementary" in the name "CMOS" implies that these digital IC logic devices contain complementary MOSFET transistors, i.e., a mix of n- and p-channel devices. Figure 23-15 shows the circuit of a typical CMOS inverter which contains one p-channel and one n-channel device. The d-s paths (i.e., channels) are connected in series with each other, while the gates are con-

nected in parallel. In essence, the CMOS inverter consists of a pair of electronic resistors in series across the dc power supply. These "resistors" are configured such that one is high resistance while the other is low resistance, and are controlled by the voltage applied to their parallel-connected gates.

Figure 23-16 shows the relationship of the channel resistances for two different situations, input-LOW and input-HIGH. Note that the output signal is the voltage between ground and the junction of R1 and R2. In each case one resistance is very high (megohms) and one is very low (about 200 ohms). Thus, the output will be either close to V− or V+. Because most CMOS devices use posi-

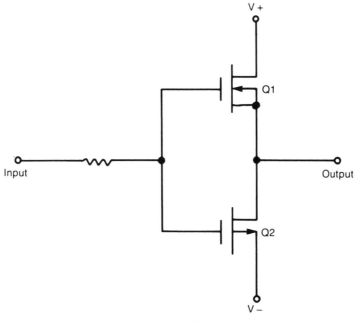

Fig. 23-15. CMOS inverter.

tive logic designations, V+ represents a HIGH output condition and V− (or ground if V− = 0) represents a LOW output condition.

The situation for input-LOW is shown in Fig. 23-16A. Because this is an inverter circuit, the output will be HIGH when the input is LOW. In this condition, R1 (i.e., the channel resistance of Q1) is very LOW, while R2 (the channel resistance of Q2) is very high. Thus, V+ is connected to the output terminal.

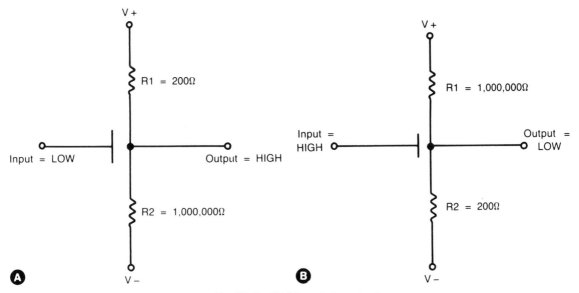

Fig. 23-16. CMOS equivalent circuit.

The opposite situation (input-HIGH) is shown in Fig. 23-16B. Here you will find that the respective roles of the resistors are reversed. In this case, R1 has a high resistance and R2 has a low resistance. Thus, the output is essentially connected to V−.

The transition point (i.e., input potential) that causes either HIGH-to-LOW or LOW-to-HIGH output changes is the midpoint voltage between V− and V+:

$$V_t = ((V+) + (V-))/2$$

In many practical situations you will find that the absolute values of V− and V+ are equal. There is no reason, however, why you can't use unequal voltages for V− and V+. In fact, in many cases CMOS devices are operated from TTL-style power supplies, especially when used with TTL devices. These CMOS devices use +5 vdc for V+ and 0 volts for V−. For reasons that are sometimes obscure, some designers set V+ to zero and then use a negative potential for V−. The normal range of ±V is 0 to ±15 vdc, with some devices operating to ±18 vdc.

SERIES A VS SERIES B CMOS DEVICES

The generic type numbers for CMOS digital logic integrated circuits are in either the 4000 range (e.g., 4049) or 4500 range (e.g., 4528). There are also special devices which carry unique house type numbers. There are two general series of CMOS devices, designated "A" and "B." The A series is the original type of CMOS device, and may or may not carry an "A" suffix on the type numbers. Thus, both "4001" and "4001A" can be assumed to be A-series devices. All B-series devices carry a "B" suffix to the type number ("4001B").

The principal difference between A and B devices is that B series CMOS ICs contain internal zener diodes that shunt high-voltage static charges around the delicate gate insulators. This reduces, but does not eliminate, potential for damage for electrostatic discharge (more on this later).

There are also other differences between A-

series and B-series devices. The B-series devices generally provide higher operating frequencies, faster rise times, and greater drive capability than A-series devices.

ELECTROSTATIC DISCHARGE (ESD) DAMAGE

All semiconductor devices can potentially be damaged by high-voltage static charges; CMOS and MOSFET devices seem particularly sensitive to electrostatic discharge (ESD) damage. Static electricity is created by the "triboelectric effect," that is, by creating excess electrons on the surface of an insulating material by mechanical rubbing. Static electricity can be generated without conscious effort. Because 35,000 volts can be generated by walking across a carpet, it's no wonder that you can get "bit" by grabbing a doorknob!

When the ESD charge hits, it punches a hole in the insulating gate. Normally, the failure is immediate and catastrophic. Metal from the gate electrode and semiconductor material from the channel fill the void and short the gate-to-channel.

At other times the failure is delayed often many months or even years. The metallization of the void is not initially sufficient to short-out the device. As time goes on, however, metallic ions and semiconductor material will migrate into the void. When this process goes on long enough, a short-circuit will develop and the IC or MOSFET will be destroyed. To the user, the result is a premature failure of the IC. In commercial equipment, where dozens or hundreds of units are made, the reliability and QA people might notice an inordinate failure rate among CMOS (or related NMOS or PMOS) and Schottky TTL devices. This fact could indicate inadequate procedures or facilities for handling ESD-sensitive devices. Obviously, users in arrid regions have to be more careful than users in high-humidity regions.

ESD PROTECTION

Damage from static electricity can be controlled or even eliminated by the use of correct methods. The aim of most techniques is to keep all pins of the device at the same potential. If excessive voltages cannot develop between pins of the IC, then damage will not occur. Other methods are just common "horsesense." For example, don't handle CMOS or other ESD-sensitive devices more than absolutely necessary. You will thereby reduce the exposure of each device to a minimum. Fortunately, few people get their kicks from fondling ICs.

CMOS devices are usually shipped in containers that serve to reduce the chance of ESD damage. Again, all of these strategies are designed to maintain all pins of the CMOS device at the same potential. In some cases, devices or groups of devices are shipped wrapped in aluminum foil. Most commercial sources, as well as those near-retail dealers who sell in quantities rather than "onesie-twosie," will ship CMOS either mounted on black conductive foam, or inside of sleeves made of anti-static plastic. Blister pack distributers often use a tiny piece of conductive foam material to hold pins of the IC at the same potential. It is prudent to keep CMOS in the containers (or on the foam) in which they were shipped until it is time to use them.

Some texts will tell you to ground yourself to prevent ESD damage. While this strategy might, in fact, prevent ESD damage to the CMOS chips, it could also KILL you! If there is any source of 115-volt ac (or other high voltage) around the bench, then it might seek ground through your ESD-grounded body—turning you into a blown fuse!

A better alternative is to ground yourself, and separately your work surface, through a very high resistance. A resistance of 2 to 10 mehohms will serve to drain off electrostatic charges while also limiting current flow to a relatively safe value in the event of an accidental contact with ac power. Use a 2-watt (or higher) resistor for this duty. We are not seeking the wattage rating in this case, but rather, the voltage rating (that's right, resistors have a voltage rating!). Lower wattage resistors have a lower voltage rating and so may short out and become either ineffective for ESD or dangerous to you.

For years, hospitals faced a similar problem in operating rooms. Explosive anesthetic agents (e.g., ether and cyclopropane) could be exploded due to sparks from ESD. To counter this problem they used conductive, high resistance-to-ground flooring, conductive high resistance shoes or shoe covers, and all-cotton garments. This may be extreme for most electronic assembly situations, but may be appropriate in arrid regions if ESD proves to be a nearly unsolvable problem.

Tools such as soldering irons should have a grounded tip in order to prevent electrostatic charge buildup. Also, the finished product (e.g., printed wiring boards that contain CMOS devices) should be handled similar to the chips. Store the PWBs on conductive foam or in static-treated plastic bags. It is a myth that all CMOS devices become safe to handle when "in the circuit;" some do, and some don't (it depends upon the circuit configuration).

DIGITAL IC DEVICE INTERFACING

One of the criteria for a "logic family" is the ability to interface members of the family with only a conductor path between inputs and outputs. For example, the TTL family is designed such that a TTL input will source (i.e., provide) 1.6 mA of current, while the TTL output will sink at least that much current. In actual practice the TTL output usually will sink enough current to provide for at least ten regular TTL inputs.

Interfacing TTL devices with each other is merely a matter of connecting the inputs of following devices to the output of the driver device, as shown in Fig. 23-17. The designer need only be mindful of the number of TTL inputs that can load the given output. When in doubt, assume that the regular TTL output will "fan-out" ten TTL inputs.

When dealing with TTL sub-families you must use somewhat modified rules. There are five major sub-families (described earlier in this chapter): regular TTL (74xx), low-power TTL (74Lxx), high-power TTL (74Hxx), Schottky TTL (74Sxx) and low-power Schottky TTL (74LSxx). These different families have different input source currents (fan-in) and output sink (fan-out) capacities.

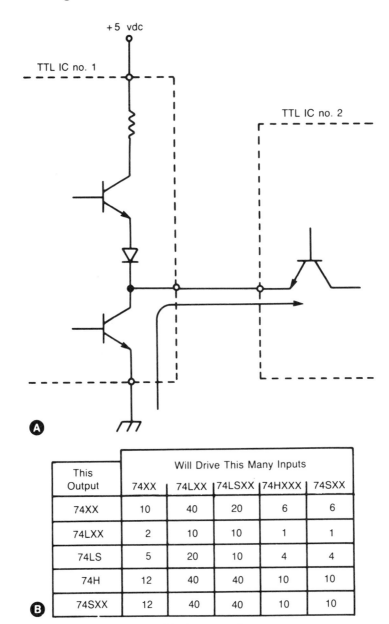

Fig. 23-17. TTL interface.

This Output	Will Drive This Many Inputs				
	74XX	74LXX	74LSXX	74HXXX	74SXX
74XX	10	40	20	6	6
74LXX	2	10	10	1	1
74LS	5	20	10	4	4
74H	12	40	40	10	10
74SXX	12	40	40	10	10

Table 23-3 gives the number of outputs each sub-family will drive, so you won't have to make calculations. Look for the family of the driver device along the left-hand vertical column, and then look to the right under the correct load device for the number of inputs that can be driven. For example, suppose you have a 74LSxx TTL device and

Table 23-3. Fan-In and Fan-Out of TTL Sub-Families

TTL Sub-Family	Fan-In (mA)	Fan-Out (mA)
74xx	16	1.6
74Lxx	3.6	0.18
74Hxx	20	2
74Sxx	20	2
74LSxx	8	0.4

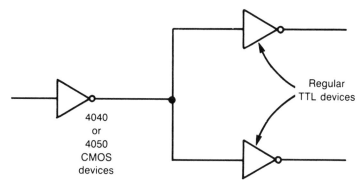

Fig. 23-18. TTL sub-families.

want to drive 74Lxx devices. Look on the left-hand column for LS-TTL, and then to the right under "L-TTL" and note that the number 20 is entered. This information tells you that a 74LSxx TTL output will drive up to 20 74Lxx inputs.

CMOS-TO-TTL INTERFACING

Most CMOS and TTL devices are not directly interfaceable due to differences in their input and output circuit structures. In this section I discuss some of the methods for making interconnections between TTL and CMOS families.

There are at least two CMOS devices that are designed for interfacing with TTL under certain circumstances. The 4049B and 4050B devices will each drive up to two regular TTL loads (Fig. 23-18) when the entire 4049B or 4050B package

is operated from TTL-compatible dc power supplies. The 4049B is a hex inverter (six inverters in one package), and the 4050B is a hex noninverting buffer. Both of these chips can be operated from the full range of CMOS dc power-supply voltages, but when V$-$ = 0 volts and V$+$ = 5 volts dc, they can be used to directly drive TTL inputs. If the package supplies are anything but 0 and +5 volts dc, however, then the package is not directly compatible with TTL inputs.

The opposite situation, that is, when a TTL device drives a CMOS input, can be accommodated most easily if the CMOS device operates from TTL power supply. If the V$-$ = 0 volts and V$+$ = 5 volts, then the TTL can be connected as in Fig. 23-19. The 2.7-kilohm resistor is used to pull up the TTL output to +5 volts in the HIGH condition.

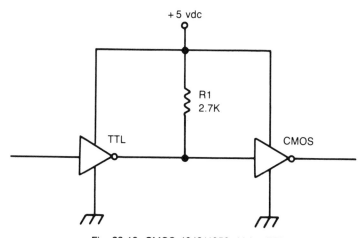

Fig. 23-19. CMOS 4049/4050 driving TTL.

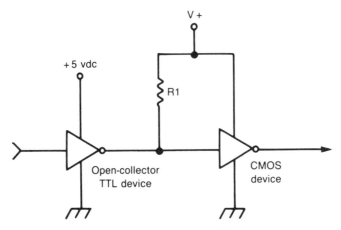

Fig. 23-20. TTL to CMOS.

The TTL output operates as a current sink, so one has to provide it with a current to sink — that's the job of the pull-up resistor. The value selected should simulate the current sourced by one regular TTL input, or about 2 to 3 kilohms.

Figure 23-20 shows a similar circuit that is used where the CMOS chip operates from other than +5 volts dc. Here you need an open-collector TTL output to sink the current from a larger value pull-up resistor. Depending upon the device selected, open-collector TTL inverters will operate at voltages up to +15 or +30 volts dc and currents to about 30 mA.

RS-232/TTL INTERFACING

Most computers have at least the I/O ports or "system bus" TTL compatible. In order to translate between TTL and RS-232 devices, you need to use certain types of circuits. One alternative is to use the Motorola MC-1488 and MC-1489 RS-232/TTL receiver/transmitter chips. Where these are not available, however, you can use discussed shown in this section.

Figure 23-21 shows how to translate the level of an incoming RS-232 signal to the TTL levels. Recall that the RS-232 has two problems. First,

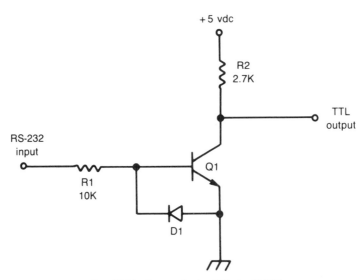

Fig. 23-21. Open-collector TTL to CMOS.

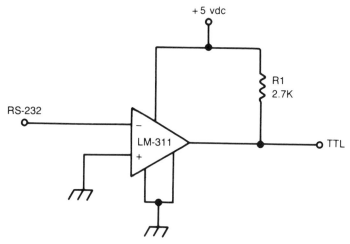

Fig. 23-22. RS-232 to TTL.

the levels are −5 to −15 volts for HIGH and +5 to +15 volts for LOW (typically: HIGH = +12 volts, and LOW = −12 volts). Thus, you need voltage-level translation. Second, RS-232 uses negative logic while TTL uses positive logic. Thus, you need an inverter in the circuit somewhere.

In Figure 23-22, the inversion function is accomplished by using an npn bipolar transistor in the common-emitter configuration. The transistor selected for Q1 can be the 2N2222 or any other medium to high beta transistor. The collector potential of the transistor is connected to a TTL-compatible +5 volt dc power supply. An npn transistor is biased on when the base is more positive than the emitter. It is biased off when the base-emitter potential is zero or negative.

In the case of Fig. 23-22, an RS-232 HIGH (−12 volts) applied to the input will be clipped to −0.7 volts by diode D1, but this potential is enough to turn off transistor Q1. Because Q1 is turned off, no collector current flows through resistor R2 and that makes the collector of Q1 HIGH at +5 volts. Thus, an RS-232 HIGH at the input makes the output TTL HIGH. Similarly, when an RS-232 LOW (+12 volts) is applied to the input, the transistor sees a positive bias (current limited by R1) and so turns on hard. In this case the transistor is saturated, so the collector is near ground potential. In other words, the RS-232 LOW produces a TTL LOW condition.

Another RS-232/TTL level translator circuit

is shown in Fig. 23-22. This circuit is based on the LM-311 voltage comparator. The LM-311 has an open-collector output that is TTL compatible when a 2.7k pull-up resistor is connected to a +5-volt dc power supply. The main chip package power is ±12 volts dc. The LM-311 compares the voltages applied to inverting and noninverting inputs and produces an output indicating their relative polarity relationship. Because the noninverting input is grounded, the transition voltage is zero volts. When the RS-232 signal is applied to the inverting input, a TTL LOW is generated in response to the RS-232 LOW and a TTL HIGH results from the RS-232 HIGH.

Methods for converting TTL signals to RS-232 levels are shown in Figs. 23-23 through 23-25. In Fig. 23-23 you see the use of an operational amplifier for this application. The noninverting input is connected to a 1.4-volt reference voltage. Because there is no negative feedback in this circuit, the operational amplifier operates as a voltage comparator. When a TTL LOW (0 to +0.8 volts) is applied to the inverting input, the inverting input is at a lower potential than the noninverting input. The operational amplifier acts as if it sees a negative voltage at the inverting input and causes its output to snap to near V+. Thus, the TTL LOW produces an RS-232 LOW. When the TTL line goes HIGH (2.4 volts or more), the inverting input is higher than the noninverting input.

The reader is left to examine the other examples of TTL to RS-232C interfacing, with the proviso that it be understood that the method of Fig. 23-26 (which uses 75188 and 75189 RS-232 receiver and transmitter chips) is the preferred method.

INTERFACING THE OUTSIDE WORLD— SOME BASIC ASSUMPTIONS

In this section I discuss matters such as interfacing LEDs, displays, relays, and other devices. Although the discussion is far from complete, the principles are similar for almost all "external device" applications other than standard peripherals. The basic assumption of this section is that the computer will have at least one input/output (I/O) port that can drive or receive data from the external devices. If the particular computer has no output port, then you can make one using the types of circuits shown elsewhere in this and other TAB books.

The output port will probably drive only one or two standard TTL loads, so you will have to be

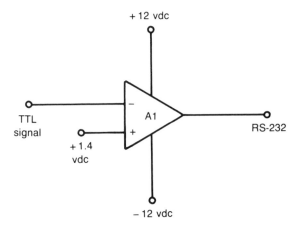

Fig. 23-23. RS-232 to TTL.

careful. If the port bits drive only one TTL load (fan-out = 1), then it will supply only 1.8 milliamperes; if it drives two TTL loads (fanout = 2) then it will drive 3.6 milliamperes. All interface circuits must be capable of being driven by these current levels. Of course, there are a few computers on the market which have higher current output capabili-

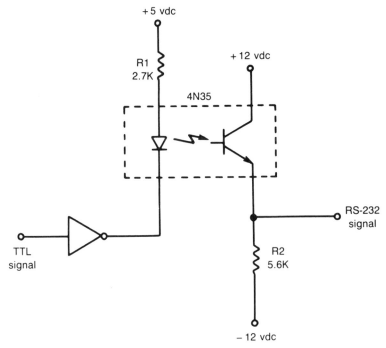

Fig. 23-24. TTL to RS-232.

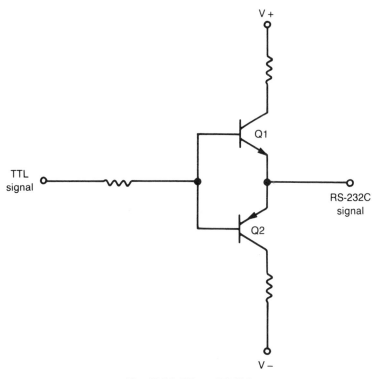

Fig. 23-25. TTL to RS-232.

ties, but it is unwise to make any assumptions in that regard unless there is specific information available either from the manufacturer or you have knowledge of the design. In the event that higher drive capacities are needed, then buffer the outputs with regular or high-current TTL devices. Most noninverting TTL buffers and inverters will drive at least ten TTL loads (18 milliamperes).

DRIVING INCANDESCENT PANEL LAMPS

The old-fashioned incandescent panel lamp has been all but eclipsed by the light-emitting diode, but it is still occasionally used. There are still enough uses for the lamp so that I will have to consider it here. The lamps specifically considered in this chapter are the little pilot or panel lamps used as indicators. For the most part, the current requirements will be 40 to 150 milliamperes. Lamps with lesser current requirements are hard to locate, while those with greater current requirements are used for panel lighting, not indicators.

If the lamp has a very low current requirement, then you could simply place it in the line between V+ and the output of an open-collector TTL device (Fig. 23-27A). But these lamps, while they exist, are not easy to come by, so I will assume the situation as shown in Fig. 23-27. Here lamp (I1) draws more current than can be accommodated by any open-collector TTL device. The lamp is turned on or off by pnp transistor switch Q1. When the transistor is turned off, there is no current return path at the "cold" end of the lamp. But when the transistor is turned on hard, i.e., it is saturated, then the collector-emitter resistance is very low (the voltage drop is only a few tenths of a volt) so the lamp has a return path and the lamp turns on.

Transistor Q1 is controlled by an open-collector TTL inverter. When the output of the inverter is HIGH (i.e., when its input is LOW), then transistor Q1 is biased on. In this case the transistor is biased on, so the lamp turns on. But when the inverter output is LOW, then the bias supply for

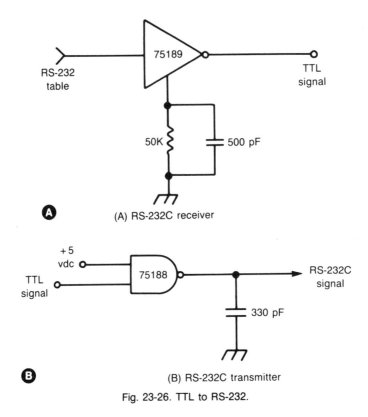

(A) RS-232C receiver

A

(B) RS-232C transmitter

B

Fig. 23-26. TTL to RS-232.

*R1 limits I to 1.8 mA

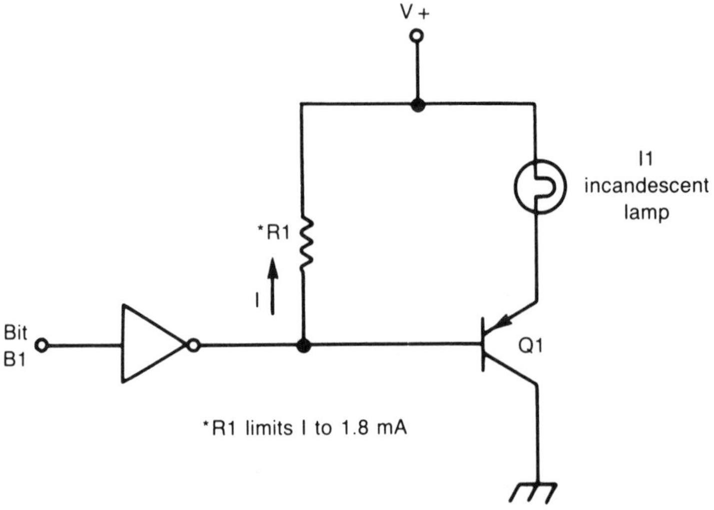

Fig. 23-27. RS-232 driver/receiver ICs.

the transistor is grounded, so the transistor turns off—and so does the lamp.

The input of the inverter is connected to bit B1 (optionally selected—any bit will do) of an output port. If this bit is LOW, then the lamp is off, but if the bit is HIGH it is on:

Bit B1	Lamp Status
0	Off
1	On

Let's see how this lamp can be turned on and off under program control. For our example I will select the Rockwell International AIM-65 microcomputer. This computer uses the popular KIM-style output port structure. The output device is a 6522 peripheral interface adapter (PIA), which is specially designed for the 6502 microprocessor used on the AIM-65. Let's assume that you are using port A of the 6522 (thus of the AIM-65). Bit PA0 (i.e., bit B0 of port A) is found on pin number 14 of the AIM-65 applications connector (which is the connector on the left rear of the SBC when viewed from the keyboard); don't forget the ground connection to the external circuitry (pin no. 1 of the applications connector).

In the AIM-65 the 6522 device is memory-mapped into the space between A000H and A00FH. The port-A output data register is found at A001H, while the port-A data direction register (DDRA) is located at A003H. In the 6522 you can configure each bit of the output ports as either input or output, depending upon the bit placed in the corresponding location in the data direction register. This "programming" can be done on a bit-for-bit basis. Thus, you may, if you want, configure B0 of port A as an output and bit B1 (or another other port-A bit) as an input. This is one of the strengths of any circuit that uses the 6522 device as an input/output port. The protocol requires a 1 for an output port, and a 0 for an input port.

As part of the initialization sequence at the first part of the program you will have to store the I/O direction commands in the DDRA. Let's assume that the plan calls for making all bits of port A as outputs. For this example, therefore, you will

Table 23-4. Initialization Sequence for the AIM-65

Memory Location	Instruction Mnemonic	Comments
0200H	LDA #N	Load accumulator with "N"
0201H	#FFH	"N = FFH = 11111111"
0202H	STA nnnn	Store accumulator data at A003H
0203H	03	NN1
0204H	A0	NN2

have to send a "00H" to DDRA. Given that AIM-65 programmers tend to start at locations 0200H, let's write a brief initialization program sequence (see Table 23-4).

This program loads FF (hex) into the accumulator and then stores that data (FFH) at memory location A003H, which is the memory-mapped location of the DDRA. Following the execution of this initialization program, the port-A data direction register DDRA will contain binary 11111111, so all eight bits are outputs.

Turning on the lamp at bit B1, as in Fig. 23-27, is then a simple matter of writing a 1 to bit B0 of port A. A program sequence such as shown below, with NN = 01H for "lamp off" and NN = 00H for "lamp on," will do the trick (see Table 23-5).

A high-current lamp driven from a 6522-type output bit may require a higher beta transistor in order to turn on fully. The problem is that some single transistors often lack the current gain required to turn on the transistor sufficiently to saturate. In this case you can use the Darlington amplifier configuration of Fig. 23-28. The Darlington configuration used for Q1 and Q2 provides a super-beta situation, and can either be built as shown or bought in the form of a Darlington transistor or

Table 23-5. Program to Turn Lamp On and Off

Memory Location	Instruction Mnemonic	Comments
0340H	LDA #N	Load accumulator with number "N"
0341H	N*	
0342H	STA nnnn	Store accumulator contents ("N") at nnnn
0343H	00H	NN1
0344H	A0H	NN2

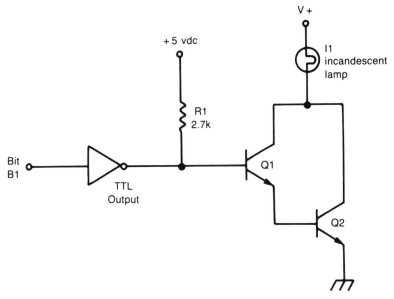

Fig. 23-28. Interfacing lamps.

"superbeta" transistor. The beta is the product of the beta ratings of the two transistors:

$$H_{fe} \text{ (total)} = H_{fe} (Q1) \times H_{fe} (Q2)$$

If the two transistors are identical, then the total beta is the individual betas squared:

$$H_{fe} \text{ (total)} = (H_{fe})^2$$

If a beta rating of 100 is used, therefore, the total beta will be:

$$H_{fe} \text{ (total)} = (H_{fe}) \times (H_{fe})$$
$$H_{fe} \text{ (total)} = (100) \times (100)$$
$$H_{fe} \text{ (total)} = 10,000$$

With a beta of 10,000 the 1.8 milliamperes available from the computer output will drive a lamp up to 1.8 × 10,000 milliamperes, or, 18 amperes (hardly a "panel lamp" level).

LIGHT-EMITTING DIODES INTERFACING

A light emitting diode, or LED, is a solid-state "lamp." It is a pn-junction diode that is specially designed to produce a light output when energized. Most common LEDs used in display applications require 15 to 20 milliamperes of current. LEDs are available in red (most common and probably the most efficient), yellow, and green.

Figure 23-29 shows the most common interfacing technique for LEDs. An open-collector output TTL inverter (U1) is connected to bit B1 of the output port. The LED is connected between V+ (in this case +5 vdc) and the output of the inverter, with a current-limiting resistor connected in series. The current limiting resistor keeps the current below 15 milliamperes, and must be scaled upward when higher V+ voltages are used. Ohm's law will determine the minimum value of R1:

R1 = V+/0.015
Thus, for a +12 vdc power supply,
R1 = 12 vdc/0.015 mA
R1 = 800 ohms

In this case an 820-ohm "standard value" resistor is sufficient.

Because this is a computer-controlled indicator lamp, you can turn the LED on with a HIGH on B1, and turn it off with a LOW on B1:

B1 Level	LED Status
1	on
0	off

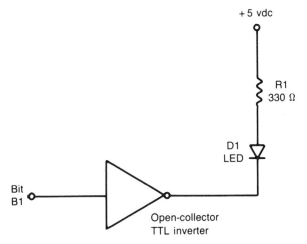

Fig. 23-29. Interfacing lamps.

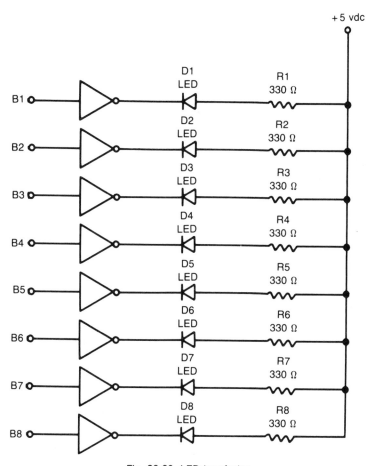

Fig. 23-30. LED interfacing.

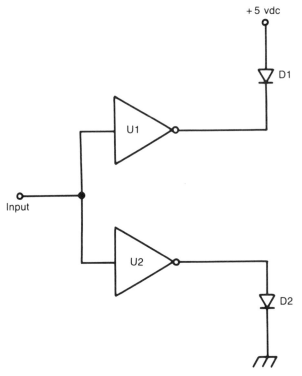

Fig. 23-31. Multiple LED interfacing.

Figure 23-30 shows an eight-bit output port on which are connected LEDs in the manner of Fig. 23-29. This circuit allows you to output various numbers to the port to turn on various numbers. If the application calls for two or more LEDs to be on at the same time, then this is the circuit to use. You can initially turn all lamps on for a test, then turn them off again by writing a FFH to the output port, holding it there for awhile, and then changing the number to 00H. You could then alter each bit as needed.

The LED interfacing circuits shown above are based on TTL devices. With certain B-series CMOS devices you can interface LEDs directly. Although inverters are used in Fig. 23-31, almost any B-series CMOS device will work, a fact which tends to open up some interesting hardware logic combinations. The circuit works because CMOS devices present a low resistance to V+ when the output is HIGH and a low resistance to V− or ground when the output is LOW. Inverter U1 in

Fig. 23-31 will turn on when the output is LOW, or when its input is HIGH. Similarly, inverter U2 will turn on when its output is HIGH and input is LOW. Some people place series current-limiting resistors in series with each LED, and that is a prudent step. About 270 ohms for +5 vdc is usually sufficient because the CMOS internal low resistance is also around 150 ohms.

INTERFACING LED NUMERICAL DISPLAYS

The seven-segment LED numerical display is shown as "LED DISPLAY" in Fig. 23-32. This device is the oldest and most common form of numerical output. By lighting selected segments a through g, you can form any of the decimal digits 0 through 9. By connecting the bits of a computer output in a circuit as previously shown (i.e., U1 and Ra), you can turn on all of our digits as needed. Given the pinouts of Fig. 23-32, the output code is given in Table 23-6.

R1-R7: 330 Ω

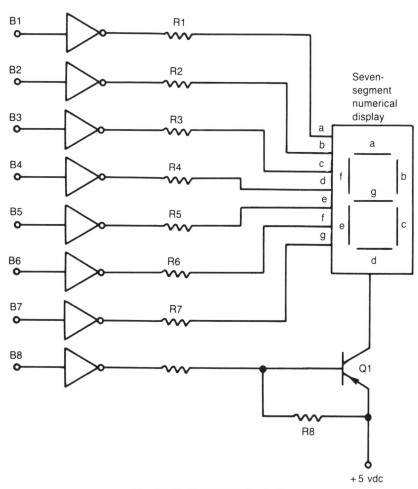

Fig. 23-32. CMOS LED interfacing.

The circuit of Fig. 23-32 assumes that the LED numerical display is of the common anode type. Transistor Q1 is a pnp bipolar type that is used as a series switch between the V+ power supply and the common anode (i.e., power terminal) of the numerical display. In some cases, Q1 may have to be either a Darlington type, or two transistors connected in the Darlington configuration shown earlier. The pnp transistor operates in exactly the opposite manner from the npn: if the base is less positive than the emitter, then the transistor is turned on. In this circuit a LOW on B7 will keep the digit on, while a HIGH will keep it off. This feature is sometimes useful in multiplexing

Table 23-6. Seven-Segment Output Codes for Decimals 0-9

1 = SEGMENT ON
2 = SEGMENT OFF

n "multiplex control" bit:
n = 1 to turn off entire display
n = 0 to turn on entire display

Decimal Digit	Segments On	Bits High	Binary
0	a,b,c,d,e,f	B0,B1,B2,B3,B4,B5	n0111111
1	b,c	B1,B2	n0000011
2	a,b,d,e,g	B0,B1,B3,B4,B6	n1011011
3	a,b,c,d,g	B0,B1,B2,B3,B6	n1001111
4	b,c,f,g	B1,B2,B5,B6	n1100110
5	a,c,d,f,g	B0,B2,B3,B5,B6	n1101101
6	a,c,d,e,f	B0,B2,B3,B4,B5	n0111101
7	a,b,c	B0,B1,B2	n0000111
8	a,b,c,d,e,f,g	B0,B1,B2,B3,B4,B5,B6	n1111111
9	a,b,c,f,g	B0,B1,B2,B5,B6	n1100111

displays and can be deleted in nonmultiplex designs. In this case you can eliminate Q1 and wire the anode directly to V+.

LED NUMERICAL DISPLAY INTERFACING WITH DECODERS

The 7447 device (see Fig. 23-33A) is a TTL IC that is designed to interface with seven-segment LED numerical displays. The 7447 accepts four-bit binary coded decimal (BCD) data on the A–D

inputs and uses it to determine which decimal digit is intended (see chart below). The corresponding seven outputs will drop LOW, thereby lighting up the correct seven-segment LED display segments (see Table 23-7).

There are several other inputs or outputs on the 7447 are of interest.

Lamp Test (pin no. 3). This terminal will turn on all segments (displaying an "8") for test purposes. It is normally kept HIGH, but when it is

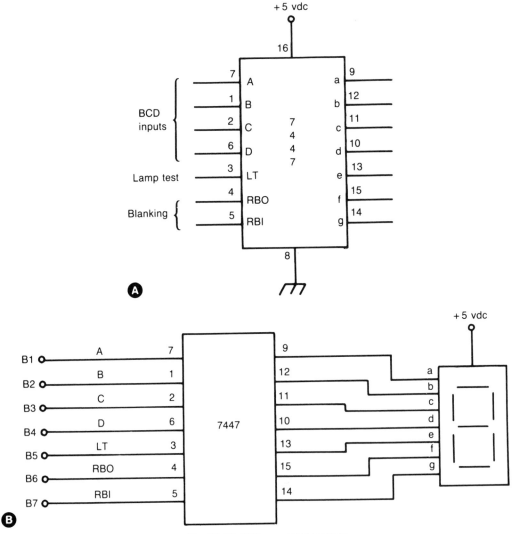

Fig. 23-33. LED numerical readout.

Table 23-7. BCD Codes Corresponding to Decimal Digits

Decimal Digit	BCD Code DCBA
0	0000
1	0001
2	0010
3	0011
4	0100
5	0101
6	0110
7	0111
8	1000
9	1001

LOW all seven segments are turned on. It is common to find instruments in which the L/T is brought LOW for about 5 or 10 seconds immediately when the instrument is turned on, or when a "TEST" pushbutton is operated on the front panel. This strategy will turn on all digits and allow the user to determine if any segments are out (the great bugaboo with the seven-segment display — a missing "g" segment makes an "8" look like a "0" — not funny!).

Ripple Blanking Input (pin no. 5) and Ripple Blanking Output (pin no. 4). The RBI terminal will cause the digit to turn off if it is zero, and the next significant digit is zero. The state of turn on all segments (displaying an "8") for test most significant RBO terminal. Thus, you can extinguish leading zeros in a multidigit display by daisy-chaining the RBO to RBI terminals in order from most significant to least significant.

There is one little bit of logic that is sometimes useful regarding the RBO. Normally one does not EVER short TTL outputs to ground. But in the case of RBO, a forced LOW on the output by a short circuit will extinquish the digit. This trick is useful in some cases, but one must remember to check the output of the computer port to determine that it can sink the current available from the RBO terminal under those conditions! Figure 23-33B shows a 7447 and seven-segment numerical display connected to a computer output port. The correct binary code (see code chart in the previous example) is output to the A — D inputs. Note that L/T and RBI are also connected. In cases where the port is configurable on a bit-by-bit basis

(e.g., when a 6522 is used, as in an AIM-65), then connect the RBO terminal and configure bit B6 as an input. Do not use the 6522 to sink RBO current.

ANALOG DISPLAYS

Analog displays are either digital voltmeters, digital current meters, or their analog equivalents. Also included are devices such as oscilloscopes or paper chart recorders. Because the output port of the computer is digital, you cannot directly drive the analog instrument with the computer outputs and so must connect a digital-to-analog converter (DAC) between the computer output port and the display device.

INTERFACING DIGITAL COUNTERS

Some users connect either a frequency, period or event counter to one bit of a parallel output port, or, to the serial output port. The "count" output from the computer will be proportional to the number to be displays. For example, let's assume that we want to display 0 to 9.99 volts. If you "scale" the voltage at 999 mS = 9.99 V (1 mS/1 mV), then you can represent any voltage in range with a specific frequency/time/number of events. If you use a simple event counter, then you could load the correct number into a register and then serially output it to the counter. Similarly, you could output a start and stop pulse to a period counter with the correct period between them. Or, you could output a frequency that is proportional to $f = 1/t$, where t is the period that will represent the value at 1 mS/1 mV).

ISOLATED CIRCUITS

When interfacing a high-voltage circuit, such as 115 volts ac (vac), it is not generally prudent to connect the computer directly to the controlled circuit. An exception to this rule is when the computer contains internal isolated circuitry such as those shown in this section. The computer manufacturer will provide specific information regarding such capability, and the lack of information is good evidence that no high voltage interfaces are provided — but you can use the methods in this chapter. Without the high-voltage methods shown in this section, the external circuit will blow out — often spectacularly — the internal circuits of the computer.

The applications of this type of interface include control of heaters, 115 vac lighting, machinery, furnaces, and other similar devices. Although homeowners have not yet discovered many of these applications, industrial users have used them for a long time.

There are basically two types of high-voltage interface that you can use. First, there is the electro-mechanical relay. These devices are electrical-mechanical switches that are actuated by an electromagnet coil. Second, there are electro-optical isolators. Such devices juxtapose inside of an IC package a light-emitting diode (LED) light source with either a photodiode or (more commonly) a phototransistor.

The connection of a low-current electro-mechanical relay to a computer output port is shown in Fig. 23-34. Note the similarity to the single LED interface discussed earlier in this chapter. Only one bit of the output port is used, and it drives the input of an open-collector TTL inverter (U1). Relay coil K1 is connected as the output load between the V+ voltage and the output terminal of inverter U1. Depending upon which of several open-collector inverters is selected, you can make V+ anything up to either +15 vdc or +30 vdc maximum.

When the output of U1 is high, both sides of coil K1 are at the same potential, so the relay is not energized. When the output of U1 drops LOW, however, one side of K1 is grounded and the coil is energized. The current flowing in the coil creates a magnetic field that pulls in the relay armature. Since U1 is an inverter circuit, a LOW on the input (i.e., at the computer output port bit) will cause the U1 output to be HIGH, and the relay remains off. Alternatively, a HIGH at the U1 input will cause the U1 output to be LOW, and the relay is energized. Because B7 is used for driving the relay circuit, any binary word that makes B7 = 1 (i.e., HIGH) will drive the relay. In most cases, I suppose, the word will be 10000000 (80 hex).

The problem with this method is that the coil current of K1 is limited to the 30 to 40 milliamperes that the open-collector inverter can handle. As you will see in a moment, an external transistor driver for the load will eliminate this problem.

The diode (D1) used in Fig. 23-34 is designed to suppress the inductive kick or high-voltage spike produced when the relay coil is de-energized. The relay coil is an inductor, and in fact can have a rather large value of inductance. As a result, de-energizing the surrounding magnetic field produces a spike of several hundreds of volts. Diode D1 is normally reverse biased, but to the counter electromotive force inductive spike it is forward biased. As a result, the CEMF spike is clamped to the

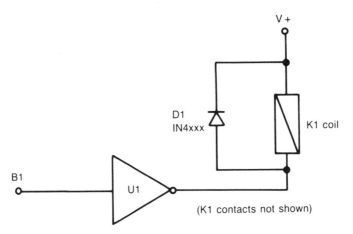

Fig. 23-34. 7447 decoder IC.

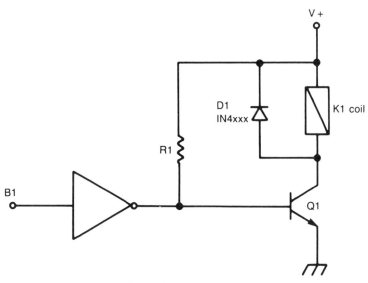

Fig. 23-35. Relay interfacing.

junction potential of the diode (0.6 to 0.7 volts). For most applications with small to moderate size coils, an ordinary 1000-volt PIV 1-ampere diode such as the 1N4007 will suffice.

Relay coil loads greater than about 30 mA cannot be handled by the method in Fig. 23-34. In Fig. 23-35 I show one means for interfacing larger relay coils to the computer output port bit. In this case a transistor switch is used to absorb the main bulk of current. The transistor (Q1) must have sufficient collector current and collector power dissipation rating to handle the load, plus a margin for safety. You must also pay attention to the collector voltage rating, but it is usually not too difficult to select the right one from the data sheets. Keep in mind that the collector power dissipation is rarely equal to or greater than the product of the maximum collector voltage and maximum collector current. Make sure that the power rating of the transistor is greater than:

$$P_{MIN} = (V+) \times I_{K1}$$

where

P is the power in watts
V+ is the supply potential in volts
I_{K1} is the actual coil current

The transistor used in this case is an npn bipolar type and so will turn on when the base is made more positive than the emitter. In this circuit, therefore, a HIGH applied to the input end of the 2.7k resistor will turn the transistor on — provided that the computer output port is the type that will source current. When the input is LOW, the transistor is turned off, so the relay is not energized. When the input is HIGH, however, a current flows in the resistor to the base of Q1, causing it to turn on and energize the relay. Because this transistor operates in a saturated mode, the relay coil is essentially grounded.

Again we use a 1N4007 diode in parallel with the relay coil. In cases of very high current, high inductance relays, however, one might want to use two or more 1N4007 diodes in series.

A potential problem with the circuit shown in Fig. 23-35 is that the beta can be very critical. Most computer output ports will sink or source only two TTL loads, i.e., about 3.2 mA. Even in the best of cases, however, only 18 mA (10 TTL loads) is available to the external circuit. In order to select the correct transistor for Q1 you must have a minimum beta gain of:

$$H_{fe} = (1.25 \times I_{K1}) / I_B$$

where

H$_{fe}$ is the minimum beta gain
I$_{K1}$ is the coil current
I$_B$ is the base current available from the computer port.

The 1.25 factor is a safety margin for transistor variation and aging losses. There will be no problems in selecting a transistor with a higher beta than the calculated value, but a lower beta can result in erratic or no operation of the circuit.

In cases where high values of load current are expected, then you might want to use a Darlington amplifier, also called a Darlington pair, as in Fig. 23-36. The Darlington circuit uses two transistors such that the emitter of Q1 drives the base of Q2, and their respective collectors are connected together. The overall gain of the Darlington circuit is the product of the individual gains:

$$H_{fe} = H_{fe(Q1)} \times H_{fe(Q2)}$$

In the case where the two transistors are identical, the total beta gain is equal to the gain of either squared. For example, if the beta of both is 100, then the total beta is $100^2 = (100)(100) = 10,000$. With this arrangement, the base current need only be 1/10,000 of the collector current. Thus, a 3.2-mA computer output current will drive a 32-ampere load through the circuit of Fig. 23-36.

The case of an optoisolator is shown in Fig. 23-37A. The optoisolator (IC1) consists of a light-emitting diode (LED) juxtaposed to a phototransistor. When the LED is turned off, the phototransistor is also turned off. But when the LED is turned on, the phototransistor is saturated and conducts current. As long as the two V+ power supplies are segregated, the two sides of the circuit are isolated to the extent of 10^{12} ohms or so.

Optoisolators can be used for ordinary digital

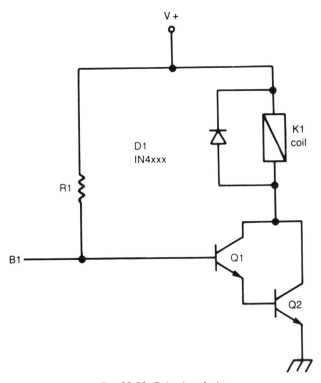

Fig. 23-36. Relay interfacing.

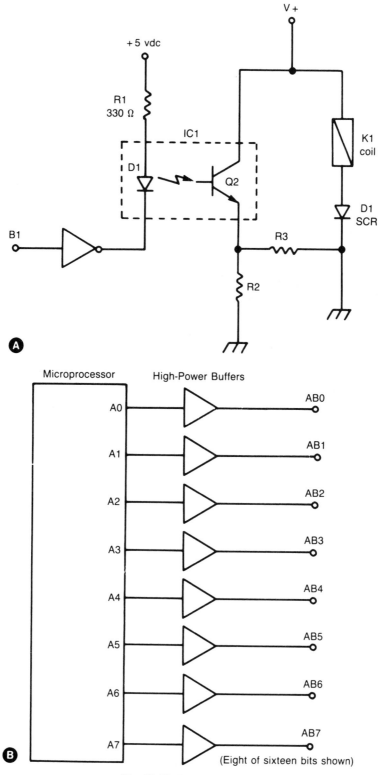

Fig. 23-37. Relay interfacing.

and analog circuits, or for high-voltage circuits. In the latter case the optoisolator is used to drive a relay or SCR/triac type of device. The relay/SCR/triac will actually control the high voltage or ac load.

PORT SELECTION AND INTERFACING TECHNIQUES

The art of interfacing involves nothing more than connecting the computer to other devices. Under the term "interfacing" falls a variety of activities that include, among others, connection of memory and I/O devices, keyboards, displays, peripherals, A/D converters, D/A converters and so forth. You have to consider certain details of interfacing. First, for example, you must learn how to increase the drive capacity of the microprocessor chip (most will only drive a couple of TTL loads!).

Second, you must learn how to decode addresses. Whether an I/O-based machine like the Z80, or a memory-mapped machine like the 6502, the address bus will carry the addresses of both memory locations and I/O ports. It is, therefore, critical to be able to decode the address when a location is called. Finally, you will learn to generate system signals such as INput and OUTput. These signals can be used to chip select I/O ports, memory chips and other devices.

BUS BUFFERING

The output lines of all microprocessors are limited in driving capacity. If you recall our earlier discussion of the TTL family of devices, then you may remember the terms "fan-in" and "fan-out." These refer to the drive requirements and drive capacity of TTL inputs and outputs respectively. The basic unit of measurement is the "standard TTL load." In TTL jargon, this means a current source of approximately 1.8 mA in the standard +2.4 to +5.0 volt range that is normal to TTL devices. A fan-in of one means that the load requirement is one standard TTL input. A typical TTL device has a fan-in of one for the inputs and a fan-out of ten for the outputs. Thus, a standard TTL device can drive up to ten standard TTL

inputs. The low-power Schottky (74LSxx devices) require less driving capacity, so have a fan-in of less than one. To keep things in perspective, therefore, let's rate microprocessor chip driving capacity in terms of TTL fan-in/fan-out figures. Some microprocessors have a fan-out of less than two, being capable of driving only two LS-TTL loads. Others, on the other hand, have a fan-out of two, which means that they can drive two standard TTL loads (or 3.2 mA at TTL logic levels). Unfortunately, this limitation places a severe constraint on the microcomputer designer.

If we want to drive more than a few TTL loads, then it is necessary to create a bus buffer, as shown in Fig. 23-37B. A bus buffer can be used on either address bus or data bus, although the one shown here is for the data bus. It consists of a set of high-power noninverting buffer stages. These devices will have a fan-out of 10 to 100, depending upon type. In some cases, they are available as six buffers in a single chip package, in which event you will need two IC packages and will have four unused stages in one of the packages. In other cases you might select a bidirectional octal bus transceiver. Such a chip will have the drive capacity, while at the same time being bidirectional — which allows input reads as well as output writes.

Some computers contain the bus buffer/driver, while others do not. In the cases where the computer does not have a buffered data/address bus, then you might want to consider constructing one.

ADDRESS DECODING

The address of any memory location within the 64K range of permissable addresses is defined by the 16-bit binary word applied to the address bus (2^{16} combinations actually results in 65,536 locations, but this is called "64K" in computerese). On Z80-based machines, the address of the I/O port will appear on the lower eight bits of the address bus, which means that the discrete I/O instructions will address 256 locations numbered from port 0 to port 255. In order to let a port or memory location know when it is being hailed by

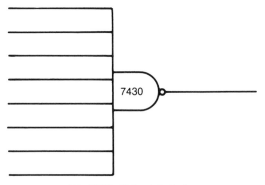

Fig. 23-38. Relay interfacing.

the CPU, therefore, you need to have some form of address decoder.

An address decoder is a circuit or device that will produce a TRUE output if and only if the correct eight- or sixteen-bit address is present on the address bus. At all other times the output of the decoder is FALSE. Please note that "TRUE" and "FALSE" are relative terms here, and are definable by you. In some cases, for example, you will define TRUE as HIGH, and in others TRUE will be LOW. In Fig. 23-38, for example, the SE-LECT signal is active-LOW, so the TRUE condition is LOW.

Figure 23-38 shows an eight-input NAND gate. Recall the rules of operation for NAND gates:

1. If any input is LOW, then the output is HIGH (i.e., "FALSE" in this case).
2. All eight inputs must be HIGH for the output to be LOW (i.e. "TRUE" in this case).

If you connect the inputs of the NAND gate to lines from the address bus, then you will generate a LOW if and only if the correct address is present on the bus. In the configuration shown, the correct address must be 11111111 (which is FF in hexadecimal, or "FFh"). Thus, only one location or I/O port (no. 255) can be addressed by this circuit. Figure 23-39 solves the problem.

The idea in using the 7430 device as an address decoder is to conspire to force all bits of the address bus HIGH when the correct address is on the line. This requirement means that you must invert all LOW's to make them HIGH's. The circuit in Fig. 23-40 has inverters on bits B3, B4 and B7. This circuit will, therefore, issue an active-LOW SELECT signal when the address present is 01100111 (or 67h). When this address is present, all inputs of the 7430 are HIGH, so the 7430

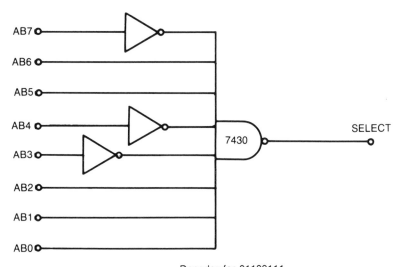

Decoder for 01100111

Fig. 23-39. 7430 8-input NAND gate.

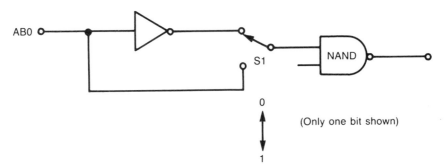

Fig. 23-40. Address decoder.

output is LOW. If any other address is present instead, then at least one bit of the group A0–A7 will be incorrect, so its corresponding input of the 7430 will be LOW — and the output remains HIGH (the inactive state).

There is still something of a problem with the circuit in Fig. 23-39: it is not easily reconfigured for different locations. In some projects you might want to reassign the memory location or port number of the decoder without having to re-wirewrap or butcher a printed wiring board. In these cases it might be nice to have a switch to select whether an inverter is present or not. There are several approaches to this problem.

First, you could place an inverter at each location, and then connect the input of the 7430 to the pole terminal or a single-pole, double-throw PC DIP switch. At one setting the switch would place the address bus bit directly on the 7430 input line; in the other setting it would place the output of the inverter on the 7430 input line (Fig. 23-40A).

Another arrangement is the circuit of Figure 23-40B. Here you see just one input of the 7430 treated with an exclusive-OR (i.e., XOR) gate (only one input is shown for sake of simplicity and to keep the artist from quitting). Recall the rules for operation of an XOR gate:

1. If either, but not both, inputs are HIGH, then the output is HIGH.
2. If both inputs are HIGH, then the output is LOW.
3. If both inputs are LOW, then the output is LOW.

In other words, as long as the two inputs see different signal levels, then the output of the XOR is HIGH, but if the same signal is applied to both then the output is LOW. In tabular form:

A	Inputs B	Output
LOW	LOW	LOW
HIGH	LOW	HIGH
LOW	HIGH	HIGH
HIGH	HIGH	LOW

In Fig. 23-40B, one input of the XOR gate is connected to a pull-up resistor and a grounding switch. When the switch (S1) is open, then the input of the XOR gate is HIGH, and when S1 is closed the input of the XOR gate is LOW. The other input of the XOR gate is connected to a bit of the address bus. Let's consider both situations.

When S1 is open, the control input of the XOR gate is HIGH. In this case a LOW applied on the A3 bit of the address bus will cause the output of the gate to be HIGH. This is because the two inputs are different. In this case the bit is inverted. Similarly, when A3 = HIGH, the XOR gate sees both inputs HIGH and so produces a LOW output; again the address bus bit was inverted.

When S1 is closed, then the control input of the XOR gate is LOW. In this case the circuit is noninverting. When A3 = HIGH, the XOR gate sees different inputs and so produces a HIGH output (no inversion). Similarly, when A3 = LOW, the XOR gate sees both inputs are the same and produces a LOW output (again, there is no inversion).

The final method is actually a modification of the first two methods. Both of the previous methods involved having an inverter on each input of the 7430 device. But this is rarely necessary. The more prudent method is to use from one to six inverters (three is a common number) on the printed circuit board and then connect them such that they can serve any bit of the address bus at will. In many cases the designers use a wire jumper system on the printed circuit board to make this work. A pair of jumpers will be used to connect the inputs and outputs of the inverter.

You could also use a 7485 (or its CMOS equivalent) to make the address decode function work. This IC is a four-bit magnitude comparator. It examines a pair of four-bit binary words ("A" and "B") and then issues outputs that indicate whether $A > B$, $A < B$, or $A = B$. You will normally use the active-HIGH $A = B$ output to indicate that the correct address is present on the bus. The address bus bits are connected to one word input (e.g., the "A" inputs), while the other inputs ("B" in our hypothetical example) are programmed with the correct address. You could use a variety of methods for programming the "B" inputs: switches such as those in Fig. 23-40B, a four-bit latch (for programmable reassignment), or hard wire jumpers. Note that the 7485 devices may be cascaded. There are cascading inputs on each 7485, which can be connected to the outputs of the lower order stages. Thus, you can create address decoder circuits in four-bit chunks.

Figure 23-41 shows the use of a 7442 chip for address decoding. This chip is a TTL device and was never intended by the original designer as an address decoder. It was, instead, a BCD-to-one-of-ten decoder. It would examine a four-bit binary coded decimal (BCD) word at its inputs, and then cause one of ten unique outputs to go LOW. Thus, if the binary word 0100 is present on the inputs, the "4" decimal output would be LOW and all others would be HIGH. The 7442 was used to decode BCD to drive lamp columns of high-voltage *Nixie*-tube switching transistors.

The decoder circuit of Fig. 23-42 uses the 7442 to select which of ten (of a possible sixteen) unique addresses are being called for by the lower four bits of the address bus (A0 – A3). Thus, on a Z80 machine you could call for I/O ports 0 through 9 with this circuit (plus an IN or OUT signal, more on this later). The outputs 0, 1, 2, 3, 4, 5, 6, 7, 8, and 9 are shown in Fig. 23-41 and Table 23-8.

COMPUTER INPUT/OUTPUT PORTS

The "port" is the means for communication by the computer, i.e., transferring data into and out of the machine. Most single-board computers have either a limited number of I/O ports built in, or a data/address bus structure that requires you to either buy add-on ports or build your own. There are two general types of ports, input and output. An input port is a circuit that allows digital data from the outside world to be input into the computer. It is unidirectional and is operated by a read instruction from either a port location or a memory location. An output port is exactly the opposite of the input port. It is a circuit that allows data from the central processing unit (CPU) of the microcomputer to be output to the external world. Out-

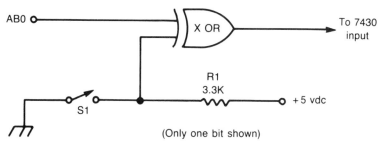

(Only one bit shown)

Fig. 23-41. Polarity switching.

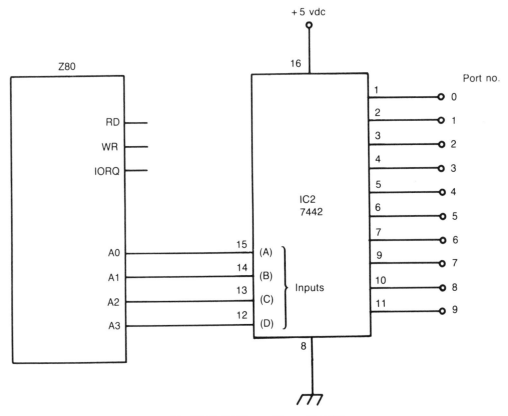

Fig. 23-42. 7447 as address decoder.

put ports are operated by a write instruction from either a port location or a memory location.

You can divide input/output ports further into I/O-based and memory-mapped I/O's. The I/O-based machine, such as the Z80 microprocessor, uses special I/O instructions for input/output operations. In the Z80 machine the address of the port is carried on the lower eight bits of the sixteen-bit address bus during I/O instruction operation. Because there is an eight-bit address for the port, there are 256 different port assignments available (numbered 0 to 255) using codes 00000000 to 11111111. Memory-mapped systems such as the 6502 machines do not have separate input/output instructions and so tend to use ordinary memory read and write instructions. Thus, an input/output port is treated as a memory location.

Another means for classifying input/output ports is according to whether they are serial or

parallel. The parallel port uses one output line for each bit and control signal. An eight-bit output port, therefore, will have at least eight separate I/O lines plus a ground. A popular example of the parallel output port is the Centronics parallel printer port. Some microcomputers, especially small machines, have at least one parallel I/O port consisting of eight TTL-compatible input lines and eight TTL-compatible output lines. Often these ports are arranged on a single sixteen-pin dual-in-line-package (DIP) integrated circuit socket and

Table 23-8. Z80 Outputs for Bits A0-A3

A0	A1	A2	A3	Active (LOW) Output	7442 Pin No.
0	0	0	0	0	1
0	0	0	1	1	2
0	0	1	0	2	3
0	0	1	1	3	4
0	1	0	0	4	5
0	1	0	1	5	6

can be connected to the external world through a ribbon cable adapter.

There are at least three types of serial port. The oldest is the 20-milliampere current loop which was developed for the original teletypewriter machines. The RS-232C is a serial voltage-oriented port that is very popular with terminal makers; this is the port that uses the popular DB-25 "D-shell" connector seen on the back of many computers. Finally, there is a simple TTL-compatible single-bit serial port. In some cases the "serial port" is simply a single bit of a parallel output port.

CONTROL SIGNALS FOR I/O PORTS

The microprocessor has three basic buses: address, data, and control. All three are needed in the creation and use of input/output ports. The address bus defines which port is being addressed, both in memory-mapped and I/O-based systems; the data bus carries the data to and/or from the central

processing unit (CPU); the control bus consists of those signals which tell external circuits what is happening.

Figure 23-43 shows the block diagram of a typical I/O port in a microcomputer that is based on the 6502 microprocessor. The address bus (A0 – A15) is passed to an address decoder that will issue an active SELECT signal (either HIGH or LOW depending upon design) if and only if the correct address is present on the address bus. Any other address on the bus will not produce an active signal.

The control signals and the SELECT signal are applied to a device select circuit. This circuit will produce either an IN or an OUT signal (either HIGH or LOW, again depending upon design) for an input or output operation respectively. The conjunction of the control signals and the address creates these system signals.

The IN signal turns on the input port, which ferries data from the outside world to the data bus,

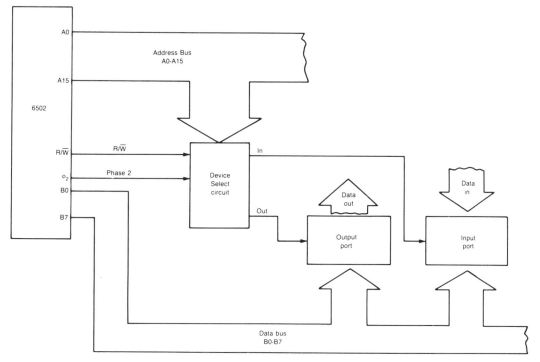

Fig. 23-43. Basic I/O system.

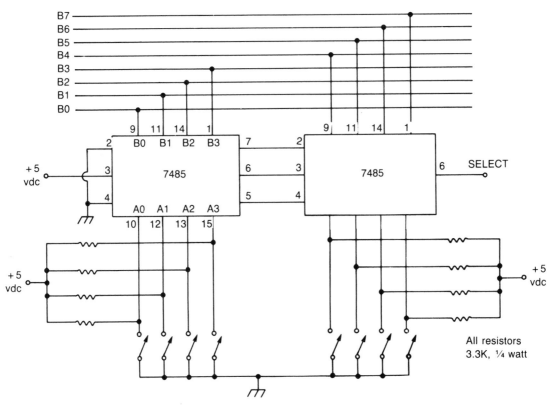

Fig. 23-44. Address selector.

and from there into the accumulator of the CPU. The OUT signal turns on the output port, which serves to deliver data less decoding; this section is representative of several of them, and is not intended to be exhaustive. The idea is to find a digital device that will produce a unique output only when all input criteria are satisfied. You can use NAND gates, NOR gates, magnitude comparators, and so forth. Figure 23-44 shows a popular address decoder device: the TTL 7430 eight-input NAND gate. The rules of operation for this device are as follows: 1) a LOW on any one input will produce a HIGH output, and 2) all inputs must be HIGH for the output to be LOW.

Thus, you must conspire to make all inputs HIGH only when the address present on the address bus is correct. The only address that will satisfy the criteria above is FF (hexadecimal), or 11111111. In all other cases you need to use one

of several strategies to make the address 11111111 when the real address is present.

Two methods for altering the LOWs present on the address bus during the correct time are shown in Fig. 23-44. In Fig. 23-44A you see the use of digital inverters. In the specific case shown, the correct address is 00010100. Because only two bits of this address (i.e., A2 and A4) are naturally HIGH, you need six inverters to make the circuit operate. The address bits A0, A1, A3, A5, A6, and A7 require an inverter to make the 7430 inputs HIGH when the address comes along. In this case only 00010100 will make the 7430 inputs 11111111. The output of the 7430 drops LOW when the correct address is present, and this output is the SELECT signal.

The other method is shown in Fig. 23-44. One of the inputs of the 7430 uses the previous method, but with a switch that allows on-board selection of the output level. Switch S1 determines whether a HIGH or LOW on address bit A0 produces the required HIGH at the input of the 7430. This method is merely an adaptation of the previous method. An alternate method is shown at bit A7. In this case we are using an exclusive-OR (XOR) gate. The rules of operation for the XOR gate are as follows:

A HIGH on either input, but not both, produces a HIGH output;

If both inputs are HIGH, then the output is LOW; and

If both inputs are LOW, then the output is LOW

In other words, the following truth table obtains:

Inputs		Output
A	B	
0	0	0
1	0	1
0	1	1
1	1	0

One input of the XOR gate in Fig. 23-44 is connected to the address bus line, while the other is connected to a switch and pull-up resistor. When switch S2 is open the XOR input is HIGH, and when it is closed the XOR input is LOW. When S2 is open and the B input is HIGH, the circuit operates as an inverter. When the A input is HIGH, then the XOR sees two equal inputs, so the output is LOW. Alternatively, when the switch is closed, the circuit operates as a noninverter. A LOW applied to the A input produces a LOW output, and a

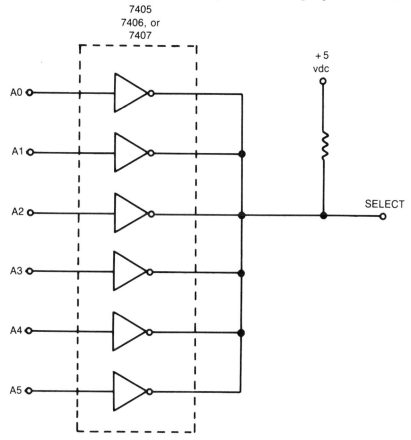

Fig. 23-45. 7405, 7406, 7407 connections.

HIGH applied to the A input produces a HIGH output.

Figure 23-45 shows another alternative address decoder. This particular circuit is based on the 7485 TTL (or CMOS 4083) four-bit magnitude comparator. This chip will examine two four-bit binary words, labelled "A" and "B," and issues a unique output that indicates which of the following conditions is true:

A is less than B
A is greater than B
A equals B

In our example we use the A = B output, pin no. 6. The 7485 device has three cascading inputs that allow several 7485 devices in cascade to accommodate higher order bit lengths. The cascade inputs are connected to the outputs of the preceding stage.

In the circuit of Fig. 23-45, the "B" inputs are connected to the address bus, and the A inputs to a bank of switches. Because the "address bus" shown in Fig. 23-46 is eight bits long you need two 7485 devices in cascade. The programming is done by switches and pull-up resistors on the A inputs. Each input is held HIGH by the pull-up resistor, and will be brought LOW if and only if the associated switch is closed. Thus, you can select HIGH or LOW for each individual input, thereby programming an input. When the address on the bus is identical to the address programmed on the A inputs, then the SELECT output goes HIGH.

SIXTEEN-BIT DECODERS

There are few applications that require an eight-bit address decoder. One of them is a Z80 input/output port, in which ports 000 through 255 carry an eight-bit address that appears on the lower eight bits of the address bus (A0–A7). Another is the case, found on some single-board computers, where no more than a small amount of RAM or ROM memory is used. But all other cases require a higher order decoder.

Figure 23-46 shows a scheme for a sixteen-bit decoder that is based on a pair of eight-bit decoder circuits such as in the previous examples. The NOR gate used in this circuit obeys the following rules: 1) a HIGH on either input will produce a LOW output, and 2) both inputs must be LOW for the output to be HIGH.

The circuit of Fig. 23-46 uses eight-bit decoders in which an active-LOW SELECT signal is generated. These circuits produce a LOW output when the correct address is on the address bus. By combining the two outputs in a NOR gate, an ac-

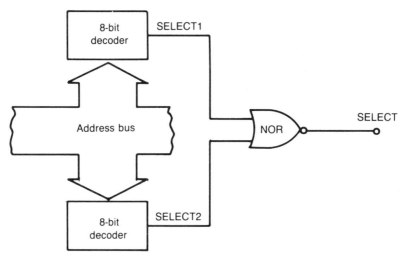

Fig. 23-46. 16-bit address decoder.

tive-HIGH SELECT output is produced from the gate.

You can also combine in cascade several four-bit magnitude comparators in order to achieve higher order decoders. The 7485 device will examine four-bit "nybbles," so two will examine eight-bits, three will examine twelve-bits, and four will examine sixteen-bits.

There are other scenarios for addressing large numbers of ports of memory locations that are based on "bank selection" concepts. You can, for example, use ten-bits to address 1024 locations. By feeding the same lower order ten lines of the address bus to all of the memory chips in several different banks, you can use the other bits as bank select switches, thereby activating each bank in succession.

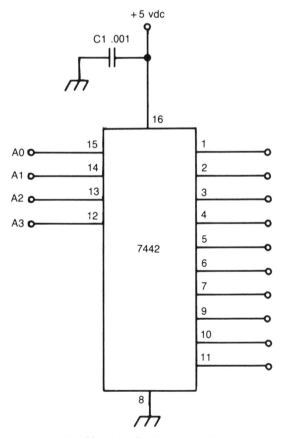

Fig. 23-47. 7442 address decoder.

GENERATING OTHER SELECT SIGNALS

There are times when we need only a few input/output ports and do not want to create a separate decoder for each one. Some small single-board computer projects fall into this category. The circuit of Fig. 23-47 shows a SELECT generator that examines the lower order four-bits of the address bus (A0–A3) and produces up to ten active-LOW outputs. The 7442 device is a TTL chip that is called a BCD-to-one-of-ten decoder. It will produce the ten outputs that correspond to the ten possible states of the BCD inputs. You could also use the 74154 to produce up to sixteen different SELECT signals, which is the maximum allowable for a four-bit binary word. The outputs are recognized by the circuit in Fig. 23-47 are given in Table 23-9.

A circuit such as Fig. 23-47 can be used in a small computer to handle all of the input/output port chores. Larger systems, where more than ten I/O ports are demanded, can use either the 74154 device, or you can add gating to the 7447 device shown in Fig. 23-37, or use the 7447 on a different set of address bus lines than A0–A3 (for example, A1–A4).

GENERATING INPUT/OUTPUT SIGNALS

In this section I will discuss the generation of input/output signals, or "read/write" signals for system use. These signals are used to tell all devices simultaneously that either an input or output (or, alternatively when memory is also addressed or the ports are memory-mapped) read/write signals. Such signals will reduce the complexity of the

Table 23-9. Output Codes for the 7442 BCD Decoder

A3	A2	A1	A0	Device Select
0	0	0	0	SELECT0
0	0	0	1	SELECT1
0	0	1	0	SELECT2
0	0	1	1	SELECT3
0	1	0	0	SELECT4
0	1	0	1	SELECT5
0	1	1	0	SELECT6
0	1	1	1	SELECT7
1	0	0	0	SELECT8
1	0	0	1	SELECT9

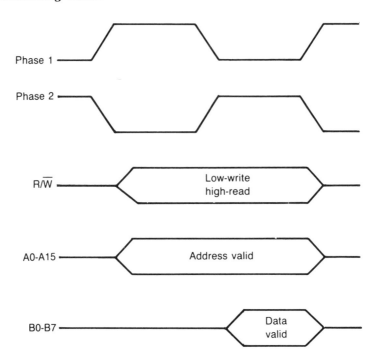

Fig. 23-48. 6502 timing.

decoding circuits at each individual port. If you have a WRITE or OUTPUT signal, for example, you could reduce output port decoding to a simple NOR (or NAND) gate with two inputs, one each for the SELECT and WRITE signals. But before going to the circuitry, let's pick a sample microprocessor to be used as the basis for our discussion, the 6502, and discuss the relevent timing signals generated by that chip.

Figure 23-48 shows the timing diagram for the 6502 microprocessor chip. There are two complementary clocks used on the 6502, and these are generated from the master clock (a 1-MHz or 2-MHz square-wave oscillator): phase 1 and phase 2. In general, the phase-1 clock period (see Fig. 23-48) is used to set up control signals and addresses so that they become stable during the phase-2 period. In Fig. 23-48, for example, the address bus and R/W signals become active during phase 1 but are actually used during phase 2. Thus, the data on those lines will be stable when it is needed (something of no small importance in a dynamic machine such as a computer). The data bus becomes active during phase 2.

There are two signals that are required in order to generate system signals in the 6502-based microcomputer: R/W and phase 2. The R/W signals will be HIGH during read (input) operations and LOW during write (output) operations. Thus, you can build an input or "READ" signal with a two-input NAND gate (Fig. 23-49A). The read condition exists only when both R/W and phase 2 are HIGH. Because a NAND gate will produce a LOW output if and only if both inputs are HIGH, the output of the 7400 shown in Fig. 23-49A will serve as an active-LOW READ signal.

Similarly, the same type of gate will be used for generation of the WRITE signal. But because the condition of R/W is opposite for write operations from read operations, you will have to invert the R/W line at the input of the WRITE NAND-gate (Fig. 23-49B). The same situation then obtains: both inputs of the NAND gate will be HIGH if and only if a WRITE operation is taking place. Thus, the output of the second NAND gate will serve as an active-LOW WRITE signal.

In some cases you will want to preserve the undecoded CPU signals from the 6502. But you

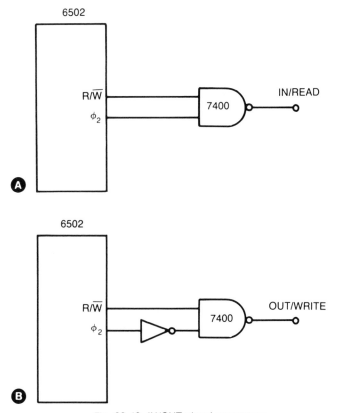

Fig. 23-49. IN/OUT signal generators.

also find that their usefulness is limited because the fan-out is LOW. A typical microprocessor output line will successfully drive only two or so "TTL inputs." You can overcome this problem as in Fig. 23-50 by buffering the lines. If you use a high fan-out buffer chip, then you will be able to drive a larger number of devices. Keep in mind that an ordinary TTL output is usually capable of driving ten TTL inputs (i.e., it has a fan-out of 10). There are other chips on the market with higher fan-out

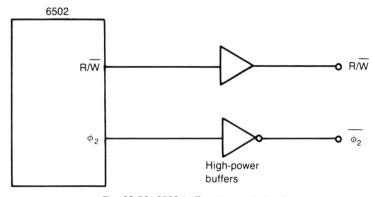

Fig. 23-50. 6502 buffered control signals.

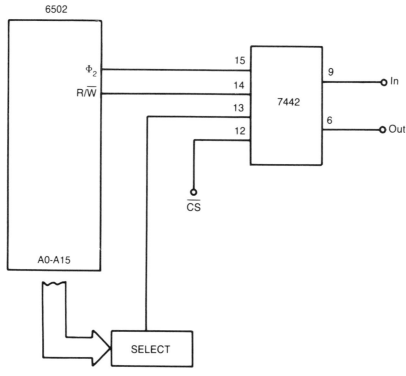

Fig. 23-51. 6502 IN/OUT signals.

potentials. The phase-2 clock line is buffered with a noninverting buffer stage in order to retain the phase information. The R/W line is treated two ways. The SYSTEM READ line is at the output of a noninverting buffer, thereby retaining its phase information. But the R/W line is of opposite phase. If you want to use this in active-HIGH circuits, than it is necessary to invert the R/W line to form an active-HIGH SYSTEM WRITE line.

You can also use any of several decoder chips to generate these signals. The ubiquitous 7442 is an example, and I will use that chip. It is an older TTL device that is a four-bit-BCD-to-1-of-10 decoder, and was intended to be a display driver. But there are more modern chips available and some of them were designed specifically for this service. Check the components catalogs of Intel and other microprocessor manufacturers for examples. In this section, however, I will examine the 7442 circuits with the understanding that similar information also applies to other chips as well.

Figure 23-51 shows a typical system signals generator circuit that is based on the 7442 device. There are ten discrete outputs on the 7442, and each one represents one of the decimal digits from 0 to 9. When the four-bit input sees binary 0110, for example, the output corresponding to decimal 6 drops LOW and all others are HIGH. Thus, by connecting the control signals from the microprocessor to the inputs of the 7442 you can generate the desired signals. The inputs of the 7442 are thus:

7442 Pin	Designation	Decimal Weight
15	A	1
14	B	2
13	C	4
12	D	8

The outputs of the 7442 are given in Table 23-10.

In the circuit of Fig. 23-51 you are connecting the phase-2 signal to the "A" input (weight = 1)

Table 23-10. Outputs for the 7442 BCD Decoder

7442 Pin	Decimal Indicated
1	0
2	1
3	2
4	3
5	4
6	5
7	6
9	7
10	8
11	9

and the R/W signal to the "B" input (weight = 2). The "C" input (weight = 4) is connected to the SELECT signal, which in this case is active-HIGH. The "D" input (weight = 8) is used as a chip select line. When this line is HIGH the 7442 will not produce a valid output regardless of the data present on the other inputs. Normally, if that feature is not needed, the "D" input, pin no. 6, is tied permanently LOW. The code for this circuit is:

Operation	D	C	B	A	Decimal	Output Pin
Read/Input	0	1	1	1	7	9
Write/Output	0	1	0	1	5	6

Note: A "1" on the "D" input invalidates the code applied to the other inputs, and so makes a dandy chip-select line.

The correct code for READ and WRITE signals appears only when 1) chip select is LOW and 2) the inputs in the above table are satisfied.

The Z80 microprocessor is another very popular type, but it uses a slightly different control signal scheme. In the Z80 there are 4 control signals:

IORQ. An active-LOW signal that indicates that either an input or an output operation is taking place.

RD. An active-LOW signal that indicates when either a memory read or input operation is taking place.

WR. An active-LOW signal that indicates when either a memory write or output operation is taking place.

Thus, you can indicate an output operation uniquely by the coincidence of the IORQ and WR signals, while an input operation is indicated by the coincidence of the IORQ and RD signals. In Fig. 23-52 we have a decoder based on the 7442 device that will generated IN/OUT (or "READ/WRITE") signals in Z80-based systems. The following codes are honored (see Fig. 23-12).

Control Signals From Z80			7442 Input	(Weight)	Pin
IORQ	A	1	15		
RD	B	2	14		
WR	C	4	13		
SELECT (active-LOW)*		D	8		12

*SELECT is created by decoding lower-order eight bits of the address bus.

The code seen and acted upon by the 7442 is

Operation	D	C	B	A	Decimal	Output Pin
Input/Read	0	1	0	0	4	5
Output/Write	0	0	1	0	2	3

In this circuit pin no. 5 will drop LOW only during the time when the Z80 is reading the input whose address is represented in decoded form by the SELECT signal, and pin no. 3 will drop LOW only when the output port at the same address is being summoned.

BUS BUFFERING

Earlier in this section I discussed buffering of controls signals. The same problem also exists for the address bus and data bus, only more so because these outputs will probably have to drive a larger number of devices. Some single-board computers have buffered outputs on these buses, especially when they go to the "outside world." There are others, however, that have absolutely no buffering, and these will usually require some sort of buffering when interfacing. Figure 23-53 shows a buffering system for the eight-bit data bus; the same scheme can also be used for the sixteen-bit address bus.

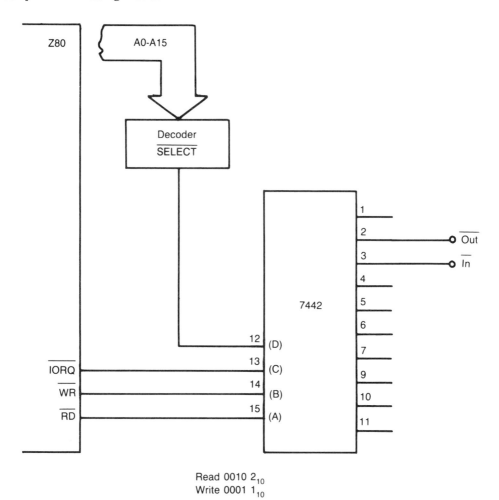

Read 0010 2_{10}
Write 0001 1_{10}

Fig. 23-52. Z80 IN/OUT signals.

The noninverting buffers are available in any of several standard TTL and special house-number chips. They are usually designated as "eight-bit bus drivers," or if they are bidirectional, "eight-bit bus transceivers." In some cases the buffers will be permanently connected, while in others a chip-select pin is available for turning the external or "system" bus on and off. The buffer chip(s) should be located close to the microprocessor chip.

OUTPUT PORT DEVICES

An output port must be capable of capturing transient data appearing on the data bus during output or "write" operations. The nature of any programmable digital computer is to perform operations sequentially. Thus, the microcomputer will perform an output operation (or memory write in the case of memory-mapped chips such as the 6502) and then move on to another operation. The data that is output will be very transient, and remains on the data bus for only a few microseconds. One goal of the output port, then, is to latch the data. One means for latching the data is to use a type-D flip-flop.

Figure 23-54 shows the circuit diagram for a type-D flip-flop. Let's review the operation of this circuit. There are two things you need to know:

1. Data applied to the D input is transferred to

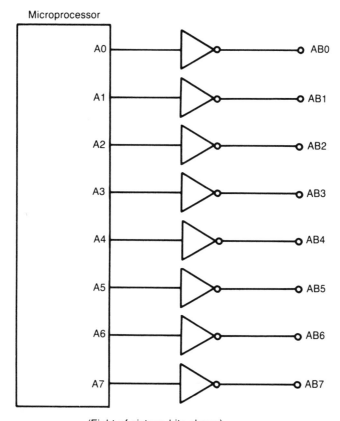

(Eight of sixteen bits shown)

Fig. 23-53. Address buffering.

the Q output only when the clock (CLK) input is HIGH; and

2. When the CLK line goes LOW the Q output will remain at the last valid D input level (HIGH or LOW) and ignores further transitions on the D input

The 74100 is a TTL chip that contains a pair of four-bit data latches. A data latch is nothing more than a type-D flip-flop renamed, and a four-bit data latch is a gang of four type-D flip-flops that all share a common CLK line. In the 74100 device the clock lines (called "strobe" when on a "latch") are pins 12 and 23, and in Fig. 23-55 these lines are connected together to form a WRITE input. The WRITE input of the output port will be con-

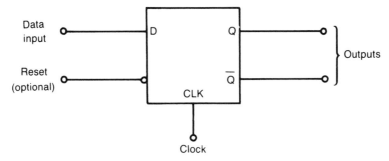

Fig. 23-54. Data latch (1 bit).

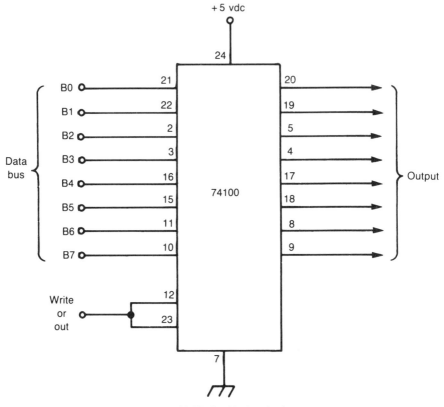

Fig. 23-55. Six-bit data latch.

nected to either a WRITE system signal, or an OUTPUT signal as described earlier in this chapter.

The circuit in Fig. 23-56 is a viable output port and can be used on most single-board computers. The WRITE line is normally held LOW, but when the microprocessor wants to output something on this port the WRITE line is made HIGH. The data on data bus lines DB0–DB7 are transferred to output port bits B0–B9.

INPUT PORT DEVICES

The input port need not latch its data, but must be able to disconnect from the data bus lines. Otherwise, a LOW on an output will permanently yank the corresponding data bus line LOW. Alternatively, when the input port line is normally held HIGH, a LOW on the data bus might damage the input port device. Thus, you need a tri-state noninverting buffer for the input port device. There are several, including 74LS244, 74125, and Intel's 8216/8226 devices. In this short section I will consider only the 74LS244 device.

The 74LS244 circuit is shown in Fig. 23-56. The input lines are connected to bits B0 through B7 of the input port (external world), while the corresponding output lines are connected to bits DB0 through DB7 of the data bus. When the READ signal is LOW, then the 74LS244 is not connected to the data bus. But an active READ signal will cause the output lines of the 74LS244 to be connected to the package pinouts and data can be transferred to the data bus.

KEYBOARD INTERFACING

Perhaps the one peripheral that is almost always associated with computers is the keyboard. The keyboard is used to allow humans to input data directly into the computer. In fact, in the popular

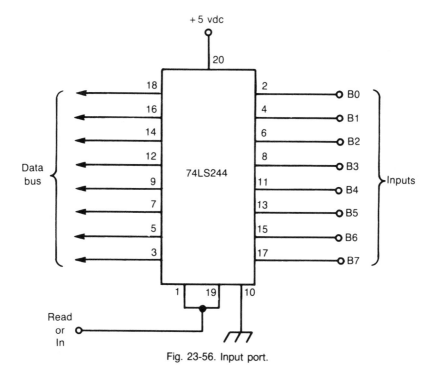

Fig. 23-56. Input port.

mind the keyboard is the computer, for the public sees little other than the keyboard and CRT. On the personal computer we find the keyboard is semi-standard and looks much like a typewriter keyboard (Fig. 23-57A). In other cases the keyboard will only be a numerical keypad. On some computers the keyboard is built-in and forms a part of the terminal and the computer together (see

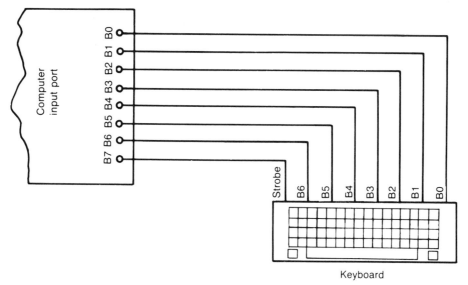

Fig. 23-57. Keyboard interfacing.

Fig. 23-57B). Computer circuitry is now so miniaturized that the computer often fits inside of the space that was once occupied solely by the computer. In still other cases, we are dealing not with a formal keyboard but a specialized terminal (e.g., the point of sale credit terminal that checks your account at the credit card company!). In some cases the "keyboard" is the front panel of an instrument that is based on a computer. I will examine all of these methods in this chapter.

STANDARD KEYBOARDS

The standard keyboard is, perhaps, the most familiar to most readers. It consists of a collection of keys, each labeled with a letter or number or symbol. When a key is pressed, a certain binary code will appear on the output. There are several different technologies used in keyboards—not all of them are contact closures (there are Hall-effect magnetic keyboards, capacitive keyboards, photoelectric keyboards, and others). For our present purposes we will concern ourselves less with the method used for the key switches, and concentrate instead on interfacing.

Figure 23-57 shows the simple method of connecting a keyboard to the computer input port. Most modern keyboards use the ASCII (American Standard Code for Information Interchange) code. Because ASCII uses a seven-bit binary word, you can easily fit it in the space allowed by an eight-bit input port. The eighth bit of the input port can be used for the strobe bit of from the keyboard. The "strobe" bit is used to tell the world when the data on the seven parallel output lines of the keyboard is valid. The strobe bit goes active following a keystroke. In most cases the strobe bit is active-HIGH, although a few are produced with active-LOW strobe lines.

In Fig. 23-58 ASCII bits A1 through A7 are connected bits B0 through B6 of the input port, while the keyboard strobe line is connected to bit B7. The computer will loop through a program such as shown in the flow chart of Fig. 23-59. The computer will input data from the keyboard input port, and then test it to see if the most significant bit (MSB) is 1 or 0. If it is 1, then it is assumed that

new data is present on the line, but if it is 0 then we assume that it is old data so the computer loops back and picks up another new data input word. Using 6502 language, the appropriate programming might look like Table 23-11.

The scheme in Fig. 23-60 is useful for both full ASCII keyboards, which require a seven-bit parallel input port, and less keyboards requiring only four bits. Examples are four-bit binary coded decimal (BCD) and four-bit hexadecimal keyboards. In the cases of the four-bit keyboards, only four bits (B0 through B3) are used, along with the strobe, so three bits are free for other uses. Figure 23-61 shows a hypothetical application. Here you see the keyboard data bits connected to bits B0–B3, and the strobe connected to bit B7 (the MSB). Bits B4, B5, and B6 are used to examine the output of a decimal counter. In this manner you can enter up to eight digits. The counter input is connected to the keyboard strobe line of the keyboard, so the counter will increment every time a key is pressed (which forces the strobe HIGH). The A, B, and C outputs (weights 1, 2, and 4) of the counter are connected to the "free" computer inputs, while the "D" output is used as an overflow indicator. When "D" goes HIGH, the external flip-flop turns on a light-emitting diode (D1) overflow indicator.

NONSTANDARD KEYBOARDS

There are many instances where the single-board computer will be used in an instrument, control system, or other application in which a custom keyboard is required. There are several ways of doing this neat trick. For some applications you will want to use an ordinary standard keyboard with custom keycaps that describe the functions represented. This approach is generally used with small hexadecimal keyboards. In still other cases you will want to make our own keyboards—Fig. 23-62 is an example of such a circuit.

One input port is reserved for the keyboard application, and the switches are connected one to each bit of the port. The switches are connected between the input port bit and ground. When the switch is closed, the represented bit is LOW, and if it is open the bit is HIGH. The "pull-up" resistors

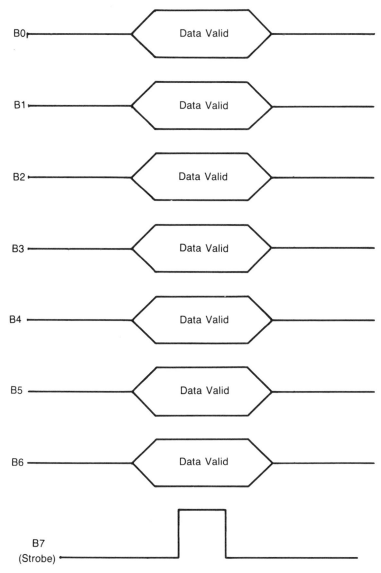

Fig. 23-58. Keyboard data signals.

are intended to keep the bits guaranteed HIGH when the switches are opened. If the switches are closed, then the "cold" end of the resistor is grounded.

Contact closure switches are not capable of making one single, solid contact but rather bounce, causing several contacts (as many as twenty). In these cases, the contacts must be "debounced." There are two ways of doing this trick: with hard-ware or with software. The hardware method is shown in Fig. 23-63. A simple monostable multivi-brator is connected to the input bit of the computer, and the switch is used to trigger the mono-stable. In general, a 5 to 10 millisecond duration will suffice to ensure that the computer sees one and only one pulse for each contact closure. The software method requires a 5 to 10 millisecond timer look to "time out" following each input.

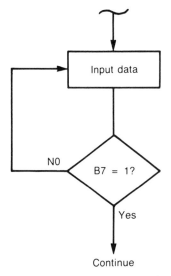

Fig. 23-59. Keyboard input routine.

Table 23-11. Keyboard Interface Program

Memory Location	Instruction Mnemonic	Comments
0200H	LDA (A000H)	Input data from
0201H	00H	keyboard
0202H	A0H	connected to port A at location A000H.
0203H	AND #80	Mask all but bit B7.
0204H	80H	
0205H	BEQ	Test bit B7 and
0206H	F9H	branch backwards if not equal 0 (F9H is two's complement for −7, the relative branch distance.

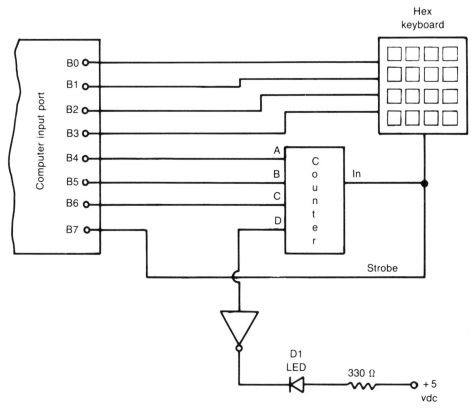

Fig. 23-60. Hexadecimal keyboard.

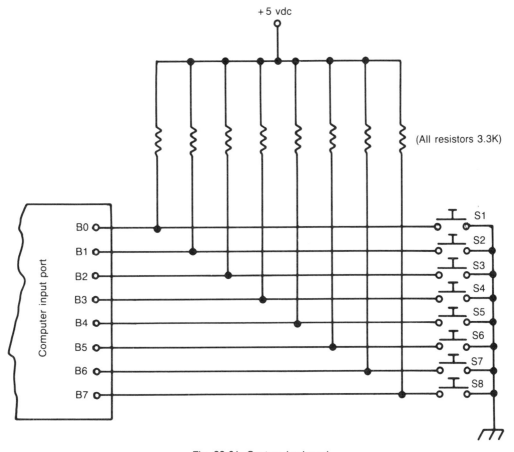

Fig. 23-61. Custom keyboard.

Almost any form of monostable multivibrator is usable for this application, provided that the output is TTL compatible. In the example shown in Fig. 23-62, a so-called "half-monostable" circuit is used based on the 4049A inverter circuit. Although this device is a CMOS chip, it has a TTL compatible output when operated from +5-vdc power supplies. The R1C1 combination is set to approximately 5 milliseconds and will produce a positive-going output pulse of the length. You could have also used a TTL type 74121 one-shot IC, 555 timer, other CMOS one-shots, and certain other "special purpose" integrated circuits.

ISOLATED KEYBOARDS

There might be several reasons for using an isolated keyboard, including safety. The circuit shown in Fig. 23-63 is used for a lot of applications, including the isolated keyboard. It consists of a light-emitting diode (LED) juxtaposed with a photosensitive transistor. When the LED light is shining on the phototransistor base region, the transistor is biased on hard and the transistor is therefore saturated. By interposing a blind between the phototransistor and LED you can turn off the transistor. It might be the blinder on a photokeyboard key, or, the paper on a printer (to form the "PAPER OUT" signal).

ON-OFF KEYS

In some cases you will want to use two buttons: one for on and one for off conditions. The circuit in Fig. 23-64 is a NAND-logic reset-set (RS) flip-flop. When switch S1 is pressed (representing

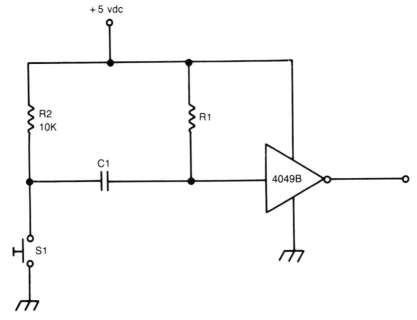

Fig. 23-62. Keyboard debouncer.

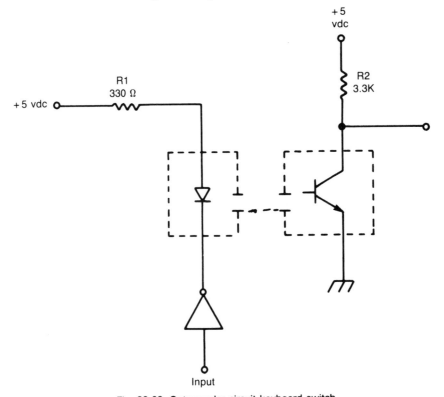

Fig. 23-63. Optocoupler circuit keyboard switch.

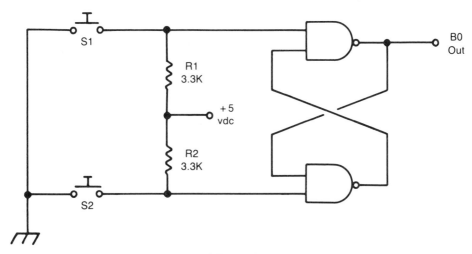

Fig. 23-64. On/Off switch.

the on state), the set input is grounded, so bit B0 snaps HIGH. If switch S2 is closed, then the RS flip-flop is reset, and B0 drops LOW. This type of circuit is used sometimes in start-stop applications.

PROGRAMMING

Any of these nonstandard circuits can be programmed in either a single operation or scanning mode. The earlier program is an example of single-shot operation. The program loops until it finds the strobe condition, and then goes off to another portion of the program to do with the input data as is necessary. The use of the strobe bit is appropriate for standard keyboards, and for nonstandard designs if all active input lines are similarly tested and acted upon. Scanning operation continually scans the input lines connected to the keyboard and searches for valid data. The scan rate must be fast enough to beat the fastest fingers jumping from key to key. In a popular radio receiver that uses digitally scanned keyboards for various functions the scan rate is 80 Hz.

SERIAL COMMUNICATIONS METHODS

There are basically two formats for transmitting digital data between computers and other machines (other computers, peripherals, telecommunications systems, and so forth). You can use either parallel or serial transmission. In parallel transmission all bits of the system are transmitted at the same time. For an eight-bit microcomputer system, parallel communications requires not less than one wire or channel per bit plus a ground and possibly additional wires for strobe, control, or handshaking signals. The parallel I/O port is very fast because the bits are banged across the system all at one time. However, there are problems.

One problem is that wired systems can only be used over distances of a few meters. The principal reason for this limitation is that wire capacitances and inductances cause the fast risetime pulses used in digital equipment to deteriorate.

Another problem is that parallel channels can be expensive. While a few feet of multiconductor wire is not terribly more expensive than two-conductor cables used in serial systems, the differential costs begin to mount up excessively when telephone or wire transmission is used between distant points. It costs a lot of money to rent or buy long distance facilities for eight or more bits of data!

Serial communications systems use only a single channel for all bits of the system, with some also adding additional channels for control signals (e.g., the RS-232 bus). When serial transmission is used, it is possible to send all data over a single pair of wires, a single telephone link, or a single radio channel.

There are several ways to catagorize serial digital communications links. On one level we can differentiate between current loops and voltage-operated systems. The 20-mA current loop used in certain teletypewriters and the 4–20 mA instrumentation system are examples of current loops. The voltage operated system uses HIGH/LOW voltages to represent the binary digits. Examples are the RS-232 and the single-bit TTL systems. There is also a pseudo-RS-232 system used in some microcomputers in which an RS-232 connector (DB-25) is used, and the RS-232 pinouts are honored, but the voltage levels are TTL compatible (i.e., 0 to 0.8 volts for LOW and +2.4 to +5 for HIGH, instead of −12 volts for HIGH and +12 volts for LOW). Still other systems are proprietary or nearly so, being used only on certain computers for certain applications. The Commodore 64 uses such a serial interface port that is neither RS-232 nor a simple single-bit TTL port.

You can also define systems as to whether they are simplex, half duplex, or full duplex. A simplex system is able to transmit data in only one direction. A half-duplex system, on the other hand, is able to transmit in both directions, but only in one direction at a time. The half-duplex system is like a CB radio channel (less interference) in which two people can communicate, but one must listen while the other transmits. When the first is finished, he listens and the other transmits. A full-duplex system is able to transmit in both directions at the same time. An example of full duplex is the telephone in which both parties can talk at the same time — unless they want to have a conversation instead of a shouting match.

Another level of discrimination of serial communications systems is synchronous versus asynchronous. These systems are diagramed in Figs. 23-65 and 23-66. The synchronous system is shown in Fig. 23-65, while the asychronous is shown in Fig. 23-66. The ability of the communications system to operate with a low or zero error rate is dependent upon the synchronization of the receiver and transmitter. If the receiver and transmitter are not synchronized, then the receiver sees little more than an unintelligible series of HIGHs and LOWs, and no sense can be made out of the data. The principal difference between synchronous and asynchronous methods lies in not whether or not synchronization takes place, but rather how it takes place.

In both systems a shift register or other system arranges the parallel data used inside the computer and converts it to a serial data stream. In the case where a shift register is used, the shift command merely forces the data to shift one place to the right for each clock pulse. This type of register is called a parallel-in, serial-out (PISO) register. The receiver uses the same sort of register but of opposite configuration: serial-in, parallel-out (SIPO).

In the synchronous system of Fig. 23-67, there are two communications channels in the transmission medium. One channel, perhaps a twisted pair of wires, is used for the data signal, i.e., the output of the PISO shift register. The

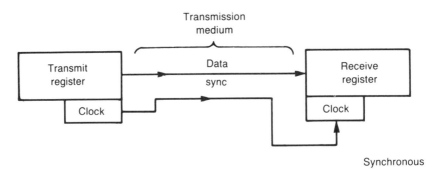

Fig. 23-65. Synchronous communications.

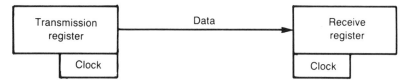

Fig. 23-66. Asynchronous communications.

other channel carries the clock signal. If there is no differential phase delay between the two channels, the clock signal can be used at the other end of the system to clock the receive register. An implication of this system is that the receive register can turn on only at certain times, that is, the time when the clock pulse arrives.

The problem with the synchronous method is that it requires a second transmission medium path, which can be expensive in radio and telephone systems. The solution to this problem is to use an asynchronous transmission system, such as that in Fig. 23-68. In this system only one transmission medium is needed. The synchronization is provided by transmitting some initial start bits that tell the receiver that the following bits are valid data bits. In most systems the data line remains HIGH when inert and signals the start of a transmission by dropping LOW.

There are two ways to keep the clock of the receiver synchronous with the transmitter. In one case an occasional sync signal will be transmitted that keeps the local receive clock on the correct frequency. This method is used in some systems but is somewhat more complex than it needs to be. The other system, used in modern equipment, is to operate the receiver and transmitter clocks very accurately, but they are locally controlled. Most system specifications call for the transmitter and receiver clock frequencies to be accurate to within 1 or 2 percent of the correct design value. As a result, it is common to find either crystal clocks or RC-timed clocks made with precision, low-temperature coefficient components.

Figure 23-68 shows how the asynchronous system works. The clock produces a train of pulses of constant duration and period. It is the constancy of the clock pulses that makes synchronization possible. When the negative-going start bit transition occurs (T1), the circuit knows that each bit will represent a specified period of time, for example, 9.09 milliseconds when 110 baud (speed) is used. The sampling clock pops up and tests the data line for HIGH or LOW at the correct intervals.

On older (and certain present low-priced) microcomputers a serial cassette was used for mass data storage. These systems used audio tones that represented the HIGH/LOW digital data. For example, one popular computer used 1050 Hz for HIGH and 1225 Hz for LOW. These tones were separated by filters and other circuitry, and turned into a serial data stream within the computer. The system was essentially asynchronous and used a method as in Fig. 23-69 for decoding the data. The

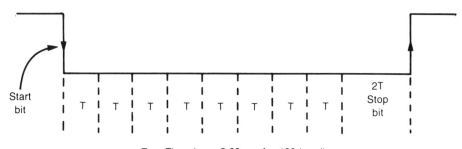

T = Time (e.g., 9.09 ms for 100 baud)

Fig. 23-67. Asynchronous data word.

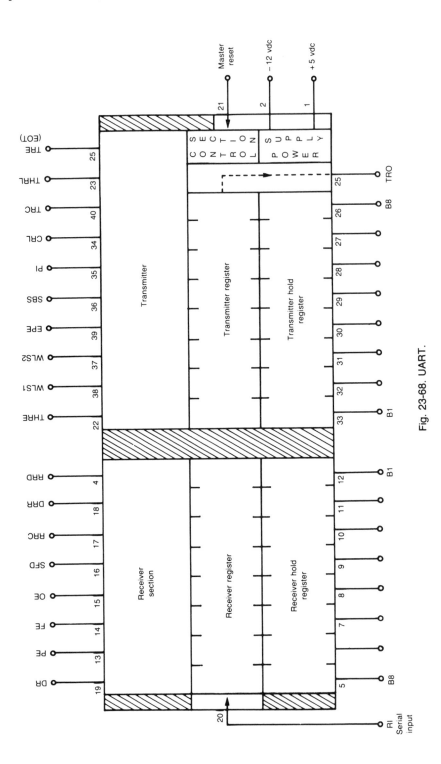

Fig. 23-68. UART.

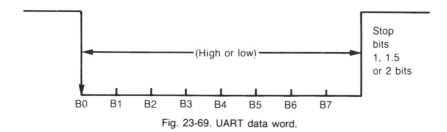

Fig. 23-69. UART data word.

reason why audio cassette recorders were so poor at reading data is that the motor speed tends to vary. This variation changes the time relationship between bits, so when the clock gate goes open to sample the data line, it finds a HIGH where a LOW was supposed to exist, or vice versa. The only cassettes that worked even half well were those that either transmitted a clock synchronizing tone to be recorded along with the data, or used a periodic stream of test data. In one system the computer attaches a leader to each data file consisting of 128 HIGH/LOW transitions. The cassette reading program measures the time required for all 128 transitions to occur, and then uses that information to set the period of the cassette read sampling clock.

The design of serial transmission circuits requires the construction of parallel-in serial-out (PISO) and serial-in parallel-out (SIPO) registers for transmitter and receiver, respectively. Fortunately, you can make use of a certain family of large scale integration (LSI) serial communications chips called the UART (universal, asynchronous receiver transmitter). Figure 23-70 shows the block diagram for a typical generic UART IC. The transmitter section has two registers: the transmitter hold register (THR) and the transmitter register (TR). The transmitter hold register is used as a buffer to the outside world, and serves as a temporary holding spot for data to be transmitted. The THR has a parallel data input system with as many bits as there are bits in the computer system. The data bit lines from outside of the UART input the data to this register. The output lines of the THR go directly to the transmitter register internally and are not accessible to the outside world. The transmitter register is of the PISO design and is used to actually transmit the

data bits to the rest of the world. The operation of the transmitter side of the UART is controlled by the transmitter register clock (TRC) input. The frequency of the clock signal applied to the TRC terminal must be 16 times the data transmission rate specified.

The receiver section is a mirror image of the transmitter section. The input is a serial line that feeds a receiver register (a SIPO type). The output register (receiver hold register, or RHR) is used to buffer the UART receiver section from the outside world. In both cases the hold registers operate semi-independently of the other registers and so can perform certain handshaking routines with other circuits to ensure that they are ready to participate in the process.

Like the transmitter, the receiver is controlled by a clock that must operate at a frequency that is 16 times the received data rate. The receiver clock (RRC) is separate from the transmitter clock (indeed, the entire receiver and transmitter circuits are separate from each other, despite sharing the same LSI IC), so the same UART IC can be used independently at the same time. Most common systems will use the UART in a half-duplex or full-duplex manner, so the receiver and transmitter clock lines will be tied together on the same 16X clock line.

The UART supports all three modes of transmission: simplex, half duplex, and full duplex. The simplex method transmits data in only one direction (see above), so a single UART will be used at the transmit end, while another UART will be used at the receive end. In the transmit end the transmitter section only is used, while on the other end of the system the receiver section only is used. In a half-duplex system, both sections of both UARTs are used, but only one UART at a time can trans-

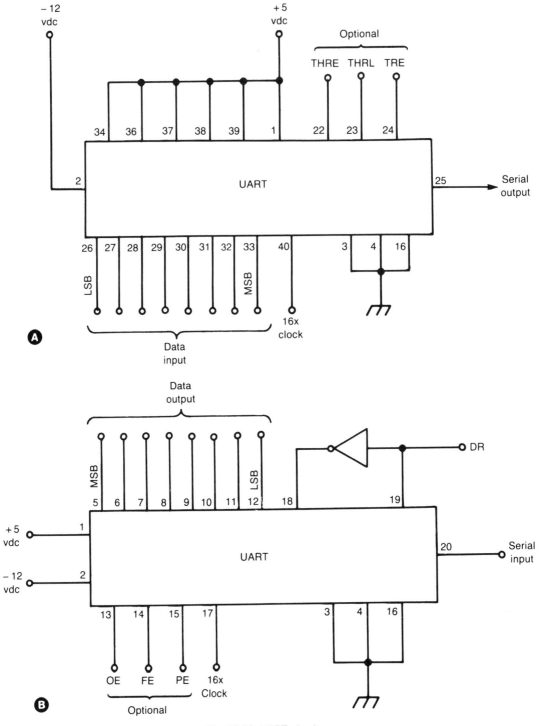

Fig. 23-70. UART circuits.

mit. An example of this method is when a single pair of wires with either current or voltage levels representing the data. Full duplex data can be accommodated when there is more than one transmission path, or where audio tones can be used to represent the data. If the transmission path is linear, then there won't be an intermodulation of the tones and they can be separated with active filters.

Several control terminals and signals are available on the UART IC, and these aid in operation of the circuit. Some of them, however, may be inactive in any given communications system. The master reset terminal is used to set all registers to zero and return all signals to their inert state. Table 23-12 shows the various signals and control input/outputs. In a section to follow I will show you how to use these in a typical interface situation with a 6502 microprocessor.

Data Received (DR). A HIGH on this terminal indicates that the data have been received and are ready for the outside world to accept.

Overrun Error (OE). A HIGH on this terminal tells the world that the data reset (DR) flag has not been reset prior to the next character coming into the internal hold register.

Parity Error (PE)[3]. Parity error indicates that the parity (odd or even) of the received data does not agree with the condition of the parity bit transmitted with that data. A lack of such a match indicates a problem in the transmission path.

Framing Error (FE). A HIGH on this line indicates that no valid stop bits were received.

B1 – B8 Receiver. Eight-bit parallel output from receiver (each line is tri-state, and so floats at a high impedance when inactive).

B1 – B8 Transmitter. Eight-bit parallel input bus to the transmitter side of the UART.

Transmitter Hold Register Empty (THRE). A HIGH on this pin indicates that the data in the transmitter hold register has been transferred to the transmitter register and that a new character may be loaded from the outside world into the transmitter hold register.

Data Receive Reset (DRR). Bringing this line LOW causes the reset (i.e., force to LOW) of the data received (DR) flag, pin no. 19.

Receiver Register Disconnect (RRD). A HIGH applied to this pin causes the data applied to the B1 – B8 transmitter input lines to be loaded into the transmitter hold register. A positive-going transition on THRL will cause the data in the

Table 23-12. Inputs and Outputs for the 6502 Microprocessor

Pin	Designation	Function
1	V_{CC}	+5 volts dc regulated power supply
2	V_{EE}	−12 volts dc regulated power supply
3	GND	Ground
4	RRD	Receiver register disconnect. A HIGH on this input tri-states (places at a high impedance) the receiver output data lines, pins 5 – 12. A LOW on this line connects the receiver data output lines to output pins 5 – 12.
5	RB8	Receiver data output line bit-8
6	RB7	Receiver data output line bit-7
7	RB6	Receiver data output line bit-6
8	RB5	Receiver data output line bit-5
9	RB4	Receiver data output line bit-4
10	RB3	Receiver data output line bit-3
11	RB2	Receiver data output line bit-2
12	RB1	Receiver data output line bit-1
13	PE	Parity error. A HIGH on this line indicates that the parity of the received data does not match the parity programmed at pin 39.
14	FE	Framing error. A HIGH on this line indicates that no valid stop bits were received.
15	OE	Overrun error. A HIGH on this line indicates that an overrun condition has occurred, which is defined as not having the DR flag (pin 19) reset before the next character is received by the internal receiver hold register.
16	SFD	Status flag disconnect. A HIGH on this line will disconnect (tri-state) the PE, FE, OE, DR, and THRE flags. This feature allows the status flags from several UARTs to be bussed together.
17	RRC	16× receiver clock input.
18	DRR	Data receive reset. Bringing this input LOW resets the data received line (DR, pin 19).
19	DR	Data received. A HIGH on this line indicates that the entire character is received and is in the receiver holding register.
20	RI	Receiver serial input (forced HIGH when no data received).

**Table 23-12. Inputs and Outputs
for the 6502 Microprocessor (*Continued*)**

Pin	Designation	Function
21	MR	Master reset. A short pulse applied to this line forces all data registers in Tx and Rx LOW, forces FE, OE, PE, and DRR LOW, and sets TRO, THRE, and TRE HIGH.
22	THRE	Transmitter hold register empty. A HIGH indicates that the data in the transmitter input register has been transferred to the TR, and allows a new data character to be loaded.
23	THRL	Transmitter holding register load. A LOW applied to this line enters the data to lines TB1 through TB8 into the transmitter holding register (THR). A positive-going level applied to this pin transfers the contents of the THR into the transmit register (TR) unless the TR is currently sending the previous word. When the transmission is finished, the THR to TR transfer will take place automatically.
24	TRE	Transmit register empty. Remains HIGH unless a transmission is taking place, in which case the TRE pin drops LOW.
25	TRO	Transmit serial output.
26	TB8	Transmitter input bit-8
27	TB7	Transmitter input bit-7
28	TB6	Transmitter input bit-6
29	TB5	Transmitter input bit-5
30	TB4	Transmitter input bit-4
31	TB3	Transmitter input bit-3
32	TB2	Transmitter input bit-2
33	TB1	Transmitter input bit-1
34	CRL	Control register load. This line can be either wired permanently HIGH, or be strobed with a positive-going pulse. It loads the programmed instructions (WLS1, WLS2, EPE,c PI, and SBS) into the internal control register.
35	PI	Parity inhibit. A HIGH disables the parity function and forces PE LOW.
36	SBS	Stop bit select. Programs the number of stop bits that are added to the data word output. A HIGH on SBS causes the UART to send 2 stop bits if the word length format is 6, 7, or 8 bits, and 1.5 stop bits if the 5-bit data word is selected. A LOW on SBS causes the UART to generate only 1 stop bit.

**Table 23-12. Inputs and Outputs
for the 6502 Microprocessor (*Continued*)**

Pin	Designation	Function		
37	WLS1	Word length select 1 (see below)		
38	WLS2	Word length select 2 (see below)		
		Word Length	WLS1	WLS2
		5 bits	LOW	LOW
		6 bits	HIGH	LOW
		7 bits	LOW	HIGH
		8 bits	HIGH	HIGH
39	EPE	Even parity enable. A HIGH applied to this line selects even parity, while a LOW selects odd parity.		
40	TRC	16× transmit clock.		

transmitter hold register to be transferred to the transmitter register, unless a data word is being transmitted at the same time. In this case the new word will be transmitted automatically as soon as the previous word is completely transmitted.

Receiver (Serial) Input (RI). Data input to the receiver section.

Transmitter Register (Serial) Output (TRO). Serial data output from the transmitter section of the UART.

Word Length Select 1 and 2 (WLS1 and WLS2). Sets the word length of the UART data word to 5, 6, 7, or 8 bits according to the protocol given above.

Even Parity Enable (EPE). A HIGH applied to this line selects even parity for the transmitted word and causes the receiver to look for even parity in the received data word. A LOW applied to this line selects odd parity.

Stop Bit Select (SBS). Selects the number of stop bits to be added to the end of the data word transmitted. A LOW on SBS causes the UART to generate only 1 stop bit regardless of the data word length selected by WLS1/WLS2. If SBS is HIGH, on the other hand, the UART will generate 2 stop bits for word lengths of 6, 7, or 8 bits and 1.5 bits for a word length of 5 bits (as selected by WLS1/WLS2).

Parity Inhibit (PI). Disables the parity function of both receiver and transmitter and forces PE LOW when PE is HIGH.

Control Register Load (CRL). A HIGH on this terminal causes the control signals WLS1,

WLS2, EPE, PI, and SBS to be transferred into the control register inside of the UART. This terminal can be treated in one of three ways: strobed, hardwired, or switch controlled. The strobed method uses a system pulse to make the transfer and is used if the parameters either change frequently or are under program control. If the parameters never change, then it can be hardwired HIGH. But if changes are made occassionally, then the control lines and CRL can be switch controlled.

The UART chip is particularly useful because it can be programmed externally for several bit lengths, baud rates, parity (odd/even, receiver verification/transmitter generation), parity inhibit, and stop bit length (1, 1.5, or 2 bits). The UART also provides six different status flags: transmission completed, buffer register transfer completed, received data available, parity error, framing error, and overrun error.

The clock speed on the common variety of UART is 320 kHz maximum for the A/B versions, 480 kHz for the AO3/BO3 versions, 640 kHz for the AO4/BO4 versions, and up to 800 kHz for the AO5/BO5 series. The receiver output lines are tri-state logic and so will float at a high impedance to both ground and the +5-volt line when inactive. The use of tri-state output allows the device to be connected directly to the data bus of a computer or other equipment or system.

The transmitter section uses an 8-bit parallel input register that will accept data to be sent serially. It will convert the 8-bit data word received in the input register to the serial format that includes the 8-bit word (also formattable to 5, 6 or 7 bits), start bit, parity bit, and stop bits.

The receiver can be viewed as simply the mirror image of the transmitter. It receives a serial input word containing start bits, data, parity, and stop bits. This serial data stream is checked for validity by comparison with parity and for the existence of the stop bits.

The UART data format (serial) is shown in Fig. 23-71. The transmitter output pin will remain HIGH unless data is being transmitted. Start bit B0 is always a HIGH-to-LOW transition, which tells the system that a new data word is about to be sent. Bits B1 through B8 are the data bits loaded into the transmitter at the sending end of the system. All 8 bits of the maximum word length format are shown in the figure, even though truncated word lengths of 5, 6, or 7 bits might be seen. The stop bit length can be programmed to be 1, 1.5, or 2 bits, according to the needs of the system.

The number of data bits, the parity, and the number of stop bits are programmed into the device using HIGH and LOW levels applied to certain pins designated for that purpose. For example, the WLS1 and WLS2 pins are used as word-length select pins, and will set the data word length according to the following protocol:

Word Length	WLS1	WLS2
5 bits	0	0
6 bits	1	0
7 bits	0	1
8 bits	1	1

Similarly, a 2-bit stop code is selected by connecting SBS HIGH, but only when the data word is 6, 7, or 8 bits. If the data word is set to 5 bits length, which is used on Baudot-encoded teletypewriters of ancient manufacture, then the 1.5-bit stop code is used. If SBS is LOW, then the stop code is 1 bit in length. The parity is set by the EPE pin and will be coded odd for a LOW and even for a HIGH.

The clock on a UART system must be stable, so you cannot generally use an RC timer-based clock circuit and expect proper performance, especially at high baud rates. The frequency of the clock must be 16 times the expected baud rate. If you want to transmit data at 300 baud, for example, the oscillator frequency must be 300×16, or 4800 Hz. While this frequency is well within the range normally competent RC oscillators can produce, it is recommended that a crystal oscillator be used to ensure the accuracy and stability of the clock. An attractive alternative is the CMOS 4060 device, which contains an internal crystal or RC oscillator and a chain of binary divider stages.

The transmitter section of the UART is shown connected in Fig. 23-72. Note that only

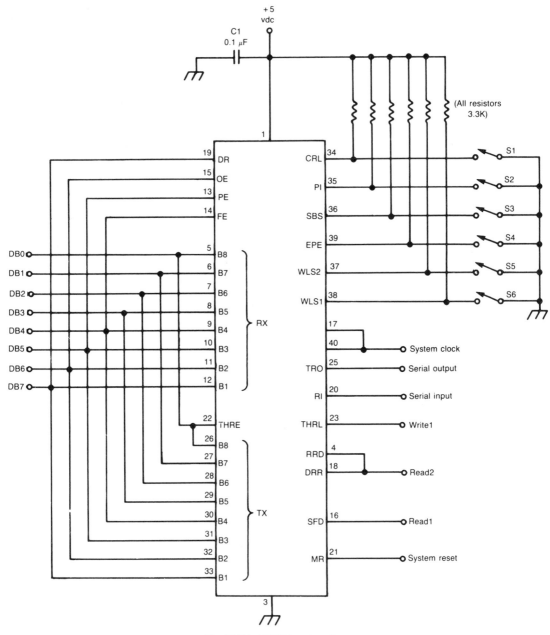

Fig. 23-71. UART bus interfacing.

the 8-bit input data, clock, and serial output are required to make this circuit operational. The TRE, THRE, and THRL signals are status flags and are optional (although in most practical applications they will be used). The status flags convey information about the status of the information transfer and are sometimes needed in the software used to control the UART. A careful review of the meaning of each flag is necessary for designers who wish to use the UART.

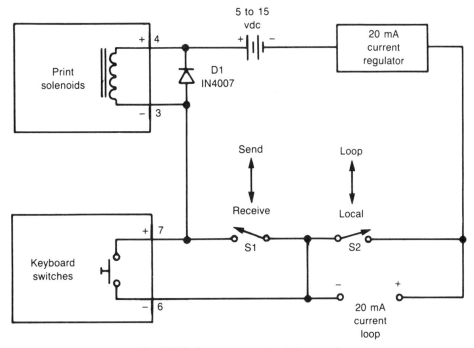

Fig. 23-72. Current loop printer/teletypewriter.

The basic receiver circuit for the UART is shown in Fig. 23-71. There is a similar simplicity in the receiver section (one of the principal attractions held by LSI devices to equipment and instrument designers). Only the clock, serial in/out, and 8-bit parallel output lines are needed. Again, however, certain signals are available that will make some applications either easier or possible; these are the DR, OE, FE, and PE flags. The table above gave the meanings of these signals and those of the transmitter section.

Note in the receiver section that an inverter is used from the data received terminal to reset the DRR terminal. This signal tells the UART to get ready for the next character and can be used to signal a distant transmitter that the UART is ready to receive another transmission.

One thing about the UART that appeals to many designers is that the two sections (receiver and transmitter) can be used either independently or in a common system. In a simplex communications channel (one direction only), a transmitter-wired UART is used on the transmitter end, while a receiver-wired UART will be installed on the receive end of the system. In a half-duplex system (bidirectional communication, but only in one direction at a time), both sections are used at each end of the system, and the status flags are used in a handshaking system to coordinate matters. Full-duplex operation is possible also but requires either a second channel (especially in radio links) or a second set of audio tones in hardwired telephone lines or coaxial cable systems. Not all telephone lines are amenable to full duplex, however, especially over long-distance lines.

In dedicated instrument applications, the programming pins will probably be hard-wired in the proper codes, but in many cases switches are used to allow the user to program as needed. You can also connect the UART control pins to an I/O port to permit programming of the UART under software control of the computer.

An example of a "standard" UART configured for use with the 6502 microprocessor is shown in Fig. 23-73. Because of the nature of the UART, you can use it directly as an I/O port and memory-map it to the 6502 without the need for any external circuitry exept device-select signals.

The transmitter input lines are high impedance and so can be connected directly to the 6502 data bus. Similarly, the receiver output lines are tri-state and so will float at high impedance (neither HIGH nor LOW) until the receiver is turned on. Therefore, you can connect both receiver and transmitter directly to the data bus (DB0–DB7). Also connected to the data bus are the DR, OE, PE, FE, and THRE signals.

The UART is programmed by the CRL, PI, SBS, EPE, WLS1, and WLS2 pins being made either HIGH or LOW. The protocols governing these control pins were covered earlier. In Fig. 23-73 the control pins are set by switches. Each input is tied HIGH through a 3.3k pull-up resistor. If the corresponding switch is open, therefore, that input is HIGH, but if the switch is closed the input is shorted to ground (and is therefore LOW).

The control input scheme of Fig. 23-73 assumes that variable control is needed over the UART programming. The use of DIP switches on the UART printed wiring board permits us to set these factors almost at will. If such a capability is not needed, however, you can also hard-wire the inputs HIGH or LOW, as needed.

A variation on the theme is to connect the control input lines to a latched output port. You could, for example, use 6 bits of a 74100 device to contain the HIGH/LOW states. The inputs of the 74100 would be connected to the data bus, while the outputs are connected to the UART control lines in place of the switches of Fig. 23-73. If you memory-map the 74100 and provide suitable device select circuitry, then simple write operations will allow you to set the UART parameters under program control.

CURRENT LOOP SERIAL PORTS

One of the earliest forms of data communications with peripherals was the teletypewriter machine. These were typewriter-like machines that used a mechanism of electrical solenoids to pull in the type bars or to position the type cylinder. The original devices used BAUDOT code and a 60-milliampere current. Later versions of the teletypewriter machine used a 20-milliampere loop and were generally more sophisticated than previous designs. Some modern teletypewriters use dot-matrix printing and contain an eight-inch floppy disk to store a magnetic copy of the data transmitted and received.

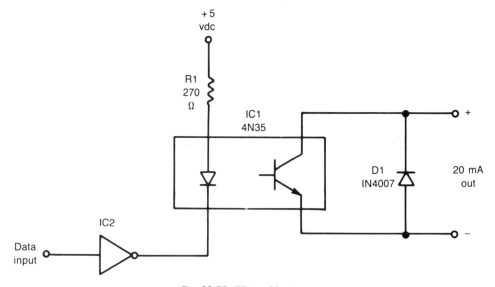

Fig. 23-73. TTL-to-20-mA loop.

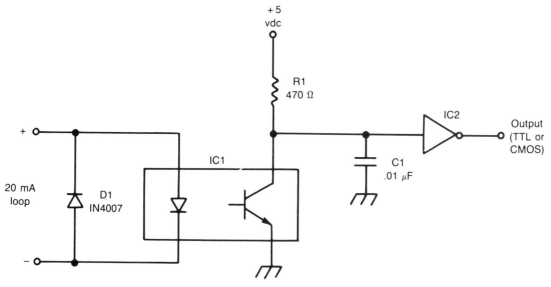

Fig. 23-74. 20-mA loop to TTL.

Figure 23-74 shows the basic elements of a teletypewriter or other printer based on the 20-milliampere current loop. The keyboard and printer are actually separate, and they usually have to be wired together if a local loop is desired (i.e., where the keystroke on the keyboard produces a printed character on the same machine. This circuit is actually grossly simplified. In a real teletypewriter there will be an encoder wheel or circuit that produces the BAUDOT code output. The keyboard consists of a series of switches (that actuate the encoder). Because these switches and their associated encoder are in series with the line, a "LOCAL" switch must be provided to bypass the transmitter section on receive.

The receiver consists of a decoder and the receive solenoids which actually operate the typebar mechanism. Note in Fig. 23-74 that a 1N4007 diode is in parallel with the receive solenoid. This is to suppress the inductive spike that will be generated when the reactive solenoids are de-energized. The diode is placed in the circuit so that it will be reverse biased under normal operation. But the counter electromotive force produced as a result of the "inductive kick" forward biases the diode. Under this condition the diode damps the spike to a harmless level. In some older machines the induc-

tive spike was safely ignored because the mass of the mechanism effectively integrated the spike to nothingness. But modern solid-state equipment does not move the mechanism directly with the 20-mA loop. The solid-state components can be damaged by the high voltage spike, so it is recommended that a 1N4007 be used even if the original design ignored it.

When the loop is closed, the circuit of Fig. 23-73 will produce a readable signal. Another similarly designed teletypewriter will be able to read the current variations produced by the machine.

Some modern printers are designed to operate with a 20-mA loop. Although most engineers would agree that the 20-milliampere loop is obsolete for modern designs, there are still large amounts of older equipment on the market, and in place at user sites, that are based on the current-loop concept. When replacing older equipment, it is prudent in most cases to simply buy a new printer that operates from a 20-milliampere loop than to redesign the whole system. Also, hobbyists and smaller users may well want to take advantage of older 20-mA loop equipment that comes on the surplus market at low cost when larger users upgrade their systems and no longer need the old machine.

Figures 23-73 and 23-74 shows how to interface 20-milliampere equipment to TTL-compatible serial outputs from computers. The circuit in Fig. 23-74 shows the transmitter arrangement. The assumption is that there is a single TTL-compatible bit from either a serialized-parallel output, or a UART IC. The TTL level is applied to an open collector TTL inverter, which has as its collector load an LED inside an optoisolator. When the LED is turned on, the phototransistor is turned on hard. Because this transistor operates as an electronic switch in series with the 20-milliampere current loop. Thus, when the TTL bit is HIGH, the LED is on and the transistor is saturated. In this condition the current loop transmit a "MARK" sign (equivalent to a logical 1 in binary).

The receive end of the current loop-to-TTL interface is shown in Fig. 23-74. In this case the optoisolator is still used, but in reverse. Here the LED is connected in series with the current loop. Thus, when a MARK is transmitted, the LED will be turned on; when a SPACE is transmitted, the LED is turned off. During the MARK periods, the optoisolator phototransistor is saturated, so the input to the TTL inverter is LOW. This condition results in a HIGH on the output to the computer. Again, a MARK is a logical 1 (HIGH) and a SPACE is a logical 0 (LOW). The 0.01 μF capacitor is used for noise suppression.

RS-232 SERIAL PORTS

Serial data communications requires only one channel (i.e., pair of wires, radio, or telephone channel), so it is ultimately less costly than parallel data transmission. The benefit is especially noticeable on long-line systems where the extra cost of wire and/or telecommunications channels becomes most apparent. The current loops discussed in the previous section were the earliest form of interface to peripherals and are still in use (albeit declining in popularity). In this section I will discuss what is probably the most common form of voltage-operated serial communications port, the RS-232C.

The Electronic Industries Association (EIA) RS-232 standard concerns itself with serial data transmission using voltage levels, as opposed to current levels. The standard calls for the use of a 25-pin D-shell standard connector (the DB-25), always wired in exactly the same manner and using the same voltage levels. If all signals are defined according to the standard, then it is possible to interface any two RS-232 devices without any problems.

A large collection of peripherals (modems, printers, video terminals) are fitted with DB-25 RS-232 connectors. Unfortunately, there are some manufactures who also use the DB-25 series of connectors in exactly the same gender as RS-232, but not in the RS-232 manner. Thus, the existence of an RS-232 connector is not adequate proof of RS-232 in use.

The RS-232 standard is older than most of our present-day digital devices and so uses an obsolete set of voltages for the levels. In RS-232 format the logical 1 (HIGH) is represented by a potential between -5 and -15 volts, while a logical 0 (LOW) is represented by a potential between $+5$ and $+15$ volts. Because most digital equipments today are based on TTL-compatible formats, some level translation is needed. Perhaps the most common method for doing this neat trick is to use the Motorola MC-1488 line driver and MC-1489 line receiver chips.

The circuit converts an RS-232 input to a TTL output. The active element is an npn transistor connected in a common-emitter configuration. When this transistor is turned off, i.e., unbiased, it will produce a TTL HIGH output. But when the transistor is turned on it is saturated, so the output will be close to ground potential (TTL LOW condition). The transistor is controlled by the signal applied between the emitter and base. In the RS-232 HIGH condition, the input is at -12 volts and so the transistor is reverse biased. Under this condition the transistor is turned off, so the output is HIGH. The -12 volts is clipped by the diode (D1) so as to not damage the transistor. During the RS-232 LOW condition, the input is at $+12$ volts, so the transistor is turned on hard (saturated) and a LOW appears on the TTL output.

The circuit is used to convert TTL levels from

Table 23-13. Pin Assignments for the RS-232C Serial Interface

RS-232 Pin Assignments for DB-25 Connector
(x) = unassigned

Pin No.	RS-232 Name	Function
1	AA	Chassis ground/common
2	BA	Data from terminal
3	BB	Data received from MODEM
4	CA	Request to send
5	CB	Clear to send
6	CC	Data set ready
7	AB	Signal ground
8	CF	Carrier detection
9	(x)	
10	(x)	
11	(x)	
12	(x)	
13	(x)	
14	(x)	
15	DB	Transmitted bit clock (internal)
16	(x)	
17	DD	Received clock bit
18	(x)	
19	(x)	
20	CD	Data terminal ready
21	(x)	
22	CE	Ring indicator
23	(x)	
24	DA	Transmitted bit clock, external
25	(x)	

Table 23-14. Parallel Printer Interface Pin Standards (Amphenol 57-30360 Connector)

(x) = unassigned
* = IN on receivers (e.g., printer) and OUT on computers

Pin No.	Signal Name	IN/OUT?	Function
1	STROBE	IN	Strobe to read data in. Pulse width >0.5 μs
2	DATA1	*	Data bits
3	DATA2	*	Data bits
4	DATA3	*	Data bits
5	DATA4	*	Data bits
6	DATA5	*	Data bits
7	DATA6	*	Data bits
8	DATA7	*	Data bits
9	DATA8	*	Data bits
10	ACKNLG	OUT	Data received and printer is ready to accept more data. Pulse width >5 μS
11	BUSY	OUT	Printer cannot accept new data.
12	PE	OUT	Out-of-paper signal. Active-HIGH.
13	SLCT	OUT	Printer SELECTed
14	AUTOFEED XT	IN	Automatic paper feed signal
15	(x)	—	
16	0V	—	Logic signal common
17	CHASSISGND	—	Chassis ground (isolated from logic ground).
18	(x)		
19	(return for pin 1)		Pins 19-30 are twisted-pair returns for signal pins 1-12, and are at ground potential.
20	(return for pin 2)		
21	(return for pin 3)		
22	(return for pin 4)		
23	(return for pin 5)		
24	(return for pin 6)		
25	(return for pin 7)		
26	(return for pin 8)		
27	(return for pin 9)		
28	(return for pin 10)		
29	(return for pin 11)		
30	(return for pin 12)		
31	INIT		Reset printer to initial state. Pulse width >50 μS
32	ERROR		LOW if "paper end" state, "off-line" state, or "error" state.

a computer to RS-232 levels. This circuit is based on the optoisolator. The driver shown in the LED side of the circuit is made from a pair of TTL open-collector inverters in cascade or from a single CMOS noninverting output (B-series CMOS only).

The TTL-to-RS-232C converter circuit is based on the popular 741 operational amplifier. The −12 vdc and +12 vdc power supplies used for the operational amplifier are the source of the RS-232 levels. The TTL input is biased for noise-band immunity by a reference voltage (V_{REF}) of 1.4 vdc.

The DB-25 connector pin assignments for the RS-232C are shown in Table 23-13.

The RS-232 connectors and cables are readily available on the market. Some companies offer complete cables, including connectors, while other sources offer only the connectors with the view that you roll your own. Keep in mind that not all of

Table 23-14. Parallel Printer Interface Pin Standards (Amphenol 57-30360 Connector) (*Continued*)

(x) = unassigned
* = IN on receivers (e.g., printer) and OUT on computers

Pin No.	Signal Name	IN/OUT?	Function
33	GND		(same as pins 19-30)
34	(x)		
35	—		Pulled up to +5 vdc through 3300 ohms
36	SLCTIN		Data entry to printer possible in active-LOW.

the assigned contacts are always used. There are some printers, for example, that will work fine with only seven of the above connections in place; the others are irrelevant and so are not used.

Some printers use "similar to RS-232C" interfaces without bothering to tell you that a special cable is needed. First, though, let's define the two forms of device that can be accommodated: DTE and DCE. The acronym "DTE" stands for "data terminal equipment," and a DTE device is capable of acting as either (or both) data source and data destination. "DCE" means "data communications equipment," and this type of equipment provides the functions required to establish, maintain, and terminate a data-transmission connection. Typically, DTE means the computer, and DCE means

the MODEM. The connection cable required for a DTE (computer) to DCE (MODEM), is "straight across." The pins of each connector are mated to the equivalent pins on the other cable. Be careful, because the connectors used on this cable are the same as for DTE-DTE, but the computers won't work.

Note that certain signals are "crossed over," that is, they are switched in order to make the computers think they are talking to MODEM instead of another computer.

CENTRONICS PARALLEL PORTS

The parallel interface differs from the serial in that not less than one line is required for each bit of the data bus, plus a return line or common (i.e., "ground"). The serial system transmitted all bits in sequence over the same line or other communications path, while the parallel format transmits all bits simultaneously. As a result, it is possible to make the parallel format a lot faster, even if somewhat more complex.

The simplest parallel format is the straight forward eight-bit parallel port defined elsewhere in this book. But for most parallel printer connections, the Centronics interface is used. This port uses a standard connector, the Amphenol 57-30360. The pinouts are shown in Table 23-14. All levels are TTL compatible.

Chapter 24

Basics of Data Conversion

THE DATA CONVERTER DOES ONE OF TWO JOBS; IT either: 1) converts a binary digital word to an equivalent current or voltage, or 2) converts an analog current or voltage to an equivalent binary word. The former are called digital-to-analog converters (DACs), while the latter are called analog-to-digital converters (A/D or ADC). In this chapter I discuss the basic functioning of these data-conversion building blocks.

APPROACHES

There are several approaches to converting data from electronic and scientific instruments into binary numbers required by a digital computer, but some of them are so tedious as to be ridiculous. For example, one could tabulate values directly from a display readout or strip-chart recording and then go to a keypunch machine, paper tape cutter, or CRT video terminal and manually enter the data on the keyboard. This method, however, soon proves to be too slow if more than a very few data points are involved. It is also prone to massive errors if the operator is inattentive or fatigued, creating problems in data acquisition (and cost) that nobody needs.

The next worst method is to use a digitizer table. These devices are particularly useful for old analog data that is produced on strip-chart or graph paper (X-Y) recorders in a time before data converters were available. The digitizer is an X-Y matrix that allows the operator to designate points in the Cartesian coordinate system, and then enter them into a digital computer by pressing a button. The analog voltage or current recording is placed on the machine, and the points designated. Three types are typically used: pressure pads, sound, and CRT.

The best solution is to take the data directly from the instrument, control circuit or other source, and enter it automatically into the computer in binary form. This act requires an A/D converter. The opposite type of converter (DAC) does the opposite work. In discussing data converters in general we begin with a discussion of DACs.

DIGITAL-TO-ANALOG CONVERTERS (DACs)

There are several different approaches to making DACs, but in most cases they use a weighted current or voltage system that generates binary words by appropriate switch contacts.

An example of the popular R-2R ladder method is shown in Fig. 24-1. The active element, A1, is an operational amplifier in a unity-gain inverting-follower configuration. Although you can get away with using a device from the low-cost 741-family, it is not a good practice in a precision DAC. It would be better to use a premium device or a BiMOS device such as the RCA CA-3140.

In the circuit of Fig. 24-1, the digital inputs are illustrated as switches, but in a real world data converter the switches would be replaced by a binary counter, or some other N-bit parallel data line.

A precision reference voltage source is required for accurate conversion, and for most designs this will be at a level of +2.56 volts, +5.00 volts, or +10.00 volts. The accuracy of the converter is dependent upon the precision of the reference voltage source. There are other sources of error, but if the reference is poor, then there is no hope for any other factors to be effective in improving the performance of the circuit. Although almost any precision voltage regulator can be pressed into service as the reference, it is a simple matter to use the Precision Monolithics, Inc., REF-01/REF-02CJ (or in more precision case the HJ series). These IC devices are especially designed for this application, and are easily trimmed to the required reference potential.

Returning to Fig. 24-1, let us consider the circuit action under circumstances where various bits are either HIGH or LOW. If all bits are LOW, then the output voltage will be zero. The value of the output voltage is given by $I \times R$, and when all bits are LOW this current is zero. In practical circuits, however, there might be some output voltage under these circumstances due to offsets in the operational amplifier, the R-2R ladder, and the electronic switches. These can be nulled to zero output voltage when all bits are intentionally set to zero (or ignored, if negligible).

If the most significant bit (MSB) is made 1 (i.e., HIGH), then the output voltage will be approximately $1/2$ V_{REF}. Similarly, if the most significant bit is turned on (set to HIGH) and all others are LOW, then the output will be $1/4$ V_{REF}. The least significant bit (LSB) would contribute 2^{-N} V_{REF} to the total output voltage.

ANALOG-TO-DIGITAL CONVERTERS (A/D)

Of the many techniques that have been published for performing an A/D conversion, only a few are of interest to us; so I will consider only the voltage-to-frequency, single-slope integrator, dual-slope integrator, counter (or servo), and successive approximation methods. Before unraveling the mysteries of A/D conversion, though, it would be profitable to consider what is meant by the words "analog" and "digital."

Analog Signals

The word "analog" has several meanings that connote similar things. It is interesting to note that

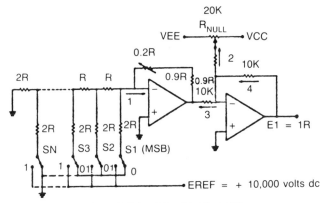

Fig. 24-1. R-2R weighted ladder DAC.

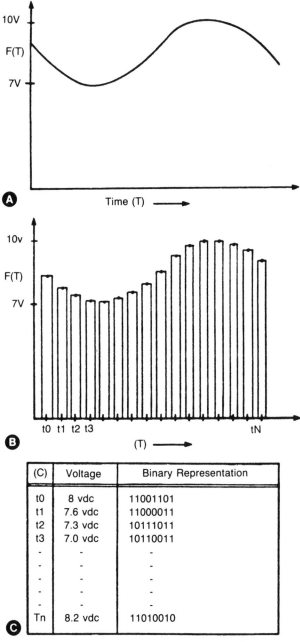

Fig. 24-2. Analog, sampled, and digital data.

(C)	Voltage	Binary Representation
t0	8 vdc	11001101
t1	7.6 vdc	11000011
t2	7.3 vdc	10111011
t3	7.0 vdc	10110011
·	·	·
·	·	·
·	·	·
·	·	·
·	·	·
Tn	8.2 vdc	11010010

it is often standard, if erroneous, practice to use the meanings interchangeably when it is not necessary to be absolutely rigorous. In the strictest sense the word is related in meaning to the word "analogous," but in daily use it connotes almost any time-varying voltage or current function (see Fig. 24-2A) that has both a continuous domain (t) and continuous range (F(t)). So any voltage or current that is allowed to take on any value within both range and domain is usually called an analog signal.

In the more rigorous sense, though, the generic term for all such signals is continuous signal. An analog signal is a subset of continuous signals and represents (i.e., is "analogous" to) some physical parameter or other quantity. Most electronic scientific instruments have an output that is either a current or voltage analog of the parameter that they are designed to measure or detect.

An example of a discrete-time signal is shown in Fig. 24-2B. Here there is a type of waveform that represents sampled data. That is to say that the time variable (t) can assume only certain values, but the variable F(t) may take on any value within its range. The sampled data signal may be obtained using a sample and hold circuit in which the clock is driven by a square-wave pulse train that has a repetition rate (i.e., frequency) equal to the number of data points required per unit of time. In this way our clocked sample and hold circuit may be called an analog/sampled data circuit.

A lot of people call sampled data by the terms "digital" or "digitized" data, but that is an error somewhat more serious than the questions over the word "analog." A true digital signal is one in which both t and F(t) are allowed to assume only certain discrete values. This is the kind of signal that is digestible in a digital computer. On paper one might represent digital signals as a table in which a series of data words corresponding to an amplitude or other feature are paired with the time signal, as in Fig. 24-2C.

In some cases you might want to use either a digital panel meter or digital multimeter both as the visual readout device and as the A/D converter for the computer. Similarly, you may wish to use a digital output of some existing instrument to provide the digital signal to the computer. Many such instruments provide parallel digital output lines on their rear panels along with control and/or strobe lines. The appeal of this approach falls down rather abruptly, though, because of two problems: relatively slow conversion speed and the fact that the instruments usually have binary-coded decimal

(BCD) coding rather than the straight binary or hexadecimal coding required by the computer.

The first of these problems really only becomes a nuisance at higher frequencies; so the approach is usable at near-dc frequencies and can be solved by either hardware of software code conversion.

Most digital panel meters (DPM) or digital multimeters (DMM) instruments use either single or dual-slope integration for the A/D conversion process. An example of a single-slope integrator A/D converter is shown in Fig. 24-3A. The single-slope integrator is simple but is limited to those applications that can tolerate accuracy of only one or two percent.

The single-slope integrator A/D converter of Fig. 24-3A consists of five basic sections: ramp generator, comparator, logic, clock, and output encoder. The ramp generator is an ordinary operational amplifier integrator with its input connected to a stable, fixed, reference voltage source. This makes the input current I_{REF} essentially constant; so the voltage at point 2 will rise in a nearly linear manner, creating the voltage ramp.

The comparator is merely another operational amplifier, but it has no feedback loop. The gain in this instance is essentially the open-loop gain of the device selected — typically very high even in low-cost operational amplifiers. When the anlog input voltage V_x is greater than the ramp voltage, then the output of the comparator is saturated at a logic HIGH level.

The logic section consists of a main AND gate, a main-gate generator, and a clock. The waveforms associated with these circuits are shown in Fig. 24-4B.

When the output of the main-gate generator is LOW, switch S1 remains closed, so the ramp voltage is zero. The main-gate signal at point 1 is a low-frequency squarewave with a frequency equal to the desired time-sampling rate. When point 1 is HIGH, S1 is open, so the ramp will begin to rise linearly. When the ramp voltage is equal to the unknown input voltage V_x, the differential voltage seen by the comparator is zero; so its output drops LOW.

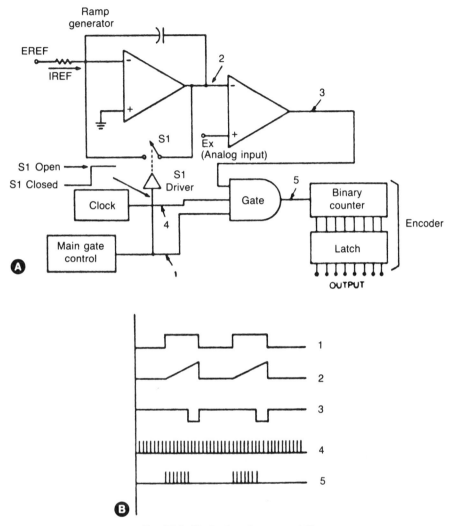

Fig. 24-3. Single-slope integrator A/D.

The AND gate requires all three inputs to be HIGH before its output can be HIGH also, from times T0 to T1, the output of the AND gate will go HIGH every time the clock signal is also HIGH.

The encoder, in this case an eight-bit binary counter, will then see a pulse train with a length proportional to the amplitude of the analog input voltage. If the A/D converter is designed correctly, then the maximum count of the encoder will be proportional to the maximum range (full-scale) value of V_x.

Several problems are found in single-slope integrator A/D converters:

- The ramp voltage may be nonlinear
- The ramp voltage may have too steep or too shallow a slope
- The clock pulse frequency could be wrong
- It may be prone to changes in apparent value of V_x caused by noise.

Many of these problems are corrected by the dual-slope integrator of Fig. 24-4. This circuit also

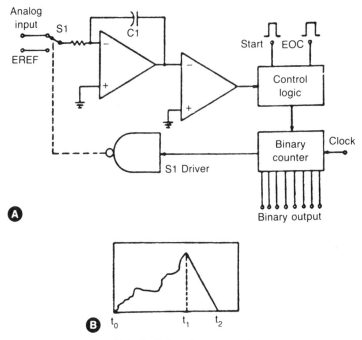

Fig. 24-4. Dual-slope integrator A/D.

consists of five basic sections: integrator, comparator, control logic section, binary counter, and a reference current or voltage source. An integrator is made with an operational amplifier connected with a capacitor in the negative feedback loop, as in the case of the single-slope version. The comparator in this circuit is also the same sort of circuit as in the previous example. In this case, though, the comparator is ground-referenced, using just one active element.

When a start command is received, the control circuit resets the counter to 00000000, resets the integrator to 0 volts (by discharging C1), and sets electronic switch S1 to the analog input. The analog voltage creates an input current to the integrator and this causes the integrator output to begin changing capacitor C1. This means that the output voltage of the integrator will begin to rise. As soon as this voltage rises a few millivolts above ground the comparator output snaps HIGH positive. A HIGH comparator output causes the control circuit to enable the counter, which begins to count pulses.

The counter is allowed to overflow and this output bit resets switch S1. The graph of Fig. 24-4B shows the integrator changing during the interval between start and the overflow of the binary counter $(t_1 - t_0)$. At time t_1 the switch changes the integrator input from the analog signal to a precision reference source. Meanwhile, at time t_1 the counter has overflowed, and again it has an output of 00000000 (maximum counter + 1 more count is the same as the initial condition). It will, however, continue to increment so long as there is a HIGH comparator output. The charge accumulated on capacitor C1 during the first time interval is proportional to the average value of the analog signal that existed between t_0 and t_1.

Capacitor C1 is discharged during the next time interval $(t_2 - t_1)$. When C1 is fully discharged the comparator will see a ground condition at its active input and so will change state and make its output LOW. Even though this causes the control logic to stop the binary counter, it does not reset the binary counter. The binary word at the counter output at the instant it is stopped is proportional to

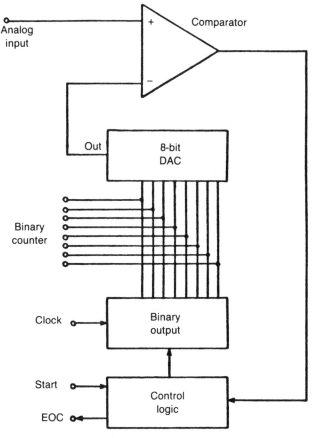

Fig. 24-5. Counter (servo/ramp) A/D converter.

the average value of the analog waveform over the interval $(t_1 - t_0)$. An end-of-conversion (EOC) signal is generated to strobe the microprocessor or other instrument so that it knows the output data is both stable and valid, and therefore ready for use.

VOLTAGE-TO-FREQUENCY CONVERTERS

These are no really genuine A/D converters in the strict sense, but there are devices that will convert analog data into a form that can be tape recorded on a machine that will not make a huge dent in your salary. Basically, it consists of only a voltage-controlled oscillator, or some variant.

COUNTER-TYPE (SERVO) A/D CONVERTERS

A counter type A/D converter (also called "servo" or "ramp" A/D converters) is shown in Fig. 24-5. It consists of a comparator, voltage output DAC binary counter, and the necessary control logic. When the start command is received, the control logic resets the binary counter 00000000, enables the clock, and begins counting. The counter outputs control the DAC inputs; so the DAC output voltage will begin to rise when the counter begins to increment. As long as analog input voltage V_{IN} is less than V_{REF}, the DAC output and then the comparator output is HIGH. When V_{IN} and V_{REF} are equal, however, the comparator output goes LOW, and this turns off the clock and stops the counter. The digital word appearing on the counter output at this time represents the value of V_{IN}.

Both slope and counter type A/D converters take too long for many applications, on the order of 2^n clock cycles (where n = number of bits). Conversion times become critical if the high-frequency component of the input waveform is to be faithfully reproduced. Nyquist's criterion requires that the sampling rate (i.e., conversions per second) be at least twice the highest frequency to be converted.

SUCCESSIVE APPROXIMATION A/D CONVERTERS

Successive approximation had been found to be best suited for many applications where speed is important. This type of A/D converter requires only n+1 clock cycles to make the conversion, and some designs allow truncation of the conversion process after fewer cycles if the final value is found prior to n+1 cycles.

The successive approximation converter operates by making several successive trials at comparing the analog input voltage with a reference generated by a DAC. An example is shown in Fig. 24-6. This circuit consists of a comparator, control logic section, a shift register, output latches, and a voltage output DAC.

When a start command is received, a HIGH is loaded into the MSB of the shift register, and this sets the output of the MSB latch HIGH. A HIGH in the MSB of a DAC will set the output voltage V_{REF} to half-scale. If the input voltage V_{IN} is greater than

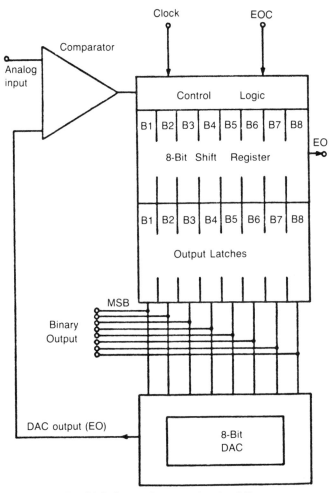

Fig. 24-6. Successive approximation A/D.

V_{REF}, the comparator output stays HIGH and the HIGH in the shift register MSB position shifts one bit to the right and therefore occupies the next most significant bit (bit 2). Again the comparator compares V_{IN} with V_{REF}. If the reference voltage from the FAC is still less than the analog input voltage, the process will be repeated with successively less significant bits until either a voltage is found that is equal in V_{IN} (in which case the comparator output drops LOW) or the shift register overflows.

If, on the other hand, the first trial with the MSB indicates that V_{IN} is less than the half-scale value of V_{REF}, the circuit will make trials below V_{REF}. The MSB latch is reset to LOW and the HIGH in the MSB shift register position shifts one bit to the right to the next most significant bit (bit 2). Here the trial is repeated again. This process will continue as before until the correct level is found or overflow occurs. At the end of the last trial (bit 8 in this case), the shift register overflows and becomes an end-of-conversion (EOC) flag to tell the rest of the world that the conversion has been completed.

This type, and most other types of A/D converters, requires a starting pulse and signals completion with an EOC pulse. This requires the computer or other digital instrument to engage in bookkeeping to repeatedly send the start command and look for the EOC pulse. If one ties the start input to the EOC output, then conversion is continuous and the computer need only look for the raising of the EOC flag.

In the chapters to follow I will examine some practical DAC and A/D circuits.

Chapter 25

Digital-to-Analog Converters

A S YOU LEARNED IN THE LAST CHAPTER, A DIGI-tal-to-analog converter (DAC) is a circuit or device that converts a binary word from a computer or other digital instrument to a proportional analog current or voltage. A number of different manufacturers offer low-cost, eight-bit, IC DACs that contain almost all of the electronics needed for the process, except possibly the reference source (which some do, indeed, contain also) and some operational amplifiers for either level shifting or current-to-voltage conversion.

For this chapter I have selected the DAC-08 device. This eight-bit DAC is now something of an industry standard and is available from several sources including Precision Monolithics, Inc., its originator. The DAC-08 is a later generation version of the old Motorola MC-1408 device. For non-PMI sources, this DAC is sometimes referred to as the LMDAC-0800.

Figure 25-1 shows the basic circuit configuration for the DAC-08. In subsequent circuits I will delete the power supply terminals for simplicity's sake, because they will be the same as shown here. The internal circuitry of the DAC-08 is the R-2R ladder discussed in the previous chapter but has two outputs: I_O and NOT-I_O. These outputs are

complementary; if the full-scale output current is 2.0 mA, then I_O + NOT-I_O = 2 mA (e.g., I_O = 1.25 mA, NOT-I_O = 0.75 mA).

Two types of input signal are required to make this DAC work: analog reference and digital. One is the reference current, I_{REF}, applied through pin no. 14. This current may be generated by a precision reference voltage source such as the REF-01 (or other) and a precision low-temperature coefficient resistor to convert V_{REF} to I_{REF}. For TTL compatibility at the binary inputs, make V_{REF} 10.000 volts, and R_{REF} 5000.00 ohms (precision).

The other type of input is the eight-bit digital word, which is applied to the IC at pins 5 through 12, as shown. The logic levels which operate these inputs can be preset by the voltage applied to pin no. 1 (for TTL operation, pin no. 1 is grounded). In the TTL-compatible configuration shown, LOW is 0 to 0.8 volts, while HIGH is +2.4 to +5 volts.

Figure 25-2 shows the connection of the DAC-08 (less power supply and reference input) required to provide the simplest form of unipolar operation over the range of approximately 0 to 10 volts. When the input word is 00000000, then the DAC output is 0 volts, ± dc offset error. A half-scale voltage (−5 volts) is given when the input

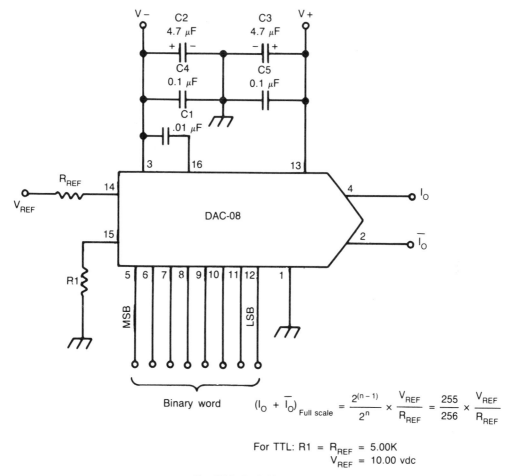

$$(I_O + \overline{I_O})_{Full\ scale} = \frac{2^{(n-1)}}{2^n} \times \frac{V_{REF}}{R_{REF}} = \frac{255}{256} \times \frac{V_{REF}}{R_{REF}}$$

For TTL: $R1 = R_{REF} = 5.00K$
$V_{REF} = 10.00$ vdc

Fig. 25-1. DAC-08 basic circuit.

word is 10000000. This situation occurs when the MSB is HIGH and all other digital inputs are LOW. The full-scale output will exist only when the input word is 11111111 (all HIGH). The output under full-scale conditions will be −9.96 volts, rather than 10-volts as might be expected (note: 9.96 volts is 1 LSB less than 10 volts).

The circuit in Fig. 25-2 works by using resistors R2 and R3 as current-to-voltage converters. When currents I_O and NOT-I_O pass through these resistors, a voltage drop of IR, or $5.00 \times I_O$ (mA), is produced. A problem with this circuit is that it has a high source impedance (5 kilohms, with the values shown for R2/R3). You can solve this problem with the circuits that follow.

Figure 25-3 shows a simple method for converting I_O to an output voltage (V_O) with a low-output impedance (less than 100 ohms) by using an inverting-follower operational amplifier. The output voltage is simply the product of the output current and the negative feedback resistor:

$$V_O = R \times I_O$$

As in the case previously described, a 5000-ohm resistor will produce a 9.96-volt output voltage when the DAC-08 is set up for TTL inputs and 2.0 I_O(max).

The frequency response of the DAC circuit can be tailored to meet certain requirements. The

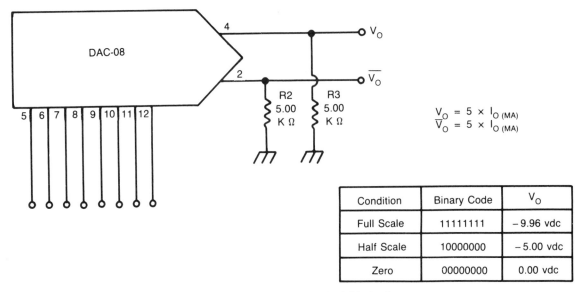

Condition	Binary Code	V_O
Full Scale	11111111	−9.96 vdc
Half Scale	10000000	−5.00 vdc
Zero	00000000	0.00 vdc

Fig. 25-2. Voltage output DAC-08 circuit.

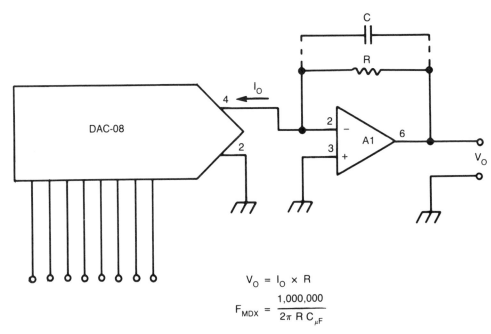

$$V_O = I_O \times R$$

$$F_{MDX} = \frac{1{,}000{,}000}{2\pi\,R\,C_{\mu F}}$$

Fig. 25-3. Low-impedance voltage-output circuit.

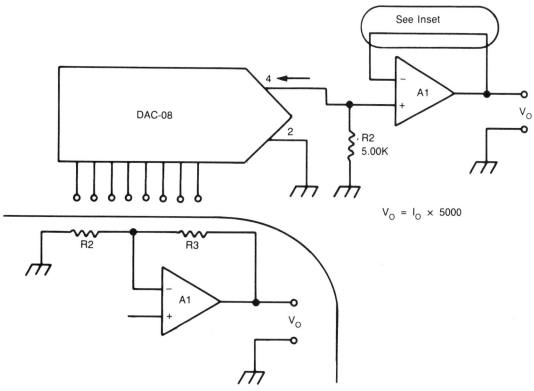

$$V_O = I_O \times 5000$$

Fig. 25-4. Low-impedance voltage-output circuit.

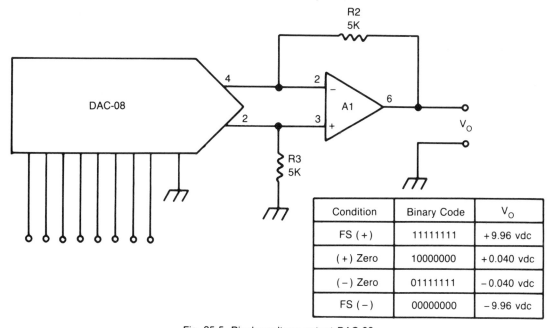

Condition	Binary Code	V_O
FS (+)	11111111	+9.96 vdc
(+) Zero	10000000	+0.040 vdc
(−) Zero	01111111	−0.040 vdc
FS (−)	00000000	−9.96 vdc

Fig. 25-5. Bipolar voltage output DAC-08.

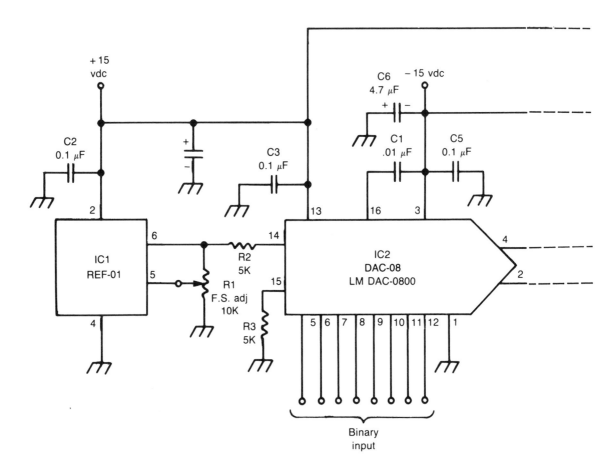

Fig. 25-6. Practical DAC-08 circuit.

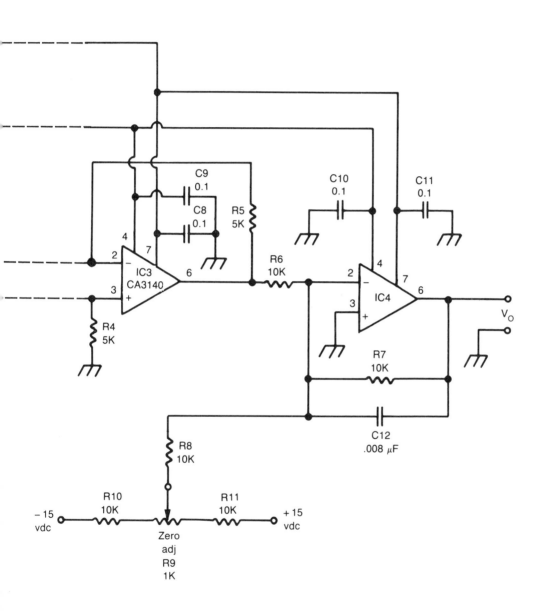

normal output waveform of the DAC is a staircase when the input ramps up from 00000000 to 11111111 in a monotonic manner. If you want to make it an actual ramp function, then you need to low-pass filter the output to remove the "stepness" of the normal waveform. An optional capacitor in shunt with feedback resistor R will offer us limited (but useful) filtering on the order of −3 dB/octave above a cut-off frequency of:

$$f = \frac{1,000,000}{6.28 \ R \ C}$$

where

f is the −3 dB frequency in hertz (Hz)
R is in ohms
C is in microfarads (μF)

In most practical circuits you will know the value of f from the application. It is the highest frequency component in the input waveform (e.g., 100-Hz in a typical pressure waveform). You then need to calculate the value of the capacitor needed to achieve this cut-off frequency, and so must swap the f and C terms in the equation above:

$$C \ (in \ \mu F) = \frac{1,000,000}{6.28 \ R \ f}$$

A related method shown in Fig. 25-4 produces an output voltage of the opposite polarity from that of Fig. 25-3. In this case you merely take the circuit of Fig. 25-2 and connect a noninverting unity-gain follower at the output. The output voltage is the product of I_0 and R2. If you need a higher output voltage, then you would use the circuit variant shown in the inset to Fig. 25-4. In this case the output amplifier has gain, so the output voltage would be:

$$V_0 = (I_0 \times R2) \times \left[\frac{R3}{R2} + 1 \right]$$

One of the ways to achieve bipolar binary operation is shown in Fig. 25-5. In this case the output amplifier is a dc differential amplifier, and both current outputs of the DAC-08 are used. The output voltage and the corresponding input codes are shown in the table in Fig. 25-5. Note that the maximum and minimum voltages are positive and negative. The zero selected can be either (+)zero (+1-LSB voltage), or (−)zero (−1-LSB voltage). It cannot be exactly zero because an even number of output codes are equally spaced around zero. In other words, the absolute value of FS(−) is equal to the absolute value of FS(+). There are also circuits that make zero = zero, but at the expense of uneven ranges for FS(−) and FS(+).

A PRACTICAL BIPOLAR DAC CIRCUIT

A practical circuit is shown in Fig. 25-6. This circuit combines the circuit fragments shown earlier to make a complete circuit that can be used in real situations. The heart of this circuit is a DAC-08, or LMDAC-0800, connected in the bipolar binary circuit discussed above.

The reference potential in Fig. 25-6 is a REF-01 10.000-volt IC reference source. Potentiometer R1 adjusts the value of the actual voltage and also serves as a full-scale adjustment for the output voltage, V_0.

The output amplifier (IC4) can be a 741-class operational amplifier, or any other form; the need is not critical. Potentiometer R9 acts as a zero adjustment for V_0. The capacitor across R7 limits the frequency response to 200 Hz (with the value shown). This limit can be changed with the equation given earlier.

Adjustment

To adjust the circuit, use the following procedure.

1. Set the binary inputs all LOW (00000000).
2. Adjust R9 for $V_0 = 0.00$ volts.
3. Set all binary inputs HIGH (11111111).
4. Adjust potentiometer R1 for $V_0 = 9.96$ volts.

Chapter 26

A/D Converters

A LARGE NUMBER OF INSTRUMENTS, TRANS-ducers, and other devices operate in what is called the "analog mode." These devices use a voltage or current to represent some physical parameter. For example, a blood pressure transducer used in medicine will produce a dc voltage output of 50 microvolts (μV) per volt of excitation per millimeter of mercury pressure (50 μV/V mmHg). This potential is amplified and displayed in the form of blood pressure. More and more, instruments such as the medical blood pressure monitor use programmable digital computers and microprocessors at their heart. The job of interfacing these instruments is the job of converting analog voltages (or currents) to a proportional binary number to be input to the digital computer. In this chapter I will examine the A/D converter function of such a system.

An analog-to-digital (A/D) converter is a device that will produce a binary word output that is proportional to the applied analog voltage. A specific A/D converter will have a certain range. Unipolar A/D converters typically have ranges of 0 to 1 volt, 0 to 2.5 volts, 0 to 5 volts, or 0 to 10 volts. In such a converter the binary word will represent voltages in a manner similar to the 0 to 10-volt case shown below:

Unipolar Voltage	Binary Word (8-BITS)	
0 volts	0 0 0 0 0 0 0 0	(Zero scale)
5 volts	1 0 0 0 0 0 0 0	(Half scale)
10 volts	1 1 1 1 1 1 1 1	(Full scale)

For the bipolar case, such as −5 volts to +5 volts, the coding is a little different:

Bipolar Voltage	Binary Word (8-BITS)
−5 volts	0 0 0 0 0 0 0 0
−zero	0 1 1 1 1 1 1 1
+zero	1 0 0 0 0 0 0 0
+5 volts	1 1 1 1 1 1 1 1

In the unipolar case the coding is relatively straightforward. Here binary 00000000 represents zero volts, and the maximum binary number b11111111 is the full-scale voltage (+2.5, +5, or +10 volts). The number of possible states is merely 2^n, where n is the bit length of the binary word. For an eight bit machine, therefore, the number of possible states is 256. Because the zero volts state is represented by 00000000, that leaves 255 different levels for voltage representation. This offset coding leaves us with a maximum full-scale voltage that is lowered by the value of voltage change caused by a change in the least significant digit (LSD) of the binary word. For a 10-volt A/D converter, the $1 - $ LSD voltage is 40 millivolts, so the maximum voltage is 10 V $-$ 0.040 volts, or +9.96 volts.

There are a number of different implementations for A/D converters, far more than can be accommodated in this chapter. For additional information, see my own book *Microprocessor Interfac-*

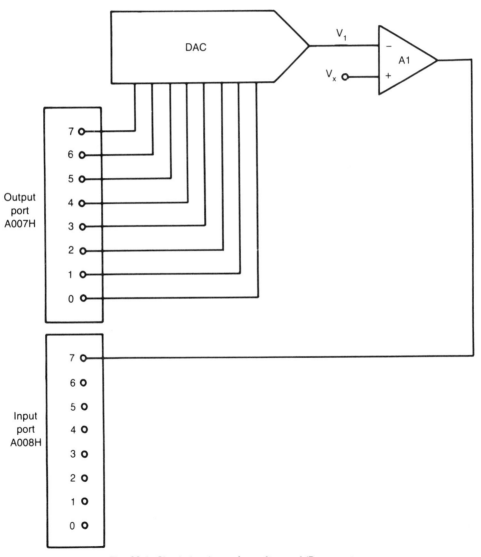

Fig. 26-1. Simple hardware for software A/D converter.

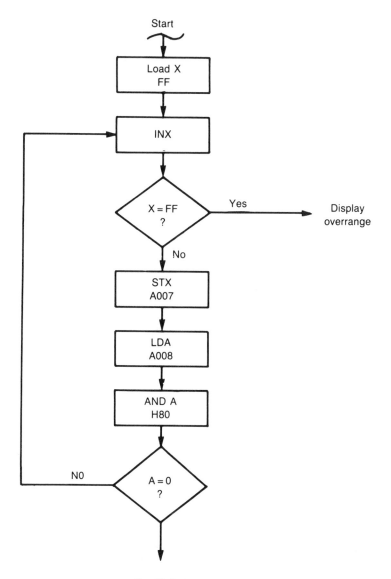

Fig. 26-2. A/D subroutine.

ing: A/D-D/A. I will, however, examine two methods that are popular with microcomputer users.

Figure 26-1 shows a simple, minimum-hardware method that uses just two integrated circuits: a digital to analog converter (DAC) and a voltage comparator (A1). The DAC produces a voltage output that is proportional to an applied eight-bit binary word. The DAC digital inputs are supplied by a computer output port.

The voltage comparator (A1) is a device that will produce an output that tells us whether or not the applied voltages are equal. In this case comparator A1 compares voltage V_1 (the DAC output) and the unknown input voltage V_x. When V1 is less than V_x, the output of A1 is HIGH. But when V_1 is equal to or greater than V_x, the output of A1 is LOW.

The output of comparator A1 is applied to a computer input port at bit 7 (MSD). When V_1 and V_x are not equal, the binary word applied to the

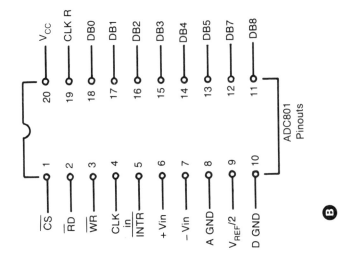

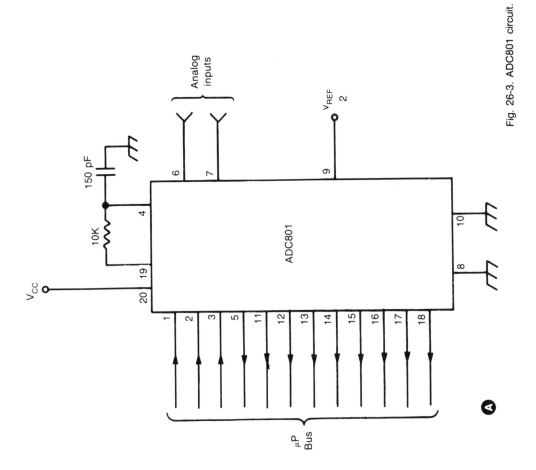

Fig. 26-3. ADC801 circuit.

port is 1xxxxxxx (where "x" denotes "don't care"), and when the two voltages are equal the word applied is 0xxxxxxx. With these facts in mind, let's take a look at Fig. 26-2 and determine how this software-controlled A/D converter will function.

Figure 26-2 shows the flow chart for a suitable A/D converter program. The idea here is to cause the output of the computer to ramp from 00000000 to 11111111, which forces the DAC voltage output to ramp from zero to full scale. In the example below we are assuming a 6502-based microcomputer in which an output port is memory mapped to address hA007 and an input is at hA008. The steps required are:

1. Load the X index register with hexadecimal number hFF (i.e. binary b11111111).

2. Increment the X register. On the initial run through the program this step brings it to h00.

3. Test for overrange. If the X register incremented to hFF, then the program must take into account that obtaining no result in the steps to follow indicates that an overrange situation occurred—and display a message to that effect.

4. Next, load the contents of the X register in memory location hA007 (which is the output port). The X register contents are thus output to the DAC, which produces a voltage proportional to the value of X register.

5. The accumulator (A register) is loaded with the value of memory location hA008, the input port. This value is 1xxxxxxx if V_1 is not equal to V_x, and 0xxxxxxx if the voltages are equal.

6. You now compare the input word from hA008 to test for the condition of the MSD. You can do this by using the AND instruction and an operand of h80 (binary b10000000). According to the rules of logical-AND operation: 0 AND 0 = 0, 0 AND 1 = 0, 1 AND 0 = 0, 1 AND 1 = 1. Thus, when 0xxxxxxx is present, the result of the AND operation is 0, and when 1xxxxxxx is present the result is 1. The test is completed by testing for zero (6502 Z-flag = 1) using the BNE instruction.

7. If the result is non-zero, then you branch back to the increment-X register point and start over until the result is zero. Alternatively, if the result is zero, the contents of the X register represent the analog voltage applied (V_x so the program falls through).

The ADC-08xx series of chips, available from several sources, are microprocessor and microcomputer compatible, and are intended specifically for the type of system discussed here. Figure 26-3 shows a basic circuit for the ADC-0801 single-channel version (ADC-0808 is an eight-channel model, and ADC-0816 is a sixteen-channel model).

Chapter 27

Selecting Data Converters

AT ONE TIME SELECTING DATA CONVERTERS WAS relatively simple: you took what was on the market, period. There was little selection unless you designed your own from scratch. Today, however, there is a wide variety of different data converters on the market in both hybrid and integrated-circuit forms. In this chapter I discuss a few of the parameters that are of interest to the largest variety of users.

The word length is the number of bits that can be accommodated by the data converter and represents the resolution of the device. The eight-bit DAC and A/D have become popular because they are 1) useful for a lot of jobs, and 2) are compatible with most modern eight-bit microprocessors and microcomputers. You can specify data converters in 4, 6, 8, 10, 11, 12, and 16-bit formats. Every time you add a bit to the word length you also double the number of gradations in the data that can be handled. The $1 - $ LSB value of a data converter is the voltage or current represented by a change in the least significant bit of the data converter. The $1 - $ LSB value is found from:

$$1 - \text{LSB} = V_{O(fs)}/2^n$$

where

$1 - $ LSB is the least significant bit voltage
$V_{O(fs)}$ is the full-scale value of V_O
n is the bit length of the binary word

For an eight-bit converter with a range of 0 to 10 volts (actually $+9.96$ volts), the $1 - $ LSB voltage is 40 mV. What the designer has to determine is whether or not the gradation represented by the eight-bit $1 - $ LSB voltage for the range given is satisfactory resolution for his or her application. If the range cannot be changed but more resolution is needed, then select a 10, 11, or 12-bit converter. Note, however, that cost increases with bit length.

Incidentally, you can always select a higher order data converter (i.e., one with more bits than is needed), and then simply ignore the lowest significant bits. For example, a 10-bit converter can be selected for an eight-bit application if the two least significant bits are ignored. This trick is sometimes done in an effort to improve accuracy.

One caution in selecting longer than eight-bit word lengths is, however, in order. The bit-length resolution should not exceed the capability of the reference supply. Put another way, selecting a long data word means that a corresponding improvement is needed in the reference supply. Buying a 12-bit A/D converter and using it with a reference supply with precision and stability specifications suitable for 8-bit operation still yields only 8-bits of resolution. When the stable reference supply drifts a substantial percentage of the $1 - $ LSB voltage, then there is no sense to using it.

Linearity is a representation of the faithfulness of the output signal in representing the input signal. Linearity is usually given in percentage of full-scale, or $1 - $ LSB, whichever is greater. Of course, the smaller the percentage the more linear the data converter.

Operating speed is another key parameter. The speed required is set by the highest Fourier (frequency) component in the input waveform. Nyquist's criterion requires a conversion speed of not less than twice the highest input frequency present. For a typical pressure waveform with Fourier components up to 100 Hz, then, the conversion speed needs to be 1/100, or 0.01 seconds (10 milliseconds). Keep in mind, however, that conversion speed is not merely the speed of the A/D converter (or DAC), but also includes the external circuitry. For example, an A/D converter has a conversion speed of two microseconds (2 μs), but is connected to an input amplifier that has a 1 μS settling time. The overall conversion time is the sum of the conversion time of the A/D and the settling time of the amplifier (1 μS in this case). Specifications to consider are: 1) A/D conversion time (or DAC settling time), 2) all amplifier settling times summed, 3) settling times of any analog switches in the circuit, and 4) one additional clock period for the ambiguity period.

Chapter 28

Interfacing Data Converters to the Computer

D ATA CONVERTERS ARE DEVICES THAT PERMIT the programmable digital computer to interface with other devices and systems. The analog-to-digital converter (ADC) converts analog voltages and currents to binary words; the digital-to-analog converter (DAC) converts binary words to analog voltages or currents. The details of these devices are covered elsewhere in this book, so here I will discuss only in the most general terms how they interface with the computer.

A/D CONVERTERS

Figure 28-1 shows a block-form generalization for the A/D converter. Connections B0–B7 are the binary output lines; in this example the A/D is an eight-bit model, so there are eight parallel lines. Some A/D converters also have a serial output line in addition to the parallel lines. Some models also "tri-state" the output lines. This means that they float at high impedance until and unless the enable line (EN) goes active (LOW in this example). When the EN line is active, the converted data are valid and so are transferred to the output lines. In still other A/D converters, the output data lines are latched; that is, they retain the last valid data presented until an update is provided by the internal circuitry.

The analog inputs shown here are differential, although some A/D converters have single-ended lines. The start line (ST) causes the conversion process to commence when made active; the end-of-conversion (EOC) line is an output signal that tells the world that new data is available. A clock (CLK) input provides timing, although some A/D converters have internal clocks (and a few types don't need it). In some cases the EOC and ST lines are replaced with a single BUSY or VALID signal.

Figure 28-2 shows a general timing diagram for A/D converters. In this example all action is synchronized by the CLOCK line. The conversion process begins when the START line goes LOW; from here on in the process is internal. The output lines B0–B7 will remain either in the high-impedance tri-state condition or latched to the formerly valid data. When the conversion process is completed, the outside world is notified by the EOC line dropping LOW. For a short time prior to EOC, and for a short time after EOC, the new data on the data bus is valid, and can be input to the computer (note: if the data is latched, then the data remains valid until the next EOC pulse is received). The

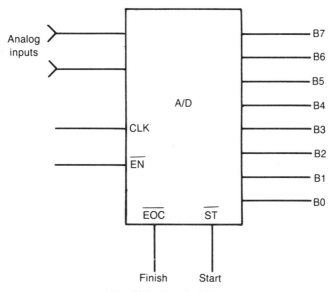

Fig. 28-1. Generic A/D.

process begins again on the next following clock pulse.

Figure 28-3 shows a simplified A/D interfacing scheme. In this case the eight output lines of the A/D converter are applied to an eight-bit parallel input on the computer (not to be confused with the parallel printer port). One bit of an output port is used to trigger the operation (the other bits remain useful for other jobs, incidentally). When a positive-going output pulse is programmed to bit B0 of the output port, it is inverted by the NOR gate, and applied to the ST input of the converter. After this initial kick in the pants, the process becomes asynchronous. The EOC pulse is also

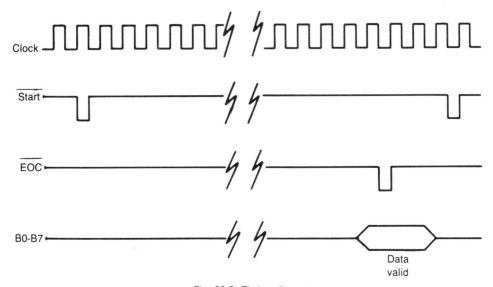

Fig. 28-2. Timing diagram.

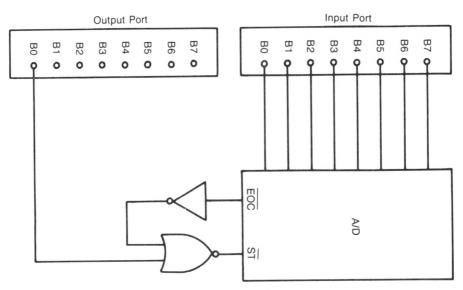

Fig. 28-3. A/D interfacing.

NOR-gated into the ST and so will restart the process when it is completed.

One danger of this method is that you must asynchronously input data and may accidentally input erroneous data during the brief DATA INVALID period at the transition from ST to EOC.

A more sophisticated variant is shown in Fig. 28-4. In this case a data latch is interposed between the A/D and the computer input port (in some ADCs, the data latch is included). An initial RESET pulse is applied to the system, and it resets the control FF, and starts the conversion. When the EOC pulse is received, it sets the control FF, causing data from the ADC to be transferred to the latch; the EOC pulse also restarts the converter. A separate input port line can be used to sample the status of the control FF to allow the computer to know whether or not the data is valid.

All data converter control is made easier if the input and output ports of the computer are programmable as either inputs or outputs on a bit-for-bit basis. Several interface chips are available that do this trick, such as the 6522 used with 6502 computers (e.g., KIM-1, AIM-65 and Apple IIe), and the 6526 used with 6510-based computers (e.g., Commodore 64).

DAC INTERFACING

The normal output circuitry of a programmable digital computer is not at all compatible with analog inputs. What is needed is a circuit to interface the binary output of the computer to the analog input of the peripherals. Figure 28-5 shows such a converter circuit.

The input of the circuit of Fig. 28-5 is an eight-bit binary word from the output port of the computer. The eight-bit word is applied directly to the digital inputs of a digital to analog converter (DAC). The DAC is a special IC or circuit that converts the binary word into a proportional voltage or current over a selected range. The DAC-08 IC is made by Precision Monolithics, Inc., National Semiconductors and others, and is very popular (it is usually available from both commercial and hobbyist suppliers). This particular DAC is designed to output a current between 0.400 milliamperes (mA) and 2 mA. The actual input current will be

$$I_0 = \frac{V_{REF} \times A}{256 \times R3}$$

where

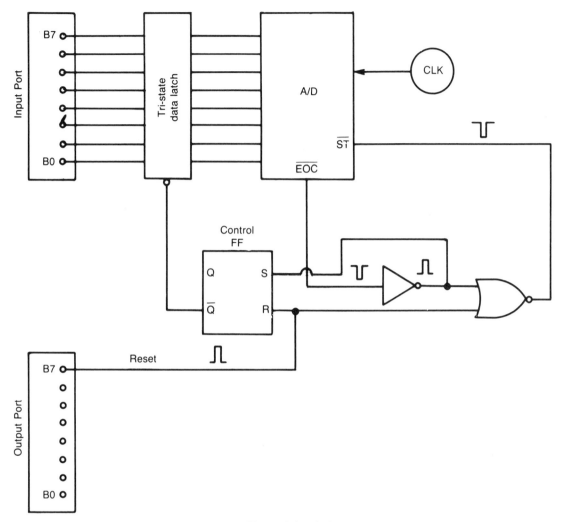

Fig. 28-4. A/D interfacing.

I_O is the output current in milliamperes

V_{REF} is the reference voltage in volts

A is the decimal equivalent of the binary input word

R3 is the precision input resistor (Fig. 28-5) in kilohms

The factor "256" is the decimal equivalent of the maximum allowable binary value (11111111 in an 8-bit machine)

In Figure 28-5 the output current from the DAC-08 is converted to a ground referenced volt-

age by operational amplifier A1. The output voltage of this amplifier is the product of the current and the feedback resistance:

$$V_O = I_O \times R_F$$

where

V_O is the output potential in volts

I_O is the DAC output current in mA

R_f is the feedback resistor in kohms

The capacitor shunting the feedback resistor at amplifier A1 is used to provide a small degree of low-pass filtering. The capacitor rolls off high-frequency components (such as the DAC steps on a continuous waveform) at a -6 dB per octave rate starting at a -3 dB frequency of

$$f = 1,000,000/6.28RC$$
$$f = 1,000,000/((6.28)\ (5000)\ (0.01\ \mu F)$$
$$f = 3,185\ Hz.$$

Because the value of the feedback resistor is fixed by other considerations (see output voltage equation above), you can vary the -3 dB roll-off point by increasing or decreasing the capacitance according to the following equation:

$$C = 159,236/Rf$$

or,

$$C = 32/f\ (if\ R3 = 5000\ ohms)$$

where

C is the capacitance in microfarads
f is the roll-off point in Hertz
R is the value of R3 in ohms

Potentiometer R1 in Fig. 28-5 is used as either a position control or a zero control. In either case you set the input binary word to zero (i.e., 00000000 in unipolar DACs as shown, or 10000000 for bipolar DACs) and then adjust the potentiometer for zero volts at the output of either A1 or A2.

Operational amplifier A2 is used as a low-pass

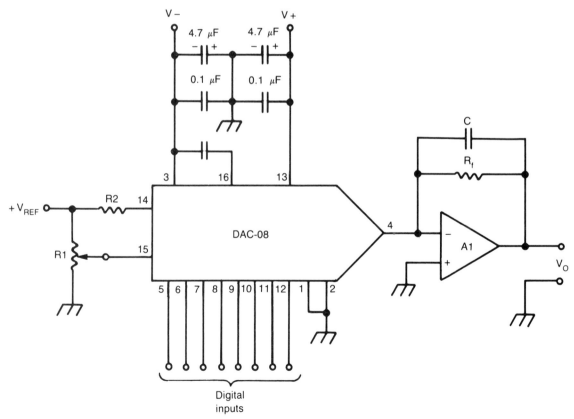

Fig. 28-5. DAC-08 interfacing.

filter. This particular design is set for a −3 dB roll-off frequency of 1000 Hz, and because it is a second order-filter it rolls off at a rate of −12 dB/octave (−20 dB/decade). A textbook on active operational-amplifier filters will discuss how to set the values of the resistors and capacitors to alter the cutoff frequencies.

The purpose of the low-pass filter and the capacitor shunting the feedback resistor is to smooth the output waveform. Because the amplifiers get their input signal from a DAC, it will be a ragged signal in which the value of each step is the voltage change equal to a change of the least significant bit of the computer output word. This value is called the "1 − LSB voltage."

Figure 28-6 shows a so-called "universal rear-end" circuit that can be used to follow the DAC

circuit shown earlier (or any other DAC circuit). This circuit uses a 1458 dual operational amplifier to control the input signal applied to the analog peripheral. Operational amplifier A1a is an inverting follower that has a gain variable over 0 to 1. A screwdriver adjustment DC BALANCE control is also part of this stage. The DC BALANCE control is used to cancel the effects of dc offset in the preceding stages. At a time when those voltages are supposed to be zero, there may be a certain dc component to the signal. Adjust the DC BALANCE control until there is no shift in the output voltage (as indicated on the oscilloscope or chart recorder) as the gain control is varied from zero to fullscale.

The second stage of the circuit in Fig. 28-6 is also a unity gain inverting follower. This stage is the output section and also contains the position

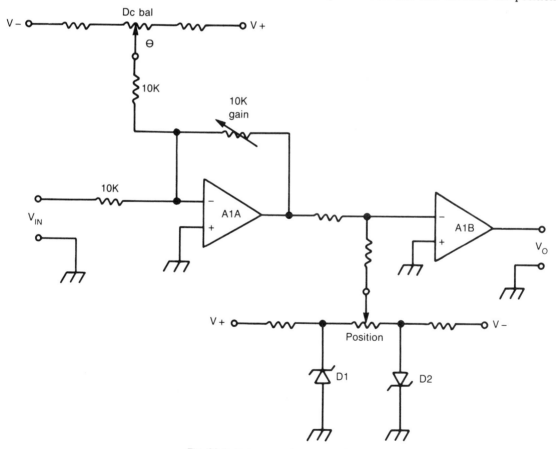

Fig. 28-6. Data-converter output circuit.

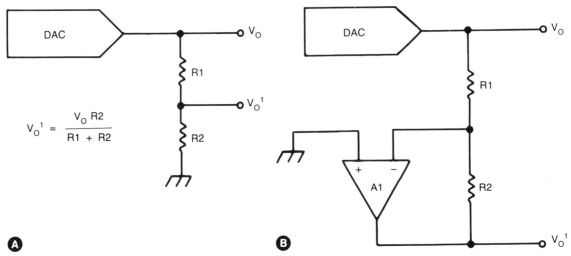

$$V_o^1 = \frac{V_o\,R2}{R1 + R2}$$

Fig. 28-7. Reduced output DACs.

control. This control allows you to position the beam or pen anywhere on the screen or chart of the display device. In some cases another DAC is used to replace the potentiometer and allows you to position the beam or pen under program control. Zener diodes D1 and D2 are used to limit the output voltage swing of the operational to a convenient level. In most cases you will want to limit the beam or pen swing to either exactly maximum scale or just off scale, but not far off scale.

There are times when the output of the DAC is not easily converted to the fixed input of the display device. For example, the digital panel meter (DPM) may require 0 to 1.999 volts, and the DAC outputs 0 to 9.96 volts. Some display oscilloscopes have few controls and a fixed input range. These oscilloscopes are intended as OEM models that other makers install in their equipment, not as general-purpose models. One popular OEM model uses 50 mV/cm deflection factor. If you want to interface with it, you have to limit the voltage to 400 mV for the standard 8-cm display screen. In still other cases you will find certain strip chart and X-Y recorders that accept only 1 mV or 10 mV maximum input voltages. You will clearly need some type of signal attenuator to accommodate these analog peripherals.

Figure 28-7A shows the output of the DAC (V_o) reduced to a lower level (V'_o) by a resistor voltage divider. The output voltage of the voltage divider is given by

$$V' = \frac{V_o \times R2}{R1 + R2}$$

Example. A certain DAC has an output of 0 to 10 volts, while the X-Y chart recorder it drives wants to see a maximum input voltage of 10 mV. Further, the value of R2 should be limited to 100 ohms. Find the value of R1 to make the attenuation required.

Rearrange the equation above to find R1:

R1 = $(V_oR2 - V'R2)/V'$
R1 = [(10 V) (100) − (0.010 V) (100)]/(0.010)
R1 = 99,900 ohms

An active attenuator is shown in Fig. 28-7B. This circuit is a simple inverting follower in which the feedback resistor is smaller than the input resistor. For example, if you need a reduction of 1/10, you could make R2 = R1/10.

Chapter 29

Computers in Data Acquisition, Instrumentation, and Control Applications

I T IS TRUE THAT THE MICROCOMPUTER IS A SUPE-rior means for implementing data acquisition, control, and instrumentation circuits. What makes the microprocessor and microcomputer so superior to other methods? After all, isn't the microprocessor just another integrated circuit? Well, I suppose you could say that is true if you are also willing to claim that a 500-lb Bengal Tiger is "just another cat!" The microcomputer is a full-fledged program-mable digital computer. In some cases, the chip inside is a microprocessor (e.g., Z80 or 6502), while in others the chip it is a "single-chip com-puter" (e.g., 8048). Microprocessors become mi-crocomputers by adding external memory and I/O circuitry.

While many readers are familiar with the busi-ness of solving instrumentation and control data acquisition problems using analog circuits or ordi-nary digital logic circuits and devices such as relays and switches, they often find that the microproces-sor is a little different. There are two main goals in microprocessor-based instrumentation:

1. Replacement of digital logic devices with software, and

2. Replacement of some (as many as possible) analog circuits with software.

Before there are too many howls of anquished protest over the second point, let me hasten to point out — and dispense with — a myth about digi-tal instrumentation that is sometimes held by ana-log circuit designers. It is often claimed that the digital implementation contains a built-in error, i.e. the so-called discrete quantization error. This error is, indeed, inherent in the digital implemen-tation of any circuit function. It is absolutely true that digitization forces us to inherit a certain basal error rate. But the argument that analog circuitry is therefore somehow "more accurate" because it permits an infinite number of discrete values be-tween limits is not supportable. Analog circuits also contain substantial errors: amplifier gain error, off-set voltage error, temperature drift, aging, and readout or display resolution, for example. The digital implementation of a circuit function often produces a superior accuracy because it eliminates the analog errors while keeping the digitization errors low and identifiable. Incidentally, the digi-talization error is reduced by a factor of two for every bit that is added to the data word length.

Another myth is the notion that, by using a computer, all of our problems are solved. It is a common fallacy that computers solve problems: this is a misconception—computers do not solve problems—a computer is merely a tool. The computer can no more "solve a problem" than a saw or hammer can "build a house." In designing electronic instrumentation using microprocessor and microcomputer techniques, you must first solve the problem at least at the flow-chart level. Only after you have produced a step-by-step plan of action (i.e., created an algorithm) that will do the job can you begin to select the components that will be used. Just as the analog-circuit designer begins with a block-diagram solution of his problem on paper, the microcomputer-oriented designer must begin with an algorithm on paper and, sometimes, a system block diagram.

You can see this process better by considering a trivial example: a simple traffic-light controller. Let's assume that the design specifications require the following operation:

1. East-west green for 30 seconds, while north-south is red;
2. East-west yellow for 7 seconds, while north-south is red;
3. East-west is red for 30 seconds, while north-south is green;
4. East-west is red for 7 seconds, while north-south is yellow;
5. Steps 1 to 4 are repeated indefinitely.

Figure 29-1 shows a proposed solution to this problem. You write a program that will cause the various on-off states for the lights to occur at appropriate times and for appropriate durations.

Next, you must allocate your resources. Suppose that you are going to use a 6502-based computer or controller, with a 6522 PIA for I/O interfacing. This arrangement will allow you to use the Rockwell AIM-65 computer as a development system (Apple-IIe will also work). This approach is especially useful if you use the same memory address allocations for the 6522 PIA as on the AIM-65 (see Table 29-1).

For those who are unfamiliar with the 6522 and the AIM-65, I will discuss the I/O protocol. The 6522 (which is used for I/O in the AIM-65 and other "similar to KIM-1 computers) has two I/O ports, designated "A" and "B." These ports are treated as memory locations because the 6502 microprocessor uses memory-mapped I/O. In the AIM-65 these ports are assigned to memory locations $A000 (port B) and $A001 (port A).

The I/O registers in the 6502 are bidirectional, meaning that they may be used as either input or output ports. The direction is controllable on a bit-for-bit basis (e.g., BA0 can be an input, while BA1 is an output). The direction of a bit in either port is set by the bit applied to the corresponding position in a data direction register (DDRA and DDRB). A HIGH written to a DDR bit will make the corresponding bit in the I/O port an output; a LOW in the DDR bit will make the port bit an input. In the AIM-65 DDRA is located at $A003, while DDRB is located at $A002. By writing $00 to $A002, you will make the entire port B an input. If you had written $FF to $A002, however, you would make all of port B an output port.

Now back to our traffic light controller. Figure 29-2 shows a circuit to connect the lights (which, in this example, are represented by LEDs) to an AIM-65 microcomputer. Each LED is driven by an open-collector TTL inverter (one section of a 7405 and 7406 hex inverter). When the input of an inverter is HIGH, then its output is LOW, which turns on the LED by completing its ground path. A HIGH applied to the corresponding port-A output bit will turn on the LED. For example, writing 00100000 ($20) to port A will make bit B5 HIGH, turning on LED D6 (N-S green). The idea is to write a program for the AIM-65 that will place a HIGH on those output bits in port A that connect to the lights that must be turned on, and a LOW to all other bits. For example, "state 1" requires R2 and G1 to be on, and all others off. The binary word for this state is 00001100 (i.e., $0C); see Figure 29-3. This state is held for 30 seconds, and then "state 2" is generated. Table 29-2 gives some programming exercises. If you have some I/O port equipped computer other than the AIM-65, then

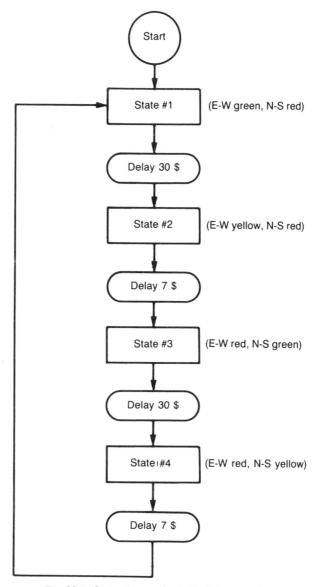

Fig. 29-1. State diagram for traffic light controller.

these exercises may be easily adapted. Figures 29-4 and 29-5 apply.

COMPUTER INTEGRATION AND DIFFERENTIATION

Some instruments or control systems require either integration or differentiation of a signal waveform. Let's briefly discuss computer solutions to these problems. The mathematical processes of integration and differentiation are fundamental to the operation of many instruments. In analog instrument designs these processes are carried out using appropriate operational-amplifier circuits. In digital instruments, however, you must use certain "numerical methods" of integration and differentiation. You must also be aware of certain anoma-

**Table 29-1. Memory
Address Allocations for the 6522 PIA.**

Location	Function
A000	Port-B output data register (ORB)
A001	Port-A output data register (ORA)
A002	Port-B data direction register (DDRB)
A003	Port-A data direction register (DDRA)
A004	T1 (write T1L-L)
A005	T1 (write T1L-H)
A006	T1 (write T1L-L)
A007	T1 (write T1L-H)
A008	T2 (write T2L-L)
A00A	T2 (write T2C-H)
A00B	Auxiliary control register (ACR)
A00C	Peripheral control register (PCR)
A00D	Interrupt flag register (FR)
A00E	Interrupt enable register (IER)
A00F	Port-A output data register (ORA)

The reader is directed to the programs in Appendix A for programs that will integrate or differentiate known functions.

The process of differentiation is used to find the instantaneous rate of change of a function. In the computerized instrument a voltage input function may be converted to a binary word by an analog-to-digital converter, and it is on the binary data that the machine operates.

Figure 29-6 shows a crude method of finding the derivative of an input signal. We can use the natural clock cycle time of the computer for T2-T1 data, or, use a software timer to create longer periods. The A/D converter will take two samples of the input signal, Y1 and Y2. The derivative is approximated by:

$$\Delta Y/\Delta T = (Y_2 - Y_1)/(t_2 - t_1)$$

provided that t (i.e., $t_2 - t_1$) is not too long.

To make a continuous derivative output, take

lies of the data that would ruin the process. In this section I will consider the more elementary numerical methods, with the understanding that the wise reader will either take a course or read a book on the subject if he needs additional information.

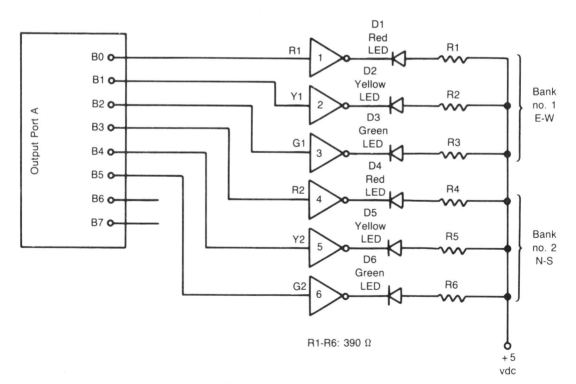

Fig. 29-2. LED driver interfacing for light problem.

Lamps Bits	— B7	— B6	G2 B5	Y2 B4	R2 B3	G1 B2	Y1 B1	R1 B0	State Code at Output	
State No			← N-S →			← E-W →			Binary Word	Hex
1	X	X	L	L	H	H	L	L	00001100	0C
2	X	X	L	L	H	L	H	L	00001010	0A
3	X	X	H	L	L	L	L	H	00100001	21
4	X	X	L	H	L	L	L	H	00010001	11

H = HIGH (i.e., logical 1)
L = LOW (i.e., logical 0)
X = don't care (assign logical 0 for convenience)

Fig. 29-3. State chart.

continuous samples of the signal Y_n, . . . Y_3, Y_2, Y_1) and make the calculations $(Y_2 - Y_1)$, $(Y_3 - Y_2)$, . . . $(Y_n - Y_{n-1})$. The time factor, t, will be the same for each period and cannot be faster than the cycle time of the A/D converter.

If an application requires an analog output that is proportional to the derivative of an analog input signal, then output the results of the calculation to a D/A converter (DAC). This circuit will produce an analog voltage that is proportional to the derivative of the analog input.

One problem faced in designing data-acquisition systems for chemistry and the life sciences is the long amounts of time required for some of the processes to take place. Changes in physiological (living) systems, for example, often take seconds,

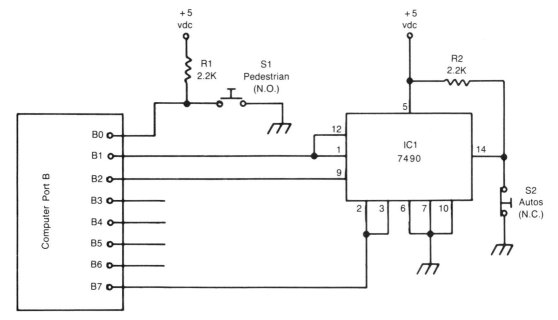

Fig. 29-4. Pedestrian light circuit.

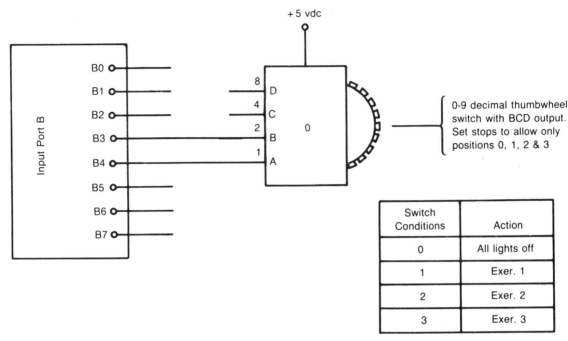

Switch Conditions	Action
0	All lights off
1	Exer. 1
2	Exer. 2
3	Exer. 3

Fig. 29-5. Thumbwheel circuit.

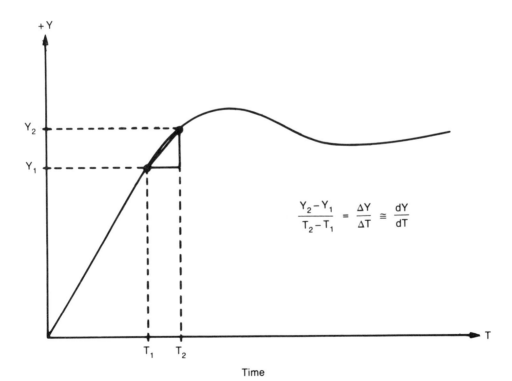

$$\frac{Y_2 - Y_1}{T_2 - T_1} = \frac{\Delta Y}{\Delta T} \cong \frac{dY}{dT}$$

Time

Fig. 29-6. Differentiation.

minutes or hours, rather than the shorter periods that electronics engineers normally see. A slow signal is difficult to differentiate in operational-amplifier differentiator circuits. In fact, differentiation is often impossible in such cases. Active differentiators require operational amplifiers to work. The input bias currents of the operational amplifier will eventually charge the input capacitor used for differentiation, causing the operational amplifier to latch up. Long before the amplifier latches up, however, the input offset bias current will create a substantial artifact in the data. The microprocessor, however, can be used to make a differentiator for slow signals. An A/D converter for the μP input, and a DAC for the output, make the instrument look like an analog device.

Integration can also be performed using a microprocessor. The "computer" version is a bit more accurate than typical analog integrators that use an operational amplifier. The problem with the analog circuit is that output bias voltages tend to charge the feedback capacitor without regard for the input signal. Figures 29-7A and 29-7B show two methods for making digital integrators.

Method 1 (Fig. 29-7A) is the more crude, but it is sufficient for many applications. The A/D converter will input a series of voltage values to the computer: V_1, V_2, V_3, V_4, V_5, etc. This will divide the area under the waveform into a series of individual rectangles. The sum of these rectangles areas is the integral of the waveform over the time of measurement. If the time periods are equal for all samples, then we can multiply each input value (V) by t. Then, by summing these values in a holding register somewhere, you will have the integral.

But there can be a significant error in the measurement if the time period is too great. Very often, you will find that some factor makes it difficult to make the time period for each rectangle sufficiently small to permit accurate integration. The shaded area of Fig. 29-7A shows the amount of error.

A more accurate method of integration is shown in Fig. 29-7B. You can include those errant shaded areas by using a slightly more complex formula for making the integration. In the case of

Fig. 29-7B, the measurement of the rectangles is made in the manner of the previous figure and added to the area of the remaining triangle.

AN INSTRUMENTATION EXAMPLE: CARDIAC OUTPUT COMPUTERS

The cardiac output computer is an instrument that measures the pumping capacity of the heart in liters per minute. In Chapter 11 I discussed the anatomy and physiology of this measurement and so will not repeat that discussion here except for a brief reiteration of the technique of the measurement. The purpose here is to show a digital-computer implementation of the cardiac output computer.

The modern way to measure CO is to inject cold saline solution into the right atrium of the heart by way of a special catheter, and then measure the temperature profile at the output of the right ventricle. The thermistor tip of the catheter is placed in the pulmonary artery by deft manipulation by the physician. The doctor does this neat trick by inserting the thin, multi-lumen thermistor-tipped catheter into the heart. The catheter is usually inserted through an opening cut into the vein in the patient's right arm, and is then threaded through the veins into the atrium, ventricle and then into the pulmonary artery. The thermistor will measure the temperature of the blood in the pulmonary artery. There is an opening in the catheter at the point where it will allow saline solution injectate from an external syringe to enter the bloodstream just before the atrium (see Fig. 29-8).

Figure 29-9 shows the concentration of cold saline in the pulmonary artery, as measured by the changing temperature of the blood. The concentration curve is merely the temperature curve upside-down.

Also shown in Fig. 29-9 is the simplified formula for calculating cardiac output. In one model, $K_1 = 60$, $K_2 = 1.08$ and $K_3 = 52.5$, so the product of these three is 3407. The V_I term is the saline injectate volume in liters. Since 10 mL is almost standard, V_I is 0.010. The term T_B is the blood temperature in degrees celsius (°C) before injection of the saline, usually around 37 degrees.

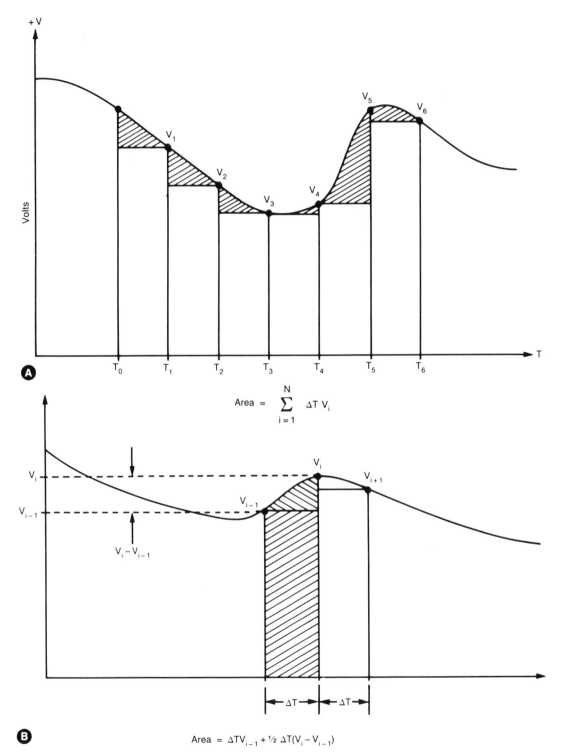

$$\text{Area} = \sum_{i=1}^{N} \Delta T\, V_i$$

$$\text{Area} = \Delta T V_{i-1} + \tfrac{1}{2}\, \Delta T (V_i - V_{i-1})$$

Fig. 29-7. Simple integration.

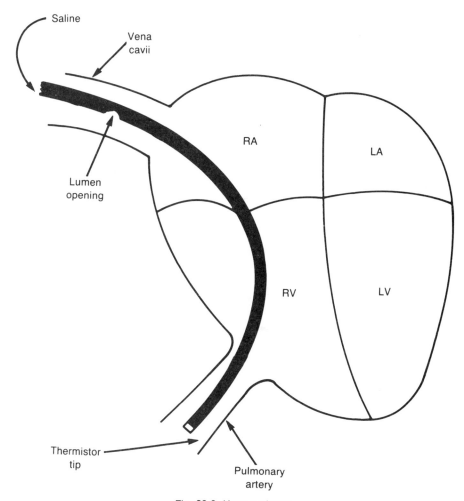

Fig. 29-8. Heart catheter.

T_I is the temperature of the injectate (saline), which is 25 degrees if room-temperature saline is used, and 0 degrees if iced saline is used. Our term T_B is the blood temperature after the saline is injected. For our simplified case, which approximates actual clinical conditions, these are sufficient to measure cardiac output.

The concentration curve of Fig. 29-9 can be broken into two segments, $(t_1 - t_2)$ and $(t_2 - t_3)$. The latter is assumed to be an exponential decay and is entered when the concentration falls off to 85 percent of the peak value. Unfortunately, some of the blood may recirculate (come around twice) before the measurement is completed, so an error term (artifact) is introduced. This defect is why you must use geometric integration for $(t_2 - t_3)$.

Now let's get down to business. The old style of cardiac output computers were nothing more than analog computers that were permanently programmed to solve some version or another of the standard cardiac-output equation of Fig. 29-9. Such a statement could be made of almost any similar analog instrument.

Typically, such an instrument consisted of several operational-amplifier circuits, including a subtractor, an integrator, and at least three analog dividers. These circuits all suffer from the same error-producing defect: thermal drift. I have seen accumulated integrator offset error (drift-caused!) of 30-percent in analog CO computers. Using a digital computer to perform these functions eliminates such errors.

Figure 29-10 shows a hypothetical CO computer implemented with a small microcomputer. The actual computer could be almost any small computer that has at least two parallel output ports and a small amount of both RAM and ROM.

The analog subsystem in this case is reduced to an A/D converter, an isolation amplifier, and a resistance Wheatstone bridge, of which the thermistor in the catheter tip is in one arm. The bridge produces a small voltage (on the order of 1 to 2 mV/°C), which is amplified by A1. This amplifier scales the bridge voltage to match the input voltage range of the A/D converter. It is an isolation amplifier in order to ensure safety—accidental electrocution of the patient is not a good trick to pull! The A/D converter produces an n-bit binary word that is proportional to the amplifier output voltage and hence the temperature of the probe.

An input port on the computer receives the A/D output data. The computer knows that new data is present because a HIGH appears on bit B1 of input port A. A single bit of an output port (B7) is used to form a start pulse. Because bit B7 is used, you can create the start pulse by writing any word to port B in which B7 is HIGH (e.g., $80, which is 10000000 in binary, fills the bill).

The computer knows to begin the process when switch S1 is closed. Normally, bit B7 of input port A is HIGH because switch S1 is open; closing S1 forces bit B7 LOW.

Let's look at a typical program sequence (Fig. 29-11) for the start button, S1. In this and other sequences I will use 6502 microprocessor mnemonics and terminology. Keep in mind that the idea is to detect a LOW on bit B7 of port A.

The first step is the "power-up routine." This program segment might be only the automatic "power-on reset pulse," or a certain amount of initialization and housekeeping chores. The purpose of the reset pulse is to force the computer to begin executing the program at the correct point. In some (e.g., Z80) the reset pulse causes the

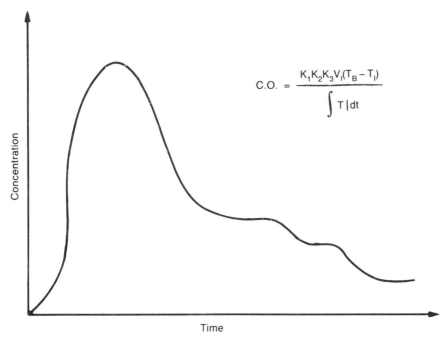

$$\text{C.O.} = \frac{K_1 K_2 K_3 V_I (T_B - T_I)}{\int T \, I \, dt}$$

Fig. 29-9. Cardiac-output waveform.

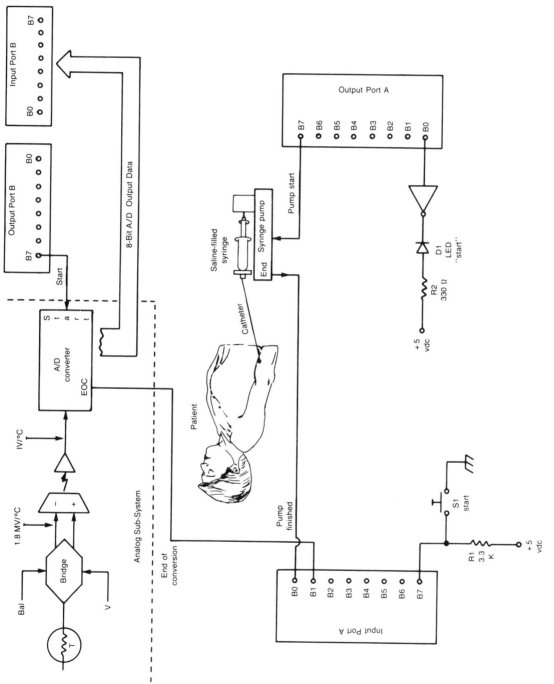

Fig. 29-10. Cardiac-output system.

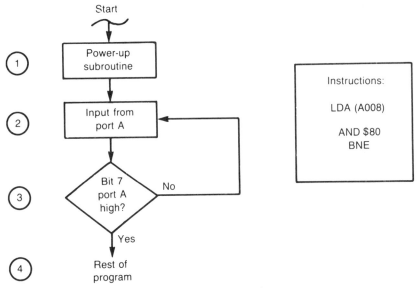

Fig. 29-11. Program routine.

computer JUMP to location $0000, while on 6502-based machines the JUMP occurs to locations $FFFC/$FFFD (this is a so-called "vectored jump"). In either case, the net result is to initialize the program at the location of the first instruction.

The next step is to input data word "A" and determine whether or not bit B7 is HIGH. If bit B7 is HIGH (i.e., 1), then the switch is open and no action takes place. Otherwise, the program "falls through" to the next step.

It is instructive to go through this program step-by-step.

1. The first step is to fetch the data from input port A. In this computer port A is memory-mapped to location $A008. Data is input to the accumulator.

2. Mask all but bit B7 by using a logical AND instruction between the contents of the accumulator with the hexadecimal number $80 (i.e., binary 10000000). According to the rules for AND operations, 1 AND 0, 0 AND 1, or 0 AND 0 always result in 0; only 1 AND 1 always yields 1. Thus, the accumulator contents will be $80 if bit B7 is 1, and $00 if bit B7 is 0.

3. The BNE instruction is "branch on result not equal to zero." In step number 2, you find that

"ANDed-A" will be $80 (non-zero) if the switch is open, and $00 if the switch is closed. If the switch is open, therefore, the program branches back to step number 1 and inputs another word from port A (the program "loops"). If, on the other hand, the switch is closed (indicating a <u>start</u> command from the user) the result of the branching instruction is zero, so the program "falls through" to the next instruction at step no. 4.

4. Step no. 4 will be whatever is next in the sequence. In this case it will be to start the syringe pump.

The next operation is to begin infusing the cool saline injectate into the patient. Because our pump is designed to work with a computer, it will turn on when it receives a TTL HIGH on the <u>pump control</u> line (i.e., bit B7 of input port A, which is located at memory location $A007). You also want to turn on a panel lamp (LED D1) to let the doctor know that the pump is computing. Because the pump requires a HIGH on B7 and the LED requires a HIGH on B0, you need to output to port A any binary word that makes these bits HIGH (10000001 in binary, or $81, fills the bill).

You must also provide a means for telling the computer when the pump is finished. In some jobs

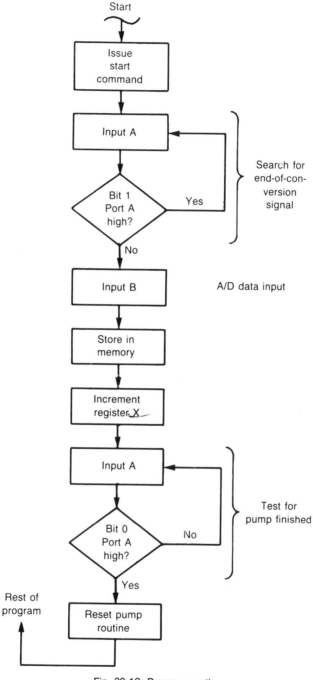

Fig. 29-12. Program routine.

you could assume a constant pump speed, and have the computer wait the correct period of time. Our pump, however, provides a "pump finished" signal to bit B0 of input port A. You need to include in the pump program a portion that looks for a HIGH on bit B0, similar to the previous program.

You must also provide a 1-millisecond start pulse to the A/D converter and arrange to look for the end-of-conversion (EOC) pulse on bit B1 of port A. When the EOC signal is received, you must input the A/D data from port B and store it. For the purposes of this discussion a 1 mS-timer subroutine at location $EF05 is assumed, and that only 256 bytes of data are to be input and stored in page $03 of memory. In a "real" product several pages of data are stored, but that complicates our simple program. The flow chart for our program is shown in Fig. 29-12, while a sample program is shown in Fig. 29-13.

Again, it is useful to go through the program step-by-step.

1. The first step is to issue a <u>start</u> command to the pump and LED by writing #$81 to output port A at memory location $A007. This is done by loading $81 into the accumulator (LDA $81) and then storing it at $A007. Because the port is the usual latching type, the data ($81) remains on port A until some program resets it.

2. This step, and steps 3 and 4, are used to generate a 1-millisecond <u>start</u> pulse to the A/D converter. The A/D start input is connected to bit B7 of output port B at location $A005. In this step bit B7 is made HIGH by loading #$80 into the accumulator (LDA #$80) and then storing it at location $A005.

3. Program "jumps to subroutine" (JSR) that generates a 1-millisecond delay.

4. Bit B7 of port B is now brought LOW in order to reset the pulse. This job is done by loading zeroes (i.e., LDA #$00) into the accumulator and then storing it at $A005.

5. Data from port A is input to the accumulator.

6. Data input from port A is masked for bit B1 by performing a logical AND operation between the accumulator and #$02 (i.e., 00000010 in binary). The result of this operation is used in the next step to determine the direction taken by program control. Until the conversion is finished, bit B1 will be 0, but when the A/D is finished to EOC line connected to bit B1 will go HIGH.

7. This step is the comparison that determines the condition generated in step no. 6. The BEQ

instruction is "branch on result equal to zero." If the result is zero, indicating that the A/D converter is not finished, then control branches backward seven lines (line 0210) to input another "A." If the result is non-zero, then the program "falls through" to step number 8.

8. Here the data input from the A/D converter is stored in memory for later use. The location is determined by an indexed-X absolute addressing scheme that will place data in a different sequential page $03 location for each iteration.

9. The INX instruction increments register X to accommodate step number 8 on the next iteration of the program. Register X requires initialization to $00 when the computer is first turned on, and will increment from $00 to $FF, one step at a time (the page $03 addresses, therefore, are $0300 to $03FF).

10. This step seeks to determine if the pump is finished by looking for a HIGH on bit B0. The program is similar to previous routines. The first step in this sequence is to input port A.

11. Port-A data is masked for bit B0 by logical-AND operation between the accumulator data and #$01. If the result is zero, then the injectate pump is still working; if the result is non-zero then the pump is finished. Program branches back to line 0205 if not finished, otherwise it falls through.

1. LDA #s81
   STA A007
2. LDA #s80
   STA A005
3. JSR EF05
4. LDA #s00
   STA A005
5. LDA (A004)
6. AND #s02
7. BEQ (−7)
8. LDA (A006)
   STA, X + s0300
9. INX
10. LDA A005
11. AND #s01
12. BEQ (−7)
13. LDA #s00
    STA A007
14. (Rest of program)

Fig. 29-13. Typical coding.

12. Branch instruction that tests data and implements the action given in step number 11.

13. Resets the pump and LED to off condition by loading $00 into the accumulator and storing it at location $A007.

14. The remainder of the program calculates and displays cardiac output.

There are other protocols that will produce the same result, of course, ours is merely the simplest. Other methods might use different means for controlling the pump. In many, perhaps most, designs, the entire injectate volume is delivered at once before the A/D converter iterations begin. Our reason for using the method we did was to demonstrate condition testing.

SIMPLE EVOKED-POTENTIALS COMPUTER

The evoked potentials computer makes a good study in simple, low-cost approaches to a formerly expensive (and therefore generally inaccessible) data-acquisition chore. It deals with data collection, signal averaging and display.

Brainwave (EEG) signals are a crude form of data by which the physician or researcher may infer a great deal by looking only at the scalp signal. Waveform elements cannot easily be attributed to specific stimulii. An analogy is the "Astrodome Metaphor." Consider a martian who wants to understand human beings. His first contact on Earth is the roof of the Houston Astrodome during a football game. By inserting a microphone just inside the hole at the peak of the dome he picks up a cacophony of shouts and screams. This signal is analogous to the scalp brainwave signal. He next lowers a directional microphone to a location only a few feet above one section of bleachers. Here the sounds resolve into a collection of different shouts, and it becomes possible to discern some individual features. This data is analogous to the medical researcher's data obtained with a microelectrode surgically embedded into a region of the brain. Finally, the martian lowers the microphone right in front of a food vendor, and notes a repetition of the data: "Hot dogs, hot dogs, hot dogs!" He then turns on a trigger device that only enables the tape recorder when the microphone picks up sound from this one source—that recording is analogous (in a rough way) to evoked potentials.

The scale signal contains elements from too many sources, so the element that is due to any specific stimulus is lost in the cacophony. But if we use evoked potentials methods, then we can discern that portion of the waveform that is due to a specific, repeated, stimulus.

In evoked potentials studies, the stimulus (e.g., a light flash) is repeated over and over (64 to 200 times, usually). Following each flash of the light, a group of 256 or more sequential voltage samples are taken. These data are coherently averaged such that each data point is either averaged with or added to the datum taken at the same instant following previous flashes. Thus, the sample taken 108 milliseconds after the present flash is either averaged with or added to data taken 108 milliseconds following all previous flashes. Interestingly enough, it doesn't matter much whether the signal is time averaged with old data, or simply added to old data. While the display will look a little different as the data is built up, the final waveform will look the same.

Evoked potentials recording is a form of data acquisition in medical and biological research in which the personal computer has made a tremendous difference in accessibility and cost. Only a decade or so ago, the laboratory that wanted an evoked potentials computer had to part with $20,000, or more, and be stuck with a single-purpose, dedicated machine that would not perform any other chores. Today, researchers and clinicians can buy simple hardware interfaces to IBM-PC and Apple II computers that allow connection to an EEG amplifier and strip-chart recorder (or oscilloscope). Most of these interfaces plug in to one of the expansion slots in the selected computer. Software will then perform the evoked potentials control and signal-averaging tasks, plus still be available for performing the statistical analysis—and then can be used as a word processor to write the doctoral dissertation or scholarly paper that results from the tests. Such a system costs only about

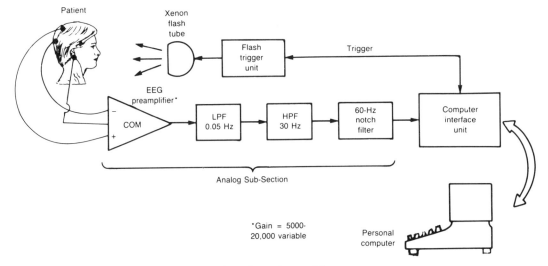

Fig. 29-14. Evoked-potentials system.

one-third what the old dedicated computers cost.

Figure 29-14 shows the block diagram for a simple evoked-potentials system based on a personal computer. The subject is placed so that he sees the xenon flash tube, and EEG electrodes are connected. For visual evoked potentials (as shown in Fig. 29-14) the common electrode (called "indifferent electrode" by life sciences people) is connected to either the earlobe or chin. The differential signal electrodes are connected to points on the top rear of the skull. If you feel that location on yourself, you will note a small indentation called the *inion*. The correct locations for the electrodes are approximately 1 centimeter either side of the inion.

The xenon flash tube trigger unit can operate in either of two ways. First, it can operate only when it is triggered from an external source, such as the evoked-potentials computer. Alternatively, it can operate in the astable mode in which it flashes automatically at a rate determined by its own internal timers. The trigger unit will issue a TTL-compatible pulse each time it triggers, so that external devices can be synchronized with the trigger unit.

The analog sub-section of this system contains amplification and filtering functions. The EEG preamplifier must increase the 50 μV EEG signal from the patient to about 1 volt. This amplification is on the order of 20,000 times. The low-pass filter (LPF) limits the low end bandpass response to 0.05 Hz, and the high-pass filter (HPF) limits the upper-end frequency response to either 30, 50, or 100 Hz, depending upon the situation.

Details of the computer interface unit are shown in Fig. 29-15. In the design of this system, it was assumed that the computer was one in which a 6522 PIA chip was used to make two ports (see discussion on cardiac output computers above). The KIM-1 and AIM-65 machines use the 6522, and certain Apple II plug-ins are available with that option. The 6522 (and certain related chips) are very useful because they can be configured as either inputs or outputs on a bit-for-bit basis. There are internal data direction registers for each port (DDRA and DDRB, respectively). Any given bit of either port will act as an input if the corresponding bit in the DDR is set to logical 0, and as an output if the bit is set to logical 1.

The A/D converter outputs an eight-bit binary word proportional to the EEG signal input, and this data is read by the computer. The DAC is used to drive analog strip-chart recorders or oscilloscopes for final presentation.

Port A of the 6522 is configured as a control port, while port B is the eight-bit data input/output

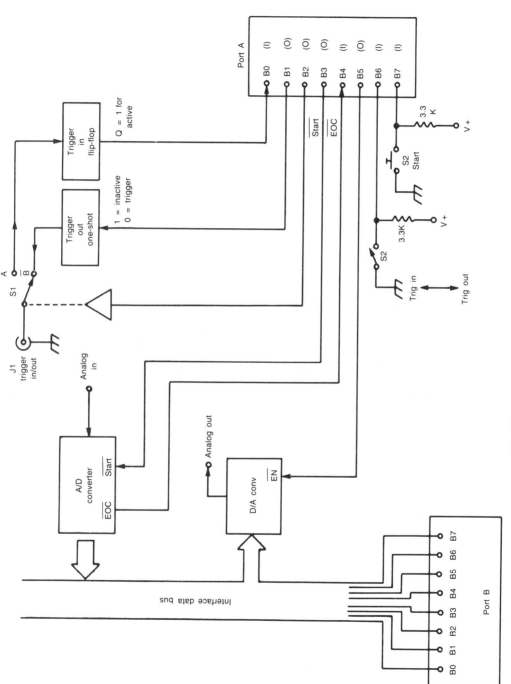

Fig. 29-15. Evoked-potentials interfacing.

port. For the protocols decided for this computer, port A will be configured as follows:

B0 = input
B1 = output
B2 = output
B3 = output
B4 = input
B5 = output
B6 = input
B7 = input

During the initialization phase of our program you must perform certain set-up actions:

1. Write the binary word 00101110 (e.g., hex 2E) to the DDRA. This action configures port A as above.

2. Set port B as an output by writing h00 to the DDRB.

3. Set port B to h00

4. Transfer port B to input of DAC, setting its output to zero volts. This is done by holding bit B5 of port A LOW for a period of time (e.g., 0.25 milliseconds).

5. Reconfigure port B as an input (write hFF to the DDRB).

6. Examine bit B6 of port A to determine whether the trigger section is an output or input (i.e., whether the computer or the xenon trigger unit controls the timing).

7. Loop for awhile, looking for a "Start" command from the operator (i.e., a LOW on bit B7 of port A). When B7 = LOW, go to the input routine.

8. In accordance with the trigger timing protocol, issue an A/D start command by holding bit B3 of port A LOW for a short period (e.g., 0.25 ms).

9. Listen for an end of conversion (e.g., NOT-EOC) by looking for a LOW on bit B4 of port A. When this signal is detected input the A/D data through port B and store it in a location where it can be found again.

It is a design choice whether to signal average (or add) the data on a byte-for-byte basis or wait until all data are input. Regardless, it is best if the data points are retained in memory and stored on disk. From that mass of data one can recreate the EEG waveforms as well as the averaged evoked potential later on. This protocol means (for 1 ms samples) more than 358K of memory for a 1.4-second flash repetition rate. For most applications in evoked potentials only the first 250 milliseconds are significant, so you can shave that memory size requirement somewhat.

Programming Exercises

Before attempting any exercise, please read through the entire list.

1. Write a program that will cause the N–S direction to flash red, and E–W to flash yellow, on one second intervals: a) alternate between E–W and N–S, and b) flash E–W and N–S simultaneously.

2. Write a program that will implement the algorithm of Fig. 29-1.

3. Write a program that will meet the following specifications:

    a) Permit E–W traffic flow at all times unless N–S traffic appears (see Fig. 29-4 for sample pedestrian and auto sensors).

    b) If pedestrian traffic appears, then switch lights to permit N–S traffic immediately.

    c) If one car appears in the N–S lane, switch traffic to permit N–S traffic after 30 seconds; if two cars appear then wait no more than 15 seconds (including time elapsed since first car arrived); and if three cars appear then switch immediately.

4. Write a program that will implement Exercises 1–3 above, depending upon the setting of the thumbwheel switch in Fig. 29-5. If you had read, and heeded, the initial instruction above, then you will have designed the programs for Exercises 1–3 as subroutines that can be meshed together nicely.

Chapter 30

Laboratory Measurement Equipment

I N SELECTING COMPONENTS FOR DATA ACQUISI-
tion systems or automatic test equipment, you
must not overlook the availability of commercial
test equipment that may or may not have originally
been intended for your application. There is a wide
variety of test equipment on the market, and much
of it is suited for a wide variety of applications.
While most of the equipment that falls into this
category was designed for either the engineering
lab, the test work station, or the troubleshooting
workshop, it can also be pressed into service for
other chores as well.

The usefulness of test equipment is enhanced
if it is capable of being used in a General-Purpose
Interface Bus (GPIB), standard IEEE-488, situation. Chapter 32 deals in some detail with this
situation, and the reader is referred there for a
discussion of the GPIB.

Some manufacturers make one or more integrated lines of equipment that are designed to
work together, often in a common frame or housing (Fig. 30-1). Tektronix, for example, makes
their TM-5xx series of equipment in which 1/8,
1/4, and 1/2 rack modules can be integrated together in a single rack/power supply housing. This
series of instruments can be configured for either

general purpose applications or for highly specialized applications.

I once had a position in a medical repair laboratory. We owned two TM-5xx racks with the test
equipment required to maintain the Tektronix 412
and 414 patient monitor sets, and because much
medical equipment is similar, a large variety of
other equipment as well. For that job, I selected
the 1/4-rack oscilloscope, a dual-polarity ±20-volt,
(500-mA) dc power supply, an 1/8-rack function
generator, a 1/8-rack digital multimeter (DMM),
and a 1/2-rack frequency counter. The frequency
counter, incidentally, was selected for two reasons:
it met all of our needs in making measurements on
medical equipment (we required frequency, period,
and events), plus (and this was a large benefit in a
hospital workshop) it met F.C.C. requirements for
measuring the output frequency of the doctor's
VHF-FM boat radios!

The modular equipment made by Tektronix is
easily configured into any of several portable,
benchtop, or rack-mounted module racks (Fig.
30-2). Some models of rack/cabinet are even designed for airline travel by field representatives
and other traveling technical people.

Some modular test equipment is not immedi-

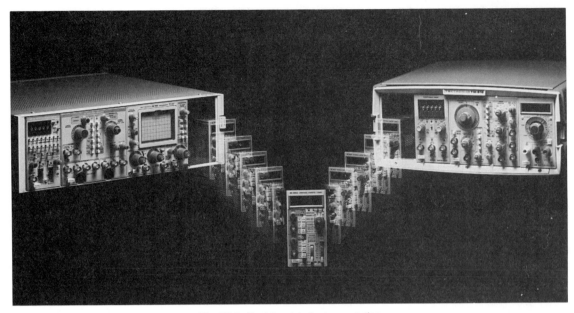

Fig. 30-1. Rack/module instrumentation.

ately suitable for use in a data-acquisition system, but can be configured as such by the enterprising engineer. For example, a collection of instruments that output data to an oscilloscope can just as easily be used to output the same analog data to an A/D converter. Almost all manufacturers of such modular equipment either offer a multiple input A/D converter module or blank modules that you can

use to construct your own plug-in. The blank module comes with a blank front panel of the same style as the rest of the series, a chassis box, and universal "prototyping" printed circuit board that matches the power supply and signals connector inside the mainframe or rack.

Today, much of the digital test equipment in the lines of modular equipment manufacturers is

Fig. 30-2. Assorted racks.

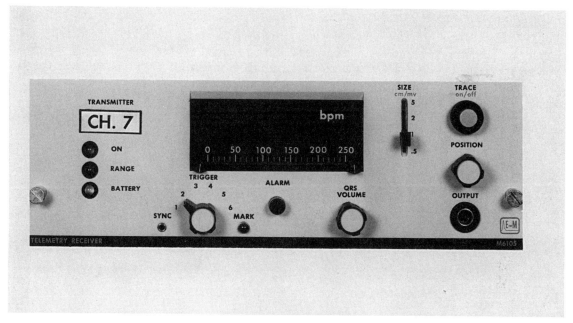

Fig. 30-3. Telemetry receiver.

already set up to output data either serially or in parallel format, either through a special connector or the backplane panel. Otherwise, some modification of the equipment may be needed.

A distinct advantage of using commercial test equipment is that the equipment can be used for other purposes. It may be the case that you will need the data-acquisition system for only a limited period of time, and will then be able to reallocate the equipment to other uses. Alternatively, you may be in a position where a single budget needs to serve two applications simultaneously.

Another advantage of using commercial test equipment is that it allows rapid system design and test-out. If custom design and construction of the instruments is required, then a large amount of engineering and fabrication dollars needs to be spent. If commercial test equipment is used, it is possible to do in a few days what might otherwise take weeks to accomplish. Besides faster times, lower cost and reconfigurability, the use of commercial test equipment also allows easy concept testing. You might not know exactly where a particular design is heading, and the test equipment

approach could give you added flexibility in that case.

Adaptability is still another advantage of commercial equipment. For example, Fig. 30-3 shows a medical telemetry receiver. This instrument is associated with a small telemetry transmitter operating in the VHF or UHF region (it is common to find these transmitters operating at powers of 1 to 10 mW, on frequencies in the guard bands between TV sound and video carrier bands). Although intended to transmit electrocardiograph waveforms from ambulatory patients a total of 100 feet or so, the same equipment can also be used to monitor any other parameter that contains Fourier-series components in the 0.05 Hz to 100 Hz (1 Hz to 45 Hz on some models) range. The transmitter is frequency modulated, and near dc coupled to the analog input, so it is easily adapted to other uses. Although the meter and alarms are designed for its original use, they can be recalibrated (or reinterpreted) for a wide variety of industrial or scientific applications.

It is not my intention to offer you a long discussion of the potential of commercial, already built, test equipment in the solution of data-acquisitions problems. The sole intent is to make you aware of its existence. You would be well-advised to review the catalogs of Tektronix, Hewlett-Packard, John Fluke, and other manufacturers to see if a suitable instrument either exists or can be adapted for your own application. While the prices in the catalog look high, it is sometimes a reasonable trade-off to buy too much instrument off the shelf, than to spend a large sum of money to custom design a specialized piece of gear that has but limited usefulness. It is, perhaps, a case of penny-wise and dollar-foolish.

Chapter 31

Plug-In Systems for Microcomputers

T HE MICROCOMPUTER REVOLUTION SPURRED IN-
terest in developing data-acquisitions systems (and many other products as well). One of the
chief benefits of selecting a computer with plug-in
slots is the ability to configure the machine as a
special-purpose instrument almost at will. Although some care is needed to prevent incompati-
bilities within a single machine (between plug-ins of
various makers), it is generally true that each slot
is memory-mapped into special space, so there is
little chance of a problem.

The utility of plug-in data-acquisition systems
was recognized early in the history of microcomputers. Figure 31-1 shows an older system made
by Burr-Brown Corporation (Tucson, AZ) for the
Motorola series of 6800-based microcomputers.
The card contains four B-B ADC-80 analog-to-digital converters and the supporting circuitry. The
analog inputs can be applied directly to the accessory connector on the top of the card, and data
read into the computer through the main bus.

Figure 31-2 shows a popular plug-in for the
IBM-PC (and those clones that have IBM-like
slots). This unit, made by Strawberry Tree Computers, is capable of handling a wide variety of
analog data. The large screw-terminal block in the
background is mounted external to the computer
and accepts the inputs from the various sensors. As
you can tell from the overlay in Fig. 31-2, the unit
will handle sixteen channels of analog data, including (but not limited to) temperatures, flow rates,
humidity, and other parameters that can be converted to either a voltage or a current.

COMPUTER PLUG-IN
INTERFACE CONNECTIONS

I have selected three main candidate computers (see Chapter 22) for data acquisition chores:
IBM-PC, Apple IIe, and AIM-65. In this section we
will take a brief look at the signals available on the
plug-in connectors.

IBM-PC CASSETTE CONNECTOR

The Cassette Connector is designed for storing and retrieving data on an audio cassette. This
method of data storage was popular when personal
computers first appeared and was a low-cost alternative to the then-expensive 5.25-inch disk drives.
Because audio-cassette storage is both time-consuming and of low reliability, and because disk
drives are a lot cheaper today, almost everyone

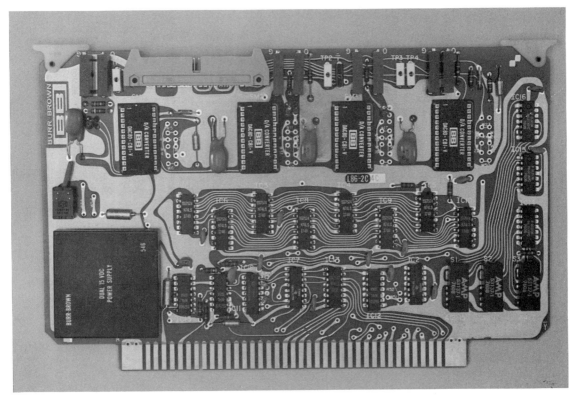

Fig. 31-1. Data acquisition plug-in.

uses disks. Nonetheless, the IBM still retains the Cassette Connector.

A five-pin DIN connector is used for the cassette port. This same connector is also used for the keyboard connector. The pinouts for the cassette port are shown in Table 31-1.

Keyboard Connector

The keyboard connector uses the same form of five-pin DIN connector as the cassette port, but the pinouts are different (see Table 31-2).

The expansion port on the read panel is a 62-pin DB-series connector that looks like a big brother of the familiar RS-232C connector. The signals on this connector are used to drive the expansion unit, hard disk drives, and other similar devices (see Table 31-3).

I/O Channel Connector on System Board

The versatility of the IBM-PC is due in large part to the ability of the user to configure the computer into a large number of different machines

Table 31-1. Pinouts for Cassette Port

Pin No.	Function
1	Motor control (common from relay)
2	Ground
3	Motor control (n.o. relay)
4	Data in (1–2 kilobaud) from earphone
5	Data output to microphone or auxiliary

Table 31-2. Pinouts for Keyboard Connector

Pin No.	Function
1	+Keyboard clock
2	+Keyboard data
3	−Keyboard reset (not used by keyboard)
4	Ground
5	+5 vdc power

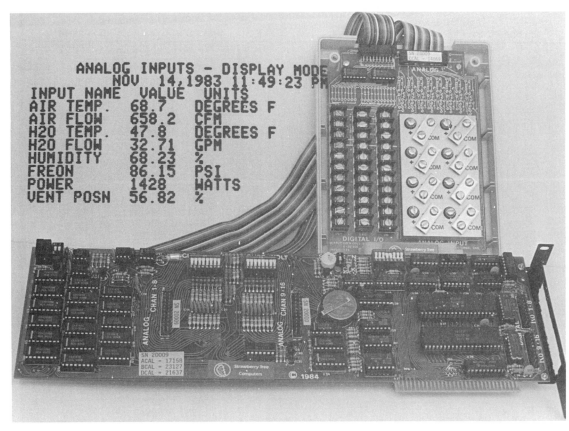

Fig. 31-2. Plug-in for the IBM-PC.

Table 31-3. Signals for the IBM PC Interface Connector

Pin No.	Function
("E" means "extended")	
1	+E IRQ6
2	+E DRQ2
3	+E DIR
4	+E ENABLE
5	+E CLK
6	−E MEM IN EXP
7	+E A17
8	+E A16
9	+E A5
10	−E DACK0
11	+E A15
12	+E A11
13	+E A10
14	+E A9
15	+E A1
16	+E A3
17	+E DACK1
18	+E A4
19	−E DACK2
20	−E IOW

Table 31-3. Signals for the
IBM PC Interface Connector (*Continued*)

Pin No.	Function
21	+E A13
22	+E D5
23	+E DRQ1
24	+E DRQ3
25	(reserved)
26	+E ALE
27	+E T/C
28	+E RESET
29	+E AEN
30	+E A19
31	+E A14
32	+E A12
33	+E A18
34	+E MEMR
35	−E MEMW
36	+E A0
37	−E DACK3
38	+E A6
39	−E IOR
40	+E A8

**Table 31-3. Signals for the
IBM PC Interface Connector (*Continued*)**

Pin No.	Function
41	+E A2
42	+E A7
43	+E IRQ7
44	+E D6
45	+E I/O CH RDY
46	+E IRQ3
47	+E D7
48	+E D1
49	−E I/O CH RDY
50	+E IRQ2
51	+E D0
52	+E D2
53	+E D4
54	+E IRQ5
55	+E IRQ4
56	+E D3
57	Ground
58	Ground
59	Ground
60	Ground
61	Ground
62	Ground

Table 31-4. Pinouts for IBM PC I/O Connector

Pin No.	Function
(Wiring side of inserted PWB)	
B1	Ground
B2	+RESET DRV
B3	+5 volts
B4	+IRQ2
B5	−5 volts
B6	+DRQ2
B7	−12 volts
B8	(reserved)
B9	+12 volts
B10	Ground
B11	−MEMW
B12	−MEMR
B13	−IOW
B14	−IOR
B15	DACK3
B16	+DRQ3
B17	−DACK1
B18	+DRQ1
B19	−DACK0
B20	CLOCK
B21	+IRQ7
B22	+IRQ6
B23	+IRQ5
B24	+IRQ4
B25	+IRQ3
B26	−DACK2
B27	+T/C
B28	+ALE
B29	+5 volts
B30	+OSC

**Table 31-4. Pinouts for
IBM PC I/O Connector (*Continued*)**

Pin No.	Function
B31	Ground
(Component side)	
A1	−I/O CH CK
A2	+D7
A3	+D6
A4	+D5
A5	+D4
A6	+D3
A7	+D2
A8	+D1
A9	+D0
A10	+I/O CH RDY
A11	+AEN
A12	+A19
A13	+A18
A14	+A17
A15	+A16
A16	+A15
A17	+A14
A18	+A13
A19	+A12
A20	+A11
A21	+A10
A22	+A9
A23	+A8
A24	+A7
A25	+A6
A26	+A5
A27	+A4
A28	+A3
A29	+A2
A30	+A1
A31	+A0

Table 31-5. Pinouts for the Apple II

Pin No.	Description	Function
1	I/O SELECT	Active-LOW signal that is LOW when the unique address (of 16 possible addresses) for that particular connector is called for in a program. All 16 of these addresses are in the range $C800 to $C8FF.
2	A0	Address bus bit 0
3	A1	Address bus bit 1
4	A2	Address bus bit 2
5	A3	Address bus bit 3
6	A4	Address bus bit 4
7	A5	Address bus bit 5
8	A6	Address bus bit 6
9	A7	Address bus bit 7
10	A8	Address bus bit 8
11	A9	Address bus bit 9
12	A10	Address bus bit 10

Table 31-5. Pinouts for the Apple II (*Continued*)

Pin No.	Description	Function
13	A11	Address bus bit 11
14	A12	Address bus bit 12
15	A13	Address bus bit 13
16	A14	Address bus bit 14
17	A15	Address bus bit 15
18	R/W	6502 read/write signal. This signal is LOW for WRITE operations, and HIGH for READ operations.
19	(no connection)	
20	I/O STR	Active-LOW signal that indicates an I/O operation is taking place. This signal is activated when any address between $C800 and $C8FF is called in the program.
21	RDY	Active-LOW input signal that will add a WAIT state to the CPU (e.g., stop program execution) if LOW during clock Phase 1.
22	DMA	Active-LOW input that allows an external device to gain control over the data bus for direct access to the memory.
23	INTOUT	Interrupt output signal that permits prioritizing of the interrupts. The INTOUT signal is connected to the INTIN signal of the next higher order plug-in printed circuit card.
24	DMAOUT	Direct-memory-access prioritization.
25	+5	Five-volt power supply from main board.
26	GND	Ground
27	DMAIN	Direct memory access prioritization input.
28	INTIN	Interrupt prioritization input
29	NMI	Nonmaskable interrupt input signal to 6502 (active-LOW).
30	IRQ	Interrupt request (active-LOW)
31	RES	Active-LOW reset line
32	INH	Active-LOW input that will disconnect on-board monitor ROMs to allow custom programmed ROMs to be used.
33	−12	−12 volt dc power supply from pc board.

Table 31-5. Pinouts for the Apple II (*Continued*)

Pin No.	Description	Function
34	−5	−5 volt-dc power supply from pc board.
35	(no connection)	
36	7 MHz	7 MHz clock from main pc board
37	Q3	2 MHz clock from main pc board
38	O1	Phase 1 clock signal
39	USER1	Disables all ROMs and locations $C800 through $C8FF.
40	O2	Phase-2 clock signal
41	DEVICESEL	Active-LOW signal that indicates that one of the 16 address of that connector is present on address bus.
42	D7	Data bus bit 7
43	D6	Data bus bit 6
44	D5	Data bus bit 5
45	D4	Data bus bit 4
46	D3	Data bus bit 3
47	D2	Data bus bit 2
48	D1	Data bus bit 1
49	D0	Data bus bit 0
50	+12	+12-volt power supply from main pc board.

Table 31-6. KIM-1 Bus Applications Connector (Numbered pins are on top of board, lettered pins are on the bottom)

Pin No.	Designation	Function
1	GND	Ground/Power Supply Common
2	PA3	Bit 3 of port A
3	PA2	Bit 2 of port A
4	PA1	Bit 1 of port A
5	PA4	Bit 4 of port A
6	PA5	Bit 5 of port A
7	PA6	Bit 6 of port A
8	PA7	Bit 7 of port A
9	PB0	Bit 0 of port B
10	PB1	Bit 1 of port B
11	PB2	Bit 2 of port B
12	PB3	Bit 3 of port B
13	PB4	Bit 4 of port B
14	PA0	Bit 0 of port A
15	PB7	Bit 7 of port B
16	PB5	Bit 5 of port B
17	KBR0	Keyboard, row 0
18	KBCF	Keyboard, column F
19	KBCB	Keyboard, column B
20	KBCE	Keyboard, column E
21	KBCA	Keyboard, column A
22	KBCD	Keyboard, column D
A	PWR5	+5 vdc from main board
B	K0	Memory bank select
C	K1	Memory bank select
D	K2	Memory bank select

Table 31-6. KIM-1 Bus
Applications Connector (Numbered pins are on top of board, lettered pins are on the bottom) (*Continued*)

Pin No.	Designation	Function
E	K3	Memory bank select
F	K4	Memory bank select
F	K4	Memory bank select
H	K5	Memory bank select
J	K7	Memory bank select
K	Decode	Memory decode signal. Used to increase memory size with external memory banks
L	AUDIN	Audio input from cassette player
M	AUDOUTL	Audio output to cassette. Low-level signal for microphone connector on recorder.
N	PWR12	+12 vdc power from main board
P	AUDOUTH	High-level audio output for use with cassette recorder that has a line input
R	TTYKBD+	Positive terminal of 20-mA TTY current loop (serial input), keyboard
S	TTYPNT+	Positive terminal of 20-mA TTY current loop (serial output), printer
T	TTYKBD−	Negative terminal of "R" above
U	TTYPNT−	Negative terminal of "S" above
V	KBR3	Keyboard, row-3
W	KBCG	Keyboard, column G
X	KBR2	Keyboard, row-2
Y	KBCC	Keyboard, column C
Z	KBR1	Keyboard, row-1

Table 31-7. KIM-1 Bus
Expansion Connector (Numbered pins on top of board, lettered pins are on bottom)

Pin No.	Designation	Function
1	SYNC	Output that is HIGH during clock phase 1
2	RDY	Active-LOW input that inserts WAIT state into program execution
3	P1	Phase-1 clock signal
4	IRQ	Active-LOW interrupt request
5	RO	Reset overflow (resets the overflow flip-flop inside CPU)
6	NM1	Active-LOW nonmaskable interrupt input
7	RST	RESET. Goes to reset line of 6502
8	DB7	Data bus bit 7
9	DB6	Data bus bit 6
10	DB5	Data bus bit 5
11	DB4	Data bus bit 4
12	DB3	Data bus bit 3
13	DB2	Data bus bit 2

Table 31-7. KIM-1 Bus
Expansion Connector (Numbered pins on top of board, lettered pins are on bottom) (*Continued*)

Pin No.	Designation	Function
14	DB1	Data bus bit 1
15	DB0	Data bus bit 0
16	K6	Active-HIGH output that indicates that the CPU addresses a location in the range $1800 to $1BFF
17	SSTOUT	Single-step output
18	(No connection)	
19	(No connection)	
20	(No connection)	
21	PWR5	+5 vdc power from main board
22	GND	Ground
A	AB0	Address bus bit 0
B	AB1	Address bus bit 1
C	AB2	Address bus bit 2
D	AB3	Address bus bit 3
E	AB4	Address bus bit 4
F	AB5	Address bus bit 5
H	AB6	Address bus bit 6
J	AB7	Address bus bit 7
K	AB8	Address bus bit 8
L	AB9	Address bus bit 9
M	AB10	Address bus bit 10
N	AB11	Address bus bit 11
P	AB12	Address bus bit 12
R	AB13	Address bus bit 13
S	AB14	Address bus bit 14
T	AB15	Address bus bit 15
U	P2	Clock phase 2
V	R/W	HIGH for read operations, LOW for write operations
W	R/W	LOW for read operations, HIGH for write operations (complement of "V" above)
X	PLLTST	Phase-locked loop test for FM-audio signal used to record data on audio cassettes.
Y	P2	Complement of clock phase 2 ("U" above)
Z	RAMRW	RAM read/write control signal to activate RAM during phase-2 clock periods

through the use of plug-in printed wiring boards. The slots for the cards are female printed-circuit card-edge connectors mounted on the system board inside of the IBM-PC. These slots are installed close to the rear panel of the IBM-PC such that interfacing connectors to the outside world can be accommodated on each plug-in card. This arrangement can be seen in Fig. 31-2. The cards are much larger than Apple II plug-in cards and so can house a lot more (which means more complex) circuitry. The pinouts of the I/O channel connector are shown in Table 31-4.

These signals and many other details of the IBM-PC are given in the IBM-PC Technical Reference Manual (IBM part number 6025005).

APPLE II-SERIES MACHINES

The pinouts of the Apple II plug-in slots are given in Table 31-5.

THE KIM-1/AIM-65 FAMILY OF COMPUTERS

Tables 31-6 and 31-7 show the signals available on the plug-in connectors of the KIM bus applications and expansion connectors.

Chapter 32

Data-Acquisition Systems Based on the IEEE-488 General-Purpose Interface Bus (GPIB)

AUTOMATIC TEST EQUIPMENT (ATE) IS NOW ONE of the leading methods for testing electronic equipment in factory production and troubleshooting situations; it has been used for this purpose for years. The data-acquisition system can also be designed around standard ATE modules and equipment.

The basic method is to use a programmable digital computer to control a bank of test instruments. The program, often in BASIC language, turns on and off the various instruments and then evaluates the results as measured by other instruments.

The bank of equipment can be configured for a special purpose or for general use. For example, you could select a particular line-up of equipment needed to test, say, a broadcast audio console, and provide a computer program to make the various measurements: gain, frequency response, total harmonic distortion, and so forth. Alternatively, you could also make a generalized test set. This is the method selected by a number of organizations who have large numbers of different electronic devices to test. There will be a main bank of electronic test equipment, adapters to make the devices under test interconnect with the system, and a special program for each type of equipment. Such an approach makes for a cost-effective system of test equipment.

Previously, the main problem in attempting to make automatic test equipment (ATE) was that it was impossible to use unmodified off-the-shelf commercial test instruments. Thus, many of the best and most useful pieces of test equipment could not be used at all, or had to be extensively modified by the ATE maker before they could be used. The problem was programmability: how to make a signal generator respond to the computer commands? In 1978, the Institute of Electrical and Electronic Engineers (IEEE) released their specification titled "IEEE Standard Digital Interface for Programmable Instrumentation," or "IEEE-488" as it is called in the trade. This specification provides details for a standard computer interface between a computer and instruments. It also calls out ASCII codes and mnemonics for program instructions. The IEEE-488 bus is also called "General Purpose Interface Bus" (GPIB). The Hewlett-Packard Interface Bus (HPIB) is a proprietary version of the IEEE-488 bus. The main purpose for the IEEE-488/GPIB is automatic test equipment, both generalized and specific.

Test instrumentation that is intended for GPIB service will have a 24-pin "blueline" connector on the rear panel. This connector is one of the Amphenol blueline series not unlike the 36-pin connector used for parallel printer interface on microcomputers. There will also be a GPIB ADDRESS DIP switch on the rear panel, usually near the connector. The purpose of the switch is to set the five-bit binary address where the instrument is located in the system, which determines whether or not the device is a listener only or a talker only and certain other details.

GPIB BASICS

The IEEE-488/GPIB specification provides technical details of the standard bus. The logic levels on the bus are generally similar to TTL: a LOW is less than or equal to 0.800 volts, while a HIGH is greater than 2.0 volts. The logic signals can be connected to the instruments through a multiconductor cable up to 20 meters (66 feet) in length, provided that an instrument load is placed every 2 meters. This specification works out to a cable length in meters of twice the number of instruments in the system. Most IEEE-488/GPIB systems operate unrestricted to 250 kilobytes per second, or faster with certain specified restrictions.

There are two basic configurations for the IEEE-GPIB system: linear and star. These configurations are created with the cable connections between the instruments and the computer. The linear configuration is basically a daisy-chain method, in which the tap-off to the next instrument is taken from the previous one in the series. In the star configuration the instruments are connected from a central point.

There are three major busses in the IEEE-488/GPIB system. Connected to the bus line are receiver and driver circuits. These similar-to-TTL logic elements provide input or output to the instrument. The driver is an output and will be a tri-state device. That is, it is inert until commanded to turn on. A tri-state output will float at high impedance until turned on. The receiver is basically a noninverting buffer with a high-impedance

input. This arrangement of drivers and receivers provides low loading to the bus.

In the basic structure of the IEEE-488/GPIB there are three busses and four different type of devices. The devices are: controllers, talkers only, listener only and talker/listener. These devices are defined as follows:

Controllers. This type of device acts as the brain of the system, and communicates device addresses and other interface messages to instruments in the system. Most controllers are programmable digital computers. Both Hewlett-Packard and Tektronix offer computers that serve this function, and certain other companies produce hardware and software that permits other computers to act as IEEE-488/GPIB controllers.

Listener. A device capable of listening will receive commands from another instrument, usually the controller, when the correct address is placed on the bus. The listener acts on the message received, and the bus does not send back any data to the controller. An example of a listener is the signal generator.

Talker. The talker responds to the message sent to it by the controller and then sends data back to the controller over the DIO data bus. A frequency counter is an example of a talker.

There is also a combination device that accepts commands from the controller to set up ranges, tasks, etc., and then returns data back over the DIO bus to the controller. An example is a digital multimeter (DMM). The controller will send the DMM commands that determines whether it is ac or dc, volts or milliamperes or ohms, and what specific range—the device is thus acting as a listener. When the measurement is made, the DMM becomes a talker and transmits the data measured back over the DIO bus to the controller.

There are three major busses in the IEEE-488/GPIB system: General Interface Management (GIM) Bus, Data Byte Transfer (DBT) Bus, and the Data Input/Output (DIO) Bus. These busses operate as described below.

DIO Bus. The data input output bus is a bidirectional eight-bit data bus that carries data and interfaces messages and device-dependent messages between the controller, talkers, and listeners. This bus sends data asynchronously in byte-serial format.

DBT Bus. The data-byte transfer bus controls the sending of data along the DIO bus. There are three lines in the DBT bus: Data Valid (DAV), Not Ready for Data (NRFD) and Not Data Accepted (NDAC). These signal lines are defined as:

1. DAV. The data valid signal indicates the availability and validity of the data on the line. If the measurement is not finished, for example, the DAV signal will be false.

2. NRFD. The Not Ready for Data signal lets the controller know whether or not the specific device addressed is in a condition to receive data.

3. NDAC. The Not Data ACcepted signal line is used to indicate to the controller whether or not the device accepted the data sent to it over DIO bus.

GIM Bus. The General Interface Management bus coordinates the system and ensures an orderly flow of data over the DIO bus; it has the following signals: Interface Clear (IFC), Attention (ATN), Service Request (SRQ), Remote Enable (REN) and End or Identify (EOI). These signals are defined as follows:

1. ATN. The attention signal is used by the controller/computer to let the system know how data on the DIO bus lines is to be interpreted, and which device is to respond to the data.

2. IFC. The interface clear signal is used by the controller to place all devices in a predefined quiescent or standby condition.

3. SRQ. The service request signal is used by a device on the system to ask the controller for attention. This signal is essentially an interrupt request.

4. REN. The remote enable signal is used by the controller to select from between two alternate sources of device programming data.

5. EOI. The end or identify signal is used by talkers for two purposes. It will follow the end of a multiple-byte sequence of data in order to indicate

Table 32-1. Pinouts for the IEEE-488/GPIB

Pin No.	Signal Line
1	DIO1
2	DIO2
3	DIO3
4	DIO4
5	EOI
6	DAV
7	NRFD
8	NDAC
9	IFC
10	SRQ
11	ATN
12	Shield
13	DIO5
14	DIO6
15	DIO7
16	DIO8
17	REN
18	Ground (6)
19	Ground (7)
20	Ground (8)
21	Ground (9)
22	Ground (10)
23	Ground (11)
24	Logic ground

that the data is now finished. It is also used in conjunction with the ATN signal for polling the system.

The signals defined above are implemented as conductors in a system interface cable. Each IEEE-488/GPIB compatible instrument will have a female 36-pin Amphenol style connector on the rear panel. The pinout definitions are shown in Table 32-1.

SUMMARY

The IEEE-488 General-Purpose Interface Bus (GPIB) can be used to marry together various pieces of standard commercial test equipment, computers, and customized equipment into a system that is capable of providing specified tests and measurements under the control of a computer program, often in BASIC. The GPIB makes either special purpose or generalized automatic test equipment (ATE) possible with a minimum of effort.

Chapter 33

Precision Voltage and Current Sources

H OW DO YOU KNOW THAT THE NEW DIGITAL MUL-
timeter that you just bought for the laboratory is accurate, that it has maintained its calibration over the years? The same question can also be
asked of oscilloscopes, laboratory power supplies,
amplifiers and other dc measuring devices. Do you
have a microcomputer with an A/D converter in it?
If the voltage reference source for an A/D or D/A
converter is not as good as the bit-length of the
converter, then it is not useful. One quick way to
get 4-bit performance out of an 8-bit data converter is to use a little voltage reference supply!
This chapter discusses how to design and build
simple but effective reference supplies.

ZENER DIODES

The simplest device that can be used for voltage regulation, and hence also for some low-preci-
sion reference applications, is the ordinary zener
(pronounced "zenner") diode. Figure 33-1 shows
the circuit symbol (inset) and transfer characteristic curve for a zener diode. The diode is little more
than a special pn junction diode, so it will behave
like any other pn diode in the forward bias region
(+V). When the applied voltage is positive, then a
forward current (+I) will flow. Below a certain

threshold voltage (V_G) the current is the reverse
leakage current I^R. Above the threshold, however,
the forward current increases in a linear ohmic
manner.

In the reverse bias region things are a little
different from ordinary pn junction diodes. The
leakage current is all that flows for applied voltages
(−V) of O to V_Z, but when V_Z is reached the diode
avalanches. In this condition the diode will regulate
the applied voltage to the value of V_Z.

The basic zener regulator circuit is shown in
Fig. 33-2. This circuit is not often used directly as
a precision reference supply, but forms the basis of
such supplies so is included here. The zener diode
(D1) is placed in parallel across the load, R_L. A
series resistor (R1) is used to limit the current to a
safe value (recall from Fig. 33-1 that −I increases
drastically at V_Z!). The output voltage will be regulated to +V_Z in this circuit.

There are some problems with the basic zener
diode, and these problems become especially acute
when it is used as a reference source (which implies
accuracy). First, the voltage is only nominal. In
other words, a "6.8-volt" zener produces a voltage
close to 6.8 vdc, but rarely is it exactly 6.8 vdc!
Another problem is that the voltage drifts some-

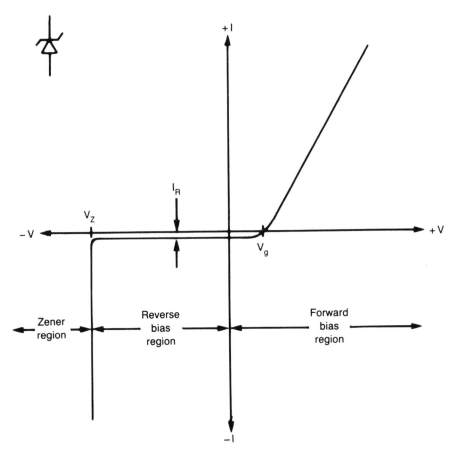

Fig. 33-1. Zener-diode I-vs-V curve.

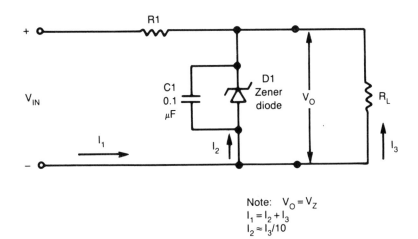

Note: $V_O = V_Z$
$I_1 = I_2 + I_3$
$I_2 \approx I_3/10$

Fig. 33-2. Zener-diode circuit.

what with temperature—hardly a desired characteristic in a reference supply.

Figure 33-3A shows a crude attempt at stabilizing the temperature drift problem. In this circuit a number of zener diodes are connected in a series-parallel arrangement. Each series string produces a voltage drop (V_1 and V_2), so the differential output voltage is ($V_1 - V_2$). The idea here is that all

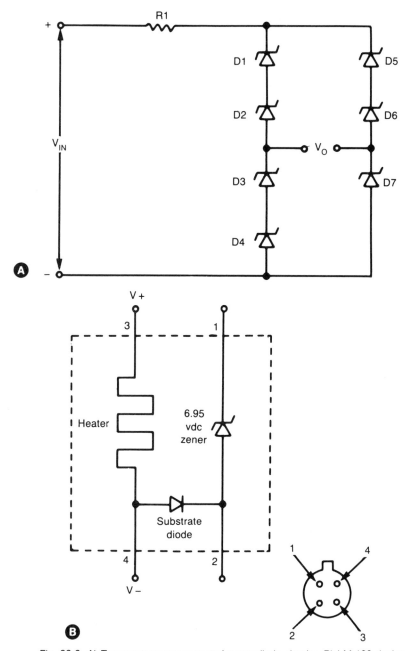

Fig. 33-3. A) Temperature-compensated zener-diode circuits, B) LM-199 device.

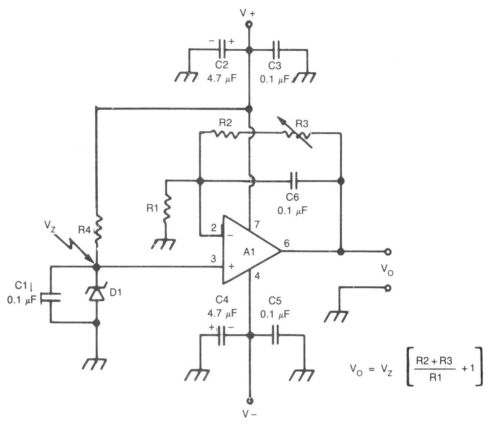

$$V_O = V_z \left[\frac{R2 + R3}{R1} + 1 \right]$$

Fig. 33-4. Precision op-amp voltage source.

diodes, assuming they are identical and in the same thermal environment, will drift approximately the same amount so the differential effects of drift are zero.

A superior idea is shown in Fig. 33-3B. The device shown here is the National Semiconductor LM-199 (or LM-399) device. It consists of a 6.95-volt zener diode embedded in an electrical heater device. One source told me that the heater was little more than a Class-A amplifier with the input shorted, and that the zener is built on the same substrate and so shares the same thermal environment. The heater acts to keep the diode at a constant temperature somewhat above ambient room temperature. With the temperature constant, the diode voltage drop will not drift. The LM-199/LM-399 devices offer some startling voltage-drift specifications. There are also other (similar) devices on the market.

OP-AMP REFERENCE SOURCE

Even the LM-199 device produces only a nominal output voltage. While that voltage remains constant, it may be a little different from the rated 6.95 volts. The circuit in Fig. 33-4 will adjust the voltage to any desired value and make it precise. In addition, the operational amplifier serves to buffer the reference supply against changes in the load conditions.

The basic circuit of Fig. 33-4 is the noninverting follower with gain op-amp configuration. The LM-199 device is used to supply the input voltage on pin no. 3, so the output voltage will be:

$$V_0 = V_z \frac{R2}{R1} + 1$$

Selection of appropriate values of R2 and R1 will produce the desired output voltage. If you

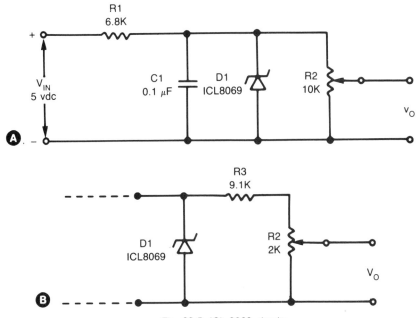

Fig. 33-5. ICL-8069 circuits.

make R1 = 1000 ohms, then a 10.00-volt power supply can be made if R2 is 438.8 ohms. In most cases R2 will be a combination of a fixed resistor (low-temperature coefficient!) and a multiturn trimmer potentiometer. The trimpot is adjusted for the desired output voltage.

INTERSIL'S ICL-8069

Another form of simple reference source is the Intersil ICL8069 band-gap zener device, shown in Fig. 33-5A. This device is a 1.2-volt temperature-compensated zener diode that will operate with low-noise at zener currents doen to 50 μA. There are several versions of the device, and they differ mostly in stability and accuracy specs.

Figure 33-5A shows the basic circuit for this diode. Note that an output potentiometer is used to pick off the correct voltage from the 1.2 volts available open-terminal. The same circuit is shown in Fig. 33-5B, with the difference that a fixed resistor is added in series with the potentiometer in order to increase the resolution of the circuit. The potentiometer selects a potential over a narrower range of total resistance.

Neither of the circuits in Fig. 33-5 offer the same benefits as the operational amplifier form of circuit. Both the earlier op-amp circuit and the circuit of Fig. 33-6 can be used to overcome some of the limitations of the simpler circuit. In Fig. 33-6 you see an operational amplifier used in a slightly unusual configuration to produce an output of 10.000 volts, adjustable with R3. The operational amplifier gain is set by the resistance of the potentiometer.

Although an LM-308 is used for the operational amplifier in Fig. 33-6, almost any quality op-amp could be used. I recommend against using the 741, however, as its drift specification might tend to decrease the accuracy of the output voltage over time.

In all of the operational amplifier circuits, incidentally, it is desirable to keep the op-amp itself from drifting. In most cases this means that the power supply voltages (V− and V+) must be kept as close as feasible to the output voltage being used. This requirement means that you must know how close to the supply voltage the output voltage will rise. In some op amps, it is several volts (as low as 0.5 volts in BiMOS devices). In one popular unit the output voltage can rise to within 1.4 volts of the power-supply voltage. For a 10.00-volt reference supply, therefore, you would want to use a standard power-supply voltage close to $(10.00 + 1.4 = 11.4)$ volts. In that this use -12 volts for V− and $+12$ volts for V+.

IC REFERENCE SOURCES

A very popular device for use in reference supplies is the integrated-circuit reference supply. Although there are many different types on the market, I will use the Precision Monolithics, Inc., REF-01 and REF-02 devices as my example (see Fig. 33-7).

The REF-02 device is a $+5.00$-volt reference source, while the REF-01 is a 10.000-volt unit. Both REF-01 and REF-02 are packaged in an 8-pin metal IC can, and use the pin-out definitions shown in Fig. 33-7. The supply voltage is applied across pins 2 and 4, while the output voltage is taken across pins 6 and 4 (pin 4 is common/ground). Pin no. 5 is used as a trim/adjust input.

The REF-02 uses pin no. 3 in a unique manner: it is an electronic thermometer transducer. The voltage appearing at pin no. 3 will have a value of 2.1 millivolts per degree Kelvin ambient temperature. It can be used to create an electronic thermometer.

Figure 33-8 shows the usual operating circuit for the REF-01 and REF-02 devices. The trimmer circuit consists of a linear taper potentiometer that selects a sample of the output voltage and inputs it to the trim circuit. This potentiometer should be a multiturn type in order to closely set the output voltage.

I have used the REF-01 device in a number of projects and found it more than satisfactory. It seems to have even better temperature stability when a Wakefield flexible IC (or TO-5 transistor) heatsink is used on the IC case. Also, keep the input voltage to a minimum required to obtain the output voltage. I have used $+12$ volts for the REF-01 and $+9$ volts for the REF-02.

CONCLUSION

The precision voltage reference source can be built using a simple zener diode-like device, such as the LM-199 or ICL-8069, or using a more complex circuit in which a zener diode is used with an

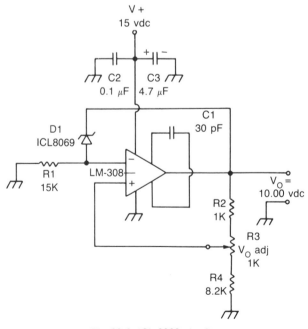

Fig. 33-6. ICL-8069 circuits.

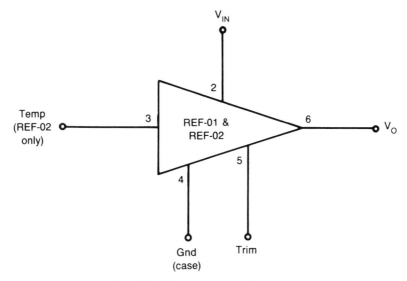

Fig. 33-7. REF-01 and REF-02 devices.

operational amplifier. You can also use a special IC reference device, such as the REF-01 or REF-02 by PMI (or one of the equivalents by other manufacturers). Check the linear and power supply IC catalogues for any of the major semiconductor makers for data sheets on other types.

Adjusting the reference source poses another problem for you. The problem is that we don't

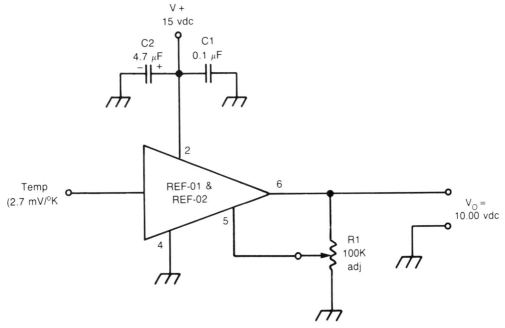

Fig. 33-8. REF-01/02 circuit.

know what to use as a reference! There are several alternatives. For those who are intrepid, take the finished reference source to someone who has a very accurate zillion digit, multi-kilobuck digital voltmeter and use that as your reference. For the most of us, however, the source can be adjusted once using a brand-new digital voltmeter (or oscilloscope), and then live with any problems that develop. In most cases you can trust your DMM (or have it commercially calibrated) so that it can be used to monitor the reference source when calibrating other dc-measuring instruments.

Chapter 34

Analog X-Y and Y-Time Recorders

T HERE ARE ANY NUMBER OF ANALOG DATA RE-corders that are used in instrumentation, data acquisition, and control systems. Although the class of "analog recorders" is large, there are only a few subclasses that I will consider in this book: oscilloscopes (Chapter 36), paper recorders (this chapter), and panel meters (Chapter 35). Note that most of these devices are displays of one sort or another. Although a completely new system might use some method other than analog, there are still large numbers of these instruments both in existing inventories and in active instrumentation systems. The common binding factor that makes these devices all similar to each other is that they are analog. In other words, they accept an analog voltage or current and display it in some format unique to the type of device.

PAPER RECORDERS

Perhaps one of the oldest yet still most common forms of electromechanical display device used in data-acquisition applications is the chart recorder. These instruments are also sometimes called "recording oscillographs." There are several different forms of instruments on the market. Figure 34-1 shows a recording made on a 50-mm strip chart recorder. The paper is 50-mm wide, and is divided into large blocks of 5-mm and small blocks of 1-mm. There are several different types of writing system. In Fig. 34-1, the paper is a paraffin-treated thermal paper that turns black when heated. A hot stylus is dragged across the paper by a permanent-magnet moving-coil galvanometer proportionally to the amplitude of the applied signal. The timebase is provided by a mechanical roller system that drags the paper under the pen tip at standard rates: 2.5, 5, 10, 15, 25, 50, and 100 millimeters/second (mm/s). Other writing systems use ink pens, ink jets, and other writing methods.

Figure 34-2 shows another type of chart recorder. In this case the paper is wider and uses edge perforations like on computer printers for the paper drive. The pen is usually an ink cartridge type. Although most quality recorders of this type use special ink cartridges, there was once a Heath-kit low-cost recorder that wrote with a user-supplied felt-tip pen.

The pen motion in the example of Fig. 34-2 is by a servomechanism system in which pen position feedback is compared with the incoming analog voltage, and a correction signal issued to null the

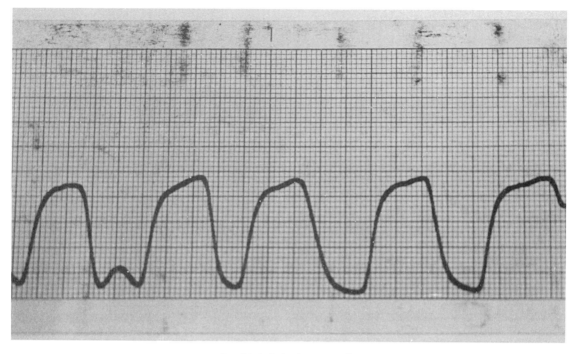

Fig. 34-1. Strip-chart recording.

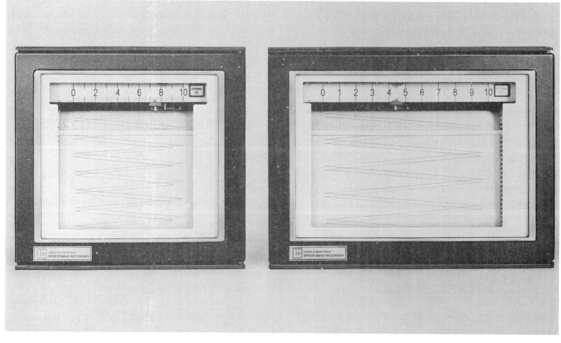

Fig. 34-2. Servo recorder.

difference. The motor that drives the pen is connected to the output of the amplifier, and the pen mechanism is connected to the motor via a string-and-pulley arrangement. The servorecorder of Fig. 34-2 is an example of a Y-time recorder, similar to the strip-chart recorder described above. The perforated paper is dragged through the machine to form the timebase, while the pen amplifier serves to supply amplitude information on the Y axis.

There are other servorecorders that use two servomechanisms to drive the pen. The paper is stationary and can be any standard 11 × 17 or smaller graph paper. Both X and Y servomechanisms are used, so these machines are called "X-Y recorders." In some cases the X axis input is supplied with a sawtooth waveform that produces a Y time mode.

One advantage of the X-Y servorecorder for computer users is that several related parameters can be recorded on a single sheet of graph paper on the same time base. If the machine uses only one pen mechanism, then successive printouts of computer-stored data will result in an overlapping picture of the interrelated events.

A variation on this theme is the X-Y plotter. These devices aren't exactly analog servomechanisms, and are seen much more frequently these days in the past. The pen mechanism is controlled by digital stepper motors. Using the pen lift feature, different elements of a drawing can be done at different times. I suspect that most future instrumentation applications that once used servorecorders of the analog variety will switch to X-Y plotters unless some compelling reason exists for using the older system (like having an analog X-Y recorder and no money).

There are only a few controls found on typical paper recorders. There will typically be position and gain controls. In many cases however, these controls will carry the labelling that makes sense to a scientist rather than a recorder designer. For example, "position" is usually labelled "zero," and "gain" or "input attenuator" is usually designated "amplitude" or "span." There may also be a timebase control labelled much like the oscilloscope controls: volts per unit of time (V/sec).

WRITING SYSTEMS

In the remainder of this chapter I will discuss writing systems used on paper recorders. Although you will learn the basics of various popular mechanisms, you should read the material with a view to the strengths and weaknesses of each system as regards data acquisition and display problems.

Permanent-Magnet Moving-Coil (PMMC) Instruments

The chart recorder used most frequently in medical applications (e.g., the ECG and EEG recorders), and many industrial or scientific applications, is the permanent magnet moving coil (PMMC) device. The PMMC galvanometer (Fig. 34-3) is very similar in concept to the D'Arsonval and Taut-Band dc meter movements. A writing pen replaces the meter pointer in paper recorders, however.

The structure of the PMMC instrument is shown in Fig. 34-3. A large permanent magnet surrounds a movable bobbin, on which is wound the coil of an electromagnet. When the current in the coil is static, the magnetic field produced by the coil is also static, so the pointer stays in one place. But if the coil current changes, or varies in any way, the magnetic field also varies in the same way. The changes in the coil electromagnetic field interact with the field of the permanent magnet to deflect the bobbin to a new equilibrium point. The amount of deflection is proportional to the degree of change in the coil current.

In a PMMC galvanometer instrument, the pen that writes on the paper is used as the pointer and is physically poised over a strip of chart paper (Fig. 34-1 is an example) that is pulled under the tip at a constant speed. This mechanism will therefore make a Y-time recording. The Y-axis data is the deflection of the pen, while the X-axis data is a timebase controlled by the paper motion. Because of this arrangement, the pen traces out the time waveform of the input signal applied to the coil.

Figure 34-4 shows the circuit of a typical solid-state PMMC galvanometer driver stage. It is basically a complementary symmetry push-pull amplifier with (usually, but not always) differential

Permanent Magnet

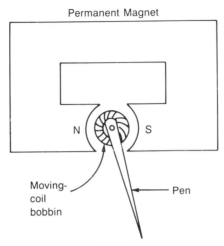

Fig. 34-3. PMMC galvanometer.

inputs. Sub-stage A1 represents the preamplifiers, while transistors Q1 and Q2 control the coil current. If you recognize this circuit as similar to certain high-fidelity audio power amplifier circuits, then you are quite correct. In older equipments vacuum tubes were used, but in all modern equipment the push-pull amplifiers are solid-state. The

state of the art now also includes power integrated circuits and hybrids for pen drivers.

Styles of Pen Action

The PMMC pen assembly sweeps a circularly arced path and so will write in a curvilinear manner (Fig. 34-5A). The pen tip travels in an arc because

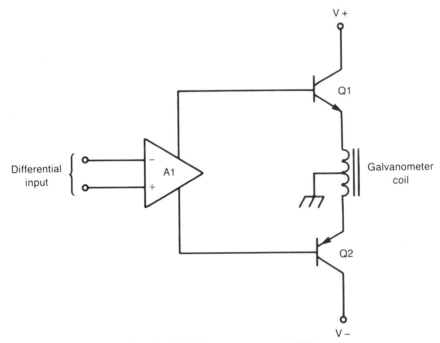

Fig. 34-4. PMMC galvanometer amplifier.

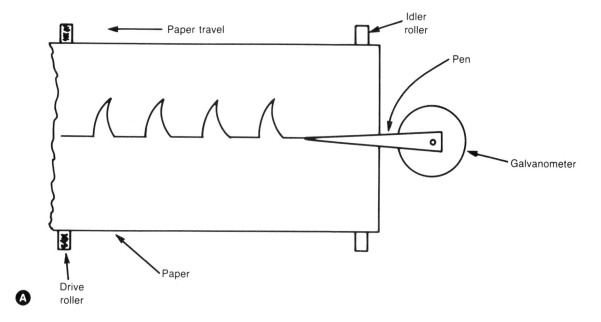

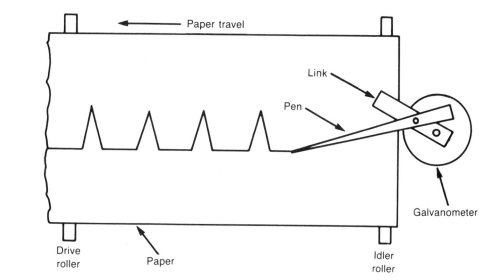

Fig. 34-5. Pen writing systems.

of the rotary motion of the moving-coil assembly. In some cases this may be tolerable, especially if the user will record on the kind of chart paper that has a curved grid. This type of recorder is all but obsolete and is only found in low-cost instruments or those intended solely for student use. It is espe-cially useful for low-cost situations where the peak amplitude, rather than the waveshape, is the data being sought.

The pivoted-pen system of Fig. 34-5B is one solution to the problem of curvilinear distortion. This method is used on almost all ECG and EEG

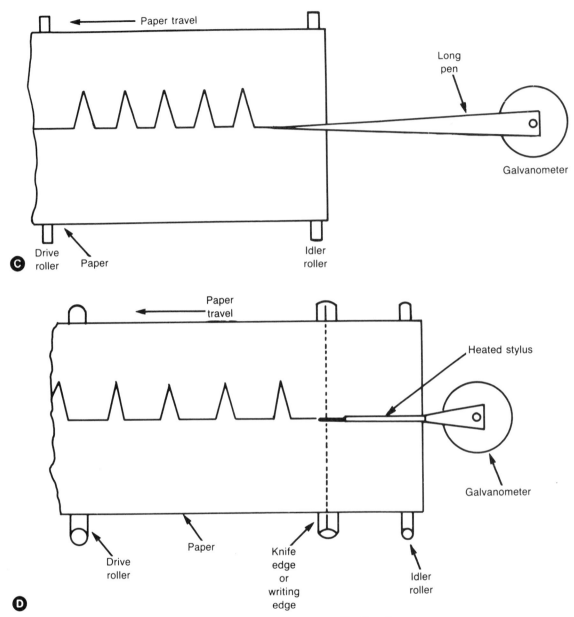

Fig. 34-5. Pen writing systems. *(Continued)*

machines, and on higher quality industrial/scientific machines. The PMMC is not connected directly to the pen, but rather through a mechanical link that is designed to translate curvilinear motion of the PMMC to nearly rectilinear motion required of the pen tip.

A pseudorectilinear system is shown in Fig. 34-5C. In this type of system, the recorder pen is very long compared with the maximum amplitude permitted on the chart paper (i.e., its width). The pen tip, therefore, travels in an arc whose length is small compared with the radius (i.e., pen length).

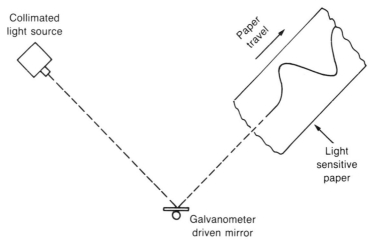

Fig. 34-6. Optical galvanometer.

The tracing will appear nearly linear (sort of, it doesn't work very well).

Our final form of pen system is shown in Fig. 34-5D. This arrangement is used with a special type of paper that turns black when heated. The pen is a heated stylus rather than an ink squirter. Note that the writing point on the stylus is not the very tip, but rather a surface immediately

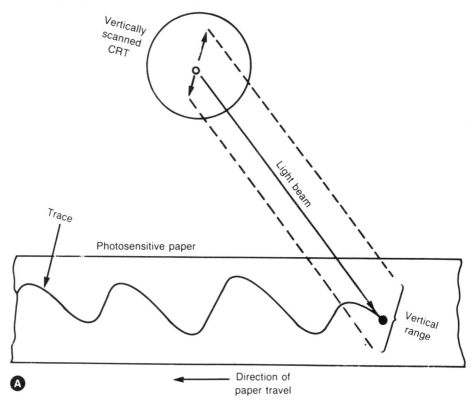

Fig. 34-7. Photo-optical camera recorder.

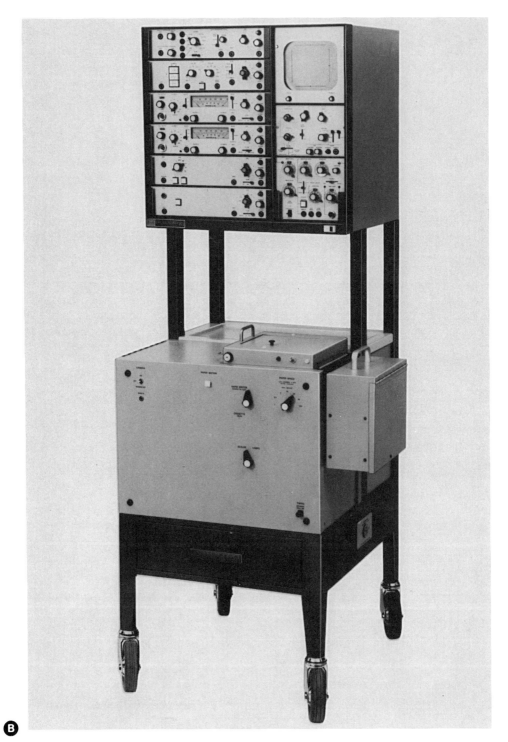

Fig. 34-7. Photo-optical camera recorder. *(Continued)*

behind the tip. This type of writer uses a special writing surface called a knife edge, or writing edge. The mark is made on the paper by the contact of the heated stylus with the paraffin-treated paper. The stylus still travels in a curvilinear path, but the resultant tracing is rectilinear because the knife-edge is straight. In maintaining a thermal recorder, incidentally, it is necessary that the pen pressure on the paper, the pen temperature, and the tautness of the paper across the knife edge be controlled in accordance with the manufacturer's instructions, or a distorted tracing will result.

WRITING METHODS

There are several different methods for writing on analog recorders; that is, for making the mark on the paper. These methods are: direct contact, thermal, ink pen, ink jet, and optical. More recently, optical methods have included lasers.

Recorders of any type, whether using a stylus (direct contact and thermal) or an ink pen, will have a very low frequency response due to the inertia of the writer assembly. Most such recorders have an upper -3 dB frequency of 100 to 200 hertz (Hz). Ink jet and optical types, on the other hand, have a higher natural frequency response because their writing assemblies are lighter. These devices have -3 dB points in the 1000 to 3000 Hz range.

Direct-contact recorders use a special type of chart paper that has a carbonized backing. When pressure is applied to the front of the paper, a black mark will appear through the sheet. Most of these instruments are in the class of "recording volt-ohm-milliammeters," and are only occasionally seen today.

The thermal recorder, discussed above also, uses a special wax or paraffin-treated paper. When this paper is heated, it turns black. Ordinary room temperatures will not make the conversion, but the heated stylus will. Storage care is needed, however, because a very hot storage room for a prolonged period of time will "gray out" the paper in a manner similar to fogging seen on improperly handled photographic films.

The stylus in a thermal system is little more than a heated resistance element connected to a low-voltage ac or dc power supply. Early models formed the stylus from a U-shaped piece of iron wire, while modern models use a resistance wire inside of a cylindrical metal housing. In both cases a low-voltage electrical power source energizes the element.

Ink pen writers (also called "ink slingers" because of their habit of spreading goo all over the shop during maintenance actions) use a hollow pen and ink reservoir to write on regular paper. In some machines the ink is pressurized in an atomizer-like reservoir, while in others a small ink cartridge is used. In some multichannel chart recorders, a single ink reservoir serves all pens via an ink manifold and pressure-pump system.

The high-velocity ink jet recorder is capable of higher frequency responses than pen-type ink recorders because the ink supply is provided through a small, low-weight nozzle mounted on the PMMC. The ink is sprayed at the paper and produces a surprisingly distinct tracing with little fuzziness. The ink-jet recording is nearly rectilinear. The popular *Mingograf* system used in Europe is an ink-jet machine.

There are two types of optical recorder in classical systems. One type is based on the PMMC galvanometer, and has a little mirror mounted on the "galvie" in place of a pen or pointer (see Fig. 34-6). The other form of optical recorder is a CRT-based design (see Fig. 34-7). In this case the vertical component of the signal is scanned across the CRT, and the photosensitive paper is pulled across the CRT face (providing the horizontal timebase). This system is shown schematically in Fig. 34-7A, and an example of such an instrument is shown in Fig. 34-7B. These instruments are capable of relatively high frequency operation, especially compared with pen-type systems.

SERVORECORDERS

The PMMC is only one form of chart recorder. In this section I will discuss the servorecorder system (archaic terms for this same class of instruments are "servopotentiometers" and "recording potentiometers"). You find these instruments in both Y-time (Fig. 34-2) and X-Y versions;

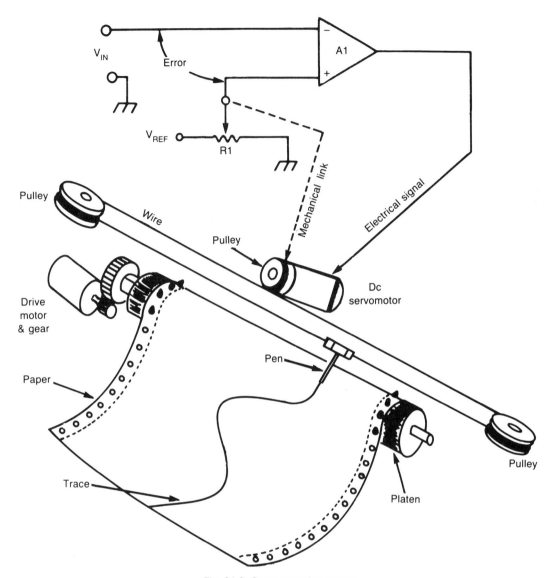

Fig. 34-8. Servo recorder system.

most X-Y versions can be made into Y-time by sweeping the X-dimension with a sawtooth timebase waveform.

The schematic presentation shown in Fig. 34-8 is a basic Y-time servorecorder, not unlike the machine shown earlier in Fig. 34-2. The paper is pulled over a platen by a motor drive system. The pen (or heated stylus) is poised over the paper, on the platen. The pen is connected to a wire and

pulley system that is driven by a dc servomotor. The motor, in turn, is driven by a DC servoamplifier. One input to this differential amplifier is the signal being recorded, V_{IN}. The other input is a reference potential supplied by a potentiometer position transducer (R1) that is mechanically driven by the dc servomotor. The motor is driven to a new position whenever the signal from the position transducer is not equal to the input signal.

The pen thus glides back and forth across the paper surface in response to the servomotor and servoamplifier reacting to the input signal.

Figure 34-9 shows the pen and position transducer system from this type of recorder. In Fig. 34-9A we see that the pen holder is attached to the drive wire but also rests on (and shorts together) a resistance element (ac) and a shorting wire (B). These elements (ABC) form the position transducer potentiometer (see Fig. 34-9B).

RECORDER SYSTEM PROBLEMS

All pen assemblies have mass, regardless of the type of drive mechanism used to propel the pen about the chart paper. Because of the inertia produced by this mass, the pen will not begin moving from rest until a certain minimum signal voltage is present. This phenomenon results in a deadband in the response characteristic (see Fig. 34-10). The deadband signal for any given recorder is the largest signal that will <u>not</u> produce a change in pen position. In most quality instruments, the deadband is 0.05 to 0.1 percent of fullscale.

The deadband can create distortion of those low-level signals that have amplitudes approximating the deadband voltage. The solution is to provide adequate preamplification to slew through the deadband as rapidly as possible. This requirement

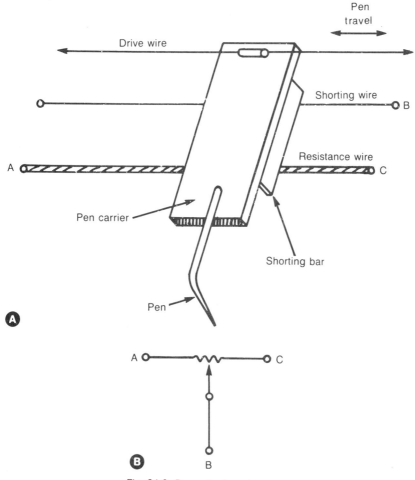

Fig. 34-9. Pen potentiometer system.

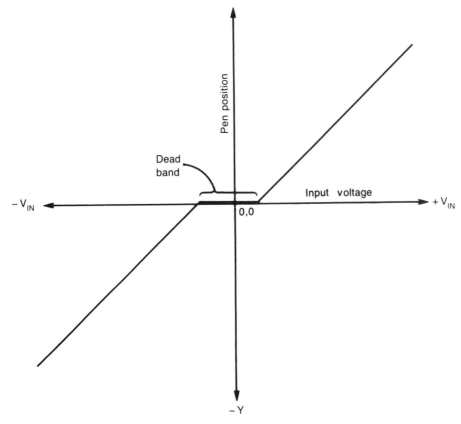

Fig. 34-10. Deadband phenomena.

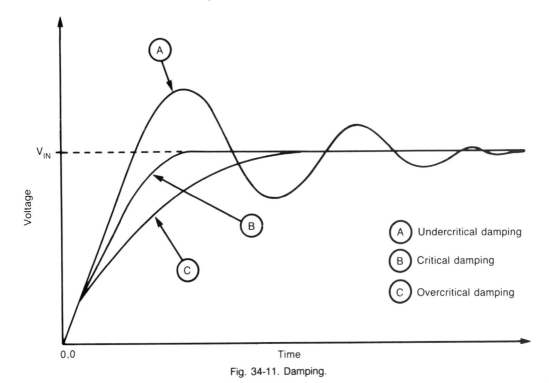

A Undercritical damping

B Critical damping

C Overcritical damping

Fig. 34-11. Damping.

forces us to use as much of the recorders amplitude range as possible.

Another problem is overshoot and undershoot in the response of the system to a step-function signal. Figure 34-11 shows three forms of response. An undercritically damped recorder (trace "A") will seriously overshoot the position dictated by the input voltage. An overcritically damped system (curve "C") is sluggish and approaches the proper limit slowly. The well-designed critically damped recorder (curve "B") approaches the limit neither too quickly nor too slowly.

Some quality recorders have a damping system to fine-tune the pen mechanism to near-critical damping. Some models, such as the Hewlett-Packard, use a capacitance position transducer in a feedback control system to assist in this damping.

Chapter 35

Analog Digital Meters

E LECTRONIC METERS HAVE LONG BEEN A PART OF data-collection efforts, and even in the computer age they find multiple uses. A meter can be either a handheld or workbench "multimeter" model, in which a variety of different measurement functions are provided. Or it can be a dedicated single-parameter model designed to be installed on an instrument panel. A meter can be either digital or analog, single scale or multimeter.

Figure 35-1 shows a digital multimeter. This type of instrument will examine the signal at the input connector and produce a digital reading indicating its voltage, current, or resistance depending upon which switch is selected. Some models available today even have a capacitance-reading capability. You will see designations such as "2-1/2 digit, 3-1/2 digit," and so forth. The "half digit" is one that can be only 0 or 1, and functions in the manner of an overflow indicator with nearly 100 percent overrange capability. The 2-1/2 digit instrument will measure 0 to 1.99 volts, the 3-1/2 digit measures 0 to 1.999 volts, and the 4-1/2 digit measures 0 to 1.9999 volts. Some digital instruments are panel meters. These instruments are typically single range and will measure from 0 to typically 1.999 volts or 19.99 volts.

There are also models available today (see Fig. 35-2) that accept the GPIB/IEEE-488 computer control connector. These instruments are computer programmable and can be used in automatic test equipment or automated data-collection systems. Although seemingly expensive compared with non-GPIB devices of the same measurement capability, the usefulness of the automated system cannot be overestimated.

The analog form of meter uses a (shown in Fig. 35-3) deflection pointer to indicate the voltage on a printed scale. Although there are now modern electronic versions of these instruments on the market, the volt-ohm-milliammeter (VOM) form of instrument dates back to before World War II . . . and is still going strong. Although modern electronics people tend to disdain the old-fashioned meters, one must be aware that there are situations where the old VOM-type of instrument is actually superior. One such situation is in the presence of strong electrical or electromagnetic fields (radio interference). Communications and broadcasting technicians often use these instruments because of their insensitivity to the strong local fields around radio transmitters.

Fig. 35-1. Digital multimeter (DMM).

USING METERS

Both analog and digital meters must be connected into the circuit correctly in order to properly measure the parameter under test. Although digital meters seem less sensitive to damage than analog meters (no pointer to break or bend), both can be damaged if connected incorrectly. For example, if a current meter (or multimeter set to a current range) is connected as a voltmeter, it will be damaged—possibly beyond repair.

Figure 35-4 shows the methods for connecting voltage and current meters. In Fig. 35-4A is a hypothetical "strawman" circuit to illustrate the principle. The voltmeter is always connected in parallel with the load, or across the points being measured. The current meter (ammeter, milliammeter, microammeter, etc.) is always connected in series with the load. Reflection on the nature of voltage and current in elementary electricity textbooks will reveal why this is so.

You can, however, sometimes use other types of meters for different tasks. It is well known, for example, that the basic meter movement is the current meter. A voltmeter is made from a current meter by simply adding a series-connected "multi-

pier" resistor to the basic meter. It is known that a given voltage is represented by a given current flow because of Ohm's law:

$$V = I \times (R_M + R_s)$$

where

V is the applied voltage
I is the current indicated on the meter
R_M is the internal resistance of the meter
R_s is the resistance of the external multiplier resistor.

Figure 35-4B shows a method for using a

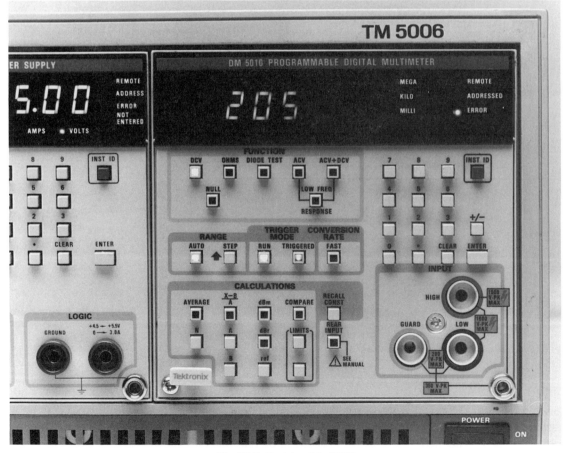

Fig. 35-2. Rack/module DMM.

Fig. 35-3. VOM.

voltmeter to measure a current flow. A physiologist I knew during my time as a biomedical engineer used this method to measure the current in an electrophoresis column. He had a collection of Datel digital voltmeters (panel meters) that would measure 0 to 1,999 millivolts. He wanted to measure currents on the order of 10 to 15 milliamperes. The circuit in Fig. 35-4B represents his

electrophoresis column: R1 is the internal resistance of the 750-volt dc regulated power supply (V), R2 is the resistance of the electrophoresis fluid column (approx. 75 kilohms), and R3 is a resistance added to make the measurement. By making R3 10 ohms, the voltage drop is $V = I \times 10$. Thus, a 10-mA current produces a 100-mV drop, for a scale factor of 10 mV/mA. As long as R3 is much

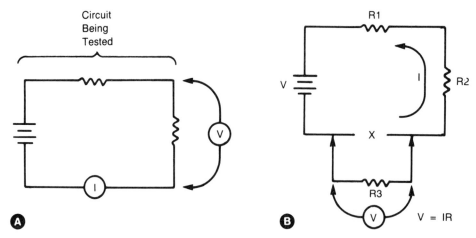

Fig. 35-4. Meter connections.

less than R1 + R2, then the accuracy of the system does not suffer.

DATA COLLECTION WITH METERS

The main purpose of a meter is to supply an
output indication that can be read by a human
operator. Unfortunately, data collection in such a
situation is notoriously difficult and totally ineffi-
cient: a human operator must read and write down
all readings. Fortunately, there is hope.

The best solution for those with money is to
supply a meter that has either an analog or digital
output, of which many examples exist. There are
also recording instruments that will somehow store
data or record it on paper. Simpson for years made

a recording VOM that found wide acceptance.
More modern instruments contain a CMOS memory (and other circuitry) and function as data
loggers.

Digital panel meters often have a digital output available on the back connector. In most of
these instruments there will be either parallel or
serial-formatted versions of the binary coded decimal (BCD) signals that actually drive the digital
display. Although the coding format of some such
instruments is a little complex, it can be decoded
easily in a computer used as a data logger. Alternatively, some digital meters can be easily altered to
bring out these lines. Although most modern instruments are based on a single-chip design, the

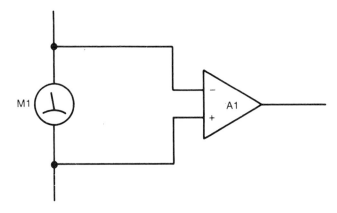

Fig. 35-5. Making an analog signal.

LED or LCD display lines are usually in multi-plexed seven-segment format and can also be used.

Figure 35-5 shows a method for making an analog signal from an analog meter. All analog meters have an internal resistance, so when current flows in the meter it produces a voltage drop. This voltage drop can be easily picked up on a differential amplifier (A1) and converted to a single-ended analog signal. By selecting the amplifier gain you can also scale the signal value. The analog signal is then applied to a computer or other data-acquisition instrument.

Chapter 36

Oscilloscopes

THE OSCILLOSCOPE USES AN ELECTRON BEAM TO write the shape of the input waveform on the phosphorous screen of a cathode-ray tube. An example of a service and laboratory oscilloscope is shown in Fig. 36-1. This particular model has a vertical bandwidth which is sufficient for anything most readers of this book might be involved except for computer repair. A timebase is provided by a sawtooth generator that deflects the beam back and forth across the screen from left to right. The characteristics of the sawtooth (its slope, frequency, etc.) determine the sweep rate.

There are two forms of basic oscilloscope as classified by sweep system. The freerunning oscilloscope is an older and lower cost method in which the sawtooth responsible for the horizontal sweep runs asynchronously. It is provided by an unsynchronized astable multivibrator. Typical of freerunning oscilloscopes is that the horizontal sweep control is marked in terms of frequency: e.g., 1 kHz, 10 kHz, etc. The other form of oscilloscope, which is the type typically used in all modern systems, is the triggered-sweep type. In this type of oscilloscope, the horizontal sweep is not allowed to run until a certain minimum amplitude signal is present in the vertical input channel. In other words, the horizontal sweep is triggered from the vertical signal. These oscilloscopes are more costly, but because of their much greater application have all but supplanted freerunning models in most areas. The model shown in Fig. 36-1 is a triggered-sweep model.

TYPICAL OSCILLOSCOPE CONTROLS

The oscilloscope in Fig. 36-1 is a Heath Model IO-4205. On the lefthand side of the oscilloscope are the vertical inputs and associated controls. The input connectors are standard BNC connectors, which are common to many different instrumentation applications. Each channel has its own input attenuator. The center control on each is a vernier and allows the input signal to be scaled as needed. But the markings on the step selector on the outer ring of each control are valid only when the vernier is in one position, designated "CAL." The calibrations on the step selector are in terms of volts (or millivolts) per centimeter. For example, when the step selector is set to "0.1 V/cm," the electron beam on the screen of the CRT will move one centimeter for each 0.1 volt of input signal.

Each channel also has its own position control. This is a continuously variable control that allows

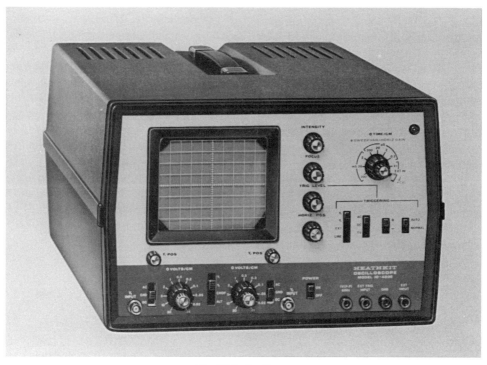

Fig. 36-1. Oscilloscope.

each of the two beams to be placed at any point on the screen of the CRT. There is no limitation on where the two beams are placed, but for the sake of consistency, most operators place the beam for the upper control (channel A) at the top of the screen, and that for channel B at the bottom of the screen.

Another control associated with each channel is the AC-GND-DC switch. These switches allow the operator to select either dc coupling or ac coupling for the input signal. In the ac-coupled mode, the dc component of the input signal is blocked by a capacitor at the input connector, and only the time-varying portion is applied to the amplifier. The GND position shorts the input of the vertical amplifier to ground (it does not short the input connector to ground!). This mode allows the operator to find the zero-volts baseline and is used when making measurements. In most cases the electron beam is positioned to either of three positions when the input selector is in the

GND mode: center screen, top line on the screen, or bottom line on the screen.

The MODE control allows the operator to select different vertical modes. For example, CH1 allows only the channel A signal to be applied to the CRT, CH2 allows only the channel B signal to be applied to the CRT, DUAL allows both to be applied to the CRT screen. There are two different submodes of dual-beam operation: chop and alternate. In the CHOP mode, the electron beam is rapidly switched back and forth between the two channels, sampling first one then another across the screen. The ALTERNATE mode is just the opposite: the oscilloscope sweeps only one channel at a time. For example, there will be one sweep of channel 1, followed by one sweep of channel 2, etc. The two beams will appear to be on the screen at the same time in both case if the sweep/sampling rates are fast enough.

You can select which channel is used to trigger the sweep of the oscilloscope beam. You can also select whether the electron beam sweeps immediately when triggered, or after a specified delay time. There is also the ability to trigger the sweep externally to the oscilloscope. The external trigger input allows the computer to instigate the sweep action. A brief TTL-compatible pulse output to the oscilloscope from a computer is applied to the EXT TRIG input, and causes the sweep to be initiated. Designers should consider this input capability when planning new systems.

Certain modern oscilloscopes are now sold with a General Purpose Interface Bus (GPIB) connector on the rear panel. The GPIB, also called the IEEE-488 bus and Hewlett-Packard Interface Bus (HPIB), allows a computer to control the operation of the oscilloscope. Designers of data-acquisition systems will find a large potential for application of GPIB-controlled instrumentation. In synopsis, however, let's point out that the oscilloscope is a "listener" device, and when its code number (0 to 15 in hexadecimal code) appears on the bus, it will perk up its ears to receive programming instructions. Using this system, you can make the "front panel" adjustments of the oscilloscope electronically, under computer program control, and thereby eliminate the need for manual adjustment between jobs. You can purchase IEEE-488/GPIB adaptors for such computers as the IBM-PC series, Apple IIe, and so forth.

Appendix A

Basic Computer Programs

PROGRAM NO. 1: ROUNDING A NUMBER

THIS PROGRAM IS INTENDED FOR USE AS A SUB-routine in a larger program. It will round off a number according to the usual rule in which a decimal fraction of 0.5 or more is rounded up, and a decimal fraction of less than 0.5 is rounded down. In the following program, a value of x is selected for test purposes. You must either enter the correct value of x, or adapt the program to the data being processed.

Program 1. Rounding a number.

```
10    REM   The name of this program is ROUND
20    REM   It is intended for use as a subroutine in a larger program
30    REM   To use this program yourself, replace line 100 with the correct
35    REM   value of X (the given value is merely an example)
100   X = 3.0864532#
120   X = X*100
130   D = X - INT(X)
140   X = INT(X)
150   IF D ¶ .5, THEN X = X+1
155   IF D = .5, THEN X = X+1
160   IF D § .5, THEN X = X
170   X = X/100
200   PRINT X
```

PROGRAM NO. 2: TRUNCATION OF A NUMBER

All decent versions of BASIC, including those of the "mini" variety intended for small computers with extremely limited memory, have an "integer" function, i.e., INT(arg). This function will truncate the number defined by the argument without rounding (some INT functions will round the nega-tive number but not the positive). Unfortunately, this is not exactly what we might want. In financial programs, for example, you really might not care about 1/1000 of a cent, or 1/10000 of a cent, but will want to express amounts only down to the penny. This requirement means two decimal points. Similarly, in scientific and engineering pro-grams, you might want to truncate a value at, for example, two decimal places because no one really believes that the pressure was 89.23476598 Torr, but might accept "89.2" as an answer — remember "significant figures" from Physics I?

This program will allow us to truncate a number to two decimal places. It uses the so-called "pull it out, cut it off, push it back in" method. You first multiply the number in line 100 by 100, then use the INT(X) function, and then divide X by 100.

A limitation on this method is that it will render zero all numbers that require three or more leading zeroes between the digit and the decimal point. Thus, "0.000470 pF" will print out "0.0 pF" much to the annoyance of the user. If you normally expect much smaller values, then either convert the units to a lower order unit, or use a different multiplier.

Program 2. Truncating a number.

```
10 REM   The name of this program is TRUNCATE
20   REM   It is intended for use as a subroutine in a larger program
30   REM    To use this program yourself, replace line X with the correct
35   REM    value of X (the given value is an example)
100 X = 3.0864532#
120 X = X*100
140 X = INT(X)
170 X = X/100
200 PRINT X
```

PROGRAM NO. 3: MONTHLY LOAN PAYMENT

This program is not generally used in scientific or engineering applications that make up the bulk of this data acquisitions text, but was included because of the immense interest (that's a pun!) generated by readers. I noticed that one of the most popular programs on the 50 kilobuck computer in our (engineering) office, after "Artillery Dual" (a video game), was the monthly loan payment program. This matter affects most of us at one time or another, whether we finance a car, a home or some dream. This program will help you determine ahead of time whether or not the monthly payment will be within your means.

I recommend that you try the same financial scenario with different loan lengths. For example, try a mortgage with 20, 25, and 30 years length. You might be surprised at how little more the 20-year mortgage payment is than the 30-year payment. Yet you only pay interest for 2/3 the

time, and thus it's much cheaper in the long run to take out the shorter mortgage.

This program will prompt you to enter several bits of information. First, you will be asked to ENTER the principal amount of the loan in dollars. Do not use commas or other symbols (e.g., "$"). For example, enter "50000" for fifty-thousand dollars instead of "50,000", "$50,000", "$50000", or some other variant. The interest rate must be entered in decimal form. For example, eleven and three-quarters percent is not "11-3/4,"but "11.75."

Finally, you will be asked to enter the loan period in MONTHS (not years). The three most popular mortgage lengths are shown in the print out of the program as an aid, but these are not the only values that you may use. There is no reason why you can't use any length in months. Of course, it is not advisable to enter a value less than one: it really doesn't make much sense to make a monthly payment on a three-day loan *(sigh)*.

Program 3. Monthly loan payment.

```
5   REM    LNPAY1
10  REM     This program computes the monthly payment on a loan
15  LET J = 30
18  GOSUB 5100
```

```
20 PRINT TAB(J);"* * * * * * * * * *"
30 PRINT TAB(J);"*   MONTHLY PAYMENT   *"
40 PRINT TAB(J);"* * * * * * * * * *"
50 GOSUB 5000
55 GOSUB 5000
60 PRINT "ENTER 1 to continue"
65 INPUT C
70 IF C = 1, THEN 90 ELSE 60
90 GOSUB 5100
100 PRINT "ENTER the amount of PRINCIPAL in dollars"
105 GOSUB 5000
110 INPUT P
120 PRINT "ENTER the INTEREST rate (APR) in percent"
125 PRINT "For example, 11.75"
130 GOSUB 5000
140 INPUT I
150 LET I = I/100
160 LET I = I/12
170 GOSUB 5000
180 PRINT "ENTER the number of MONTHS the loan will run"
185 PRINT
186 PRINT "Note: 20-years is 240 mos., 25-years is 300 mos., and"
190 PRINT "        30-years is 360 mos."
200 GOSUB 5000
210 INPUT N
300 LET Z = 1/(1+I)
320 LET Z = Z¢N
330 LET Z = 1 - Z
340 LET MP = (P*I)/Z
350 LET MP = 100*MP
360 LET MP = INT(MP)
370 LET MP = MP/100
380 GOSUB 5100
500 PRINT "Monthly payment on";P;"for";N;"Mos. is:";MP
510 GOSUB 5000
520 GOSUB 5000
530 PRINT "Do another calculation"
540 PRINT "YES = 1, NO = 2"
550 INPUT A
560 IF A =1, THEN GOTO 100
570 IF A = 2, THEN GOTO 8900
580 GOTO 510
1000 END
5000 FOR X = 1 TO 5
5010 PRINT
5020 NEXT X
5030 RETURN
5100 FOR X = 1 TO 30
5110 PRINT
5120 NEXT X
5130 RETURN
8900 GOSUB 5100
9900 PRINT "FINISHED!"
9910 GOSUB 5000
9920 GOSUB 5000
10000 END
```

PROGRAM NO. 4: TOTALING OF STRING OF NUMBERS

This is another program that is intended to be used as a subroutine in a larger program—it is a tool rather than a complete program. This program will permit you to input a string of up to 500 numbers designated (for no reason but whimsey) "X(1)" to "X(500)". Of course, the program can be modified to accept larger arrays of data if you REALLY want to sit there and fingerbone all those numbers into the computer. Just change the dimension-X statement ("DIM X(500)") argument to some number other than 500. It is necessary to have this dimension statement because the computer will not accept arrays larger than 12 entries without it. On the other hand, the program will automatically allocate one byte of memory for each element of the dimensioned array, so you need to keep the array size within some kind of reasonable bounds, or else there will be little room for anything else. Most 64K computers, for example, will hiccup if you attempt to try something like "DIM X(65536)."

The equation which this program solves is the trivial expression

$$SUM = X(i)$$

As with most of the programs in this collection, you may adopt it to line printer operation by changing all of the PRINT statements to the LPRINT statement. You may also want to delete the listing of X(i) values (see printout following listing) in your output. This can be done by eliminating lines 260 through 270 in the program. Similarly, you may want to LPRINT the x listing and result ("SUM") but not the input X(i) data. This is done by simply making the PRINT statements in lines 260, 280, and 290 LPRINT, and leaving the PRINT statements elsewhere in the program alone.

Also, this program makes rather extravagant use of spacing loops. In the event you prefer to use the line printer rather than the video screen, this feature will make your output both wasteful of overpriced computer paper and somewhat silly looking. In that event, go into the program and change several of the loop counters. Specifically, change line 310 to something a lot less than 30; try 3 for example. Also, line 450 results in a display that is pretty neat on a video display but totally ridiculous on a printer: change "30" to "2" or something else low that strikes your fancy.

Program 4. Totaling a string on numbers.

```
100   REM   The name of this program is TOTALIZE.  It will sum the values
110   REM   entered from the keyboard.
120   SUM=0
130   B=1
140   DIM X(500)
150   GOSUB 310
160   PRINT "How many values of ±X' are there?"
170   INPUT N
180   GOSUB 310
190   FOR I = B TO N
200   PRINT "Input X(";I;")";
210   INPUT X(I)
220   NEXT I
230   GOSUB 310
240   FOR S = 1 TO N
250   SUM = SUM + X(S)
260   PRINT "X(";S;") =";X(S)
270   NEXT S
280   PRINT
290   PRINT "SUM ="; SUM
```

```
300 GOTO 350
310 FOR W = 1 TO 30
320 PRINT
330 NEXT W
340 RETURN
350 PRINT
360 PRINT
370 PRINT "Finished?"
380 PRINT "1 = Yes"
390 PRINT "2 = No"
400 INPUT Z
410 IF Z = 2, THEN GOTO 100
420 IF Z = 1, THEN GOTO 440
430 GOTO 350
440 GOSUB 310
450 FOR Q = 1 TO 30
460 PRINT "********************¶¶¶¶¶¶ END OF PROGRAM §§§§§§§§*****************"
470 NEXT Q
480 END
```

PROGRAM NO. 5: GENERATING A TABLE OF VALUES FOR A MATHEMATICAL FUNCTION EMBEDDED IN THE PROGRAM

This is another program that is intended as a subroutine in a larger program. You must place the correct BASIC expression for the equation that you want to solve on line no. 320 of the program. In this case, just as an example, the expression is

$$y = 2x^2$$

You will be asked to enter the number n of x data points that will be evaluated. In the example in following the listing, the number selected is 10, so the program will evaluate the expression $y = 2x^2$ for the ten data points $x(i) = x(1), x(2), x(3), x(4), x(5), x(6), x(7), x(8), x(9)$, and $x(10)$. These data are printed out along with the y data that results. Fig. A-1 shows the quadratic formula by which the roots are initially determined.

Program 5. Evaluation of an embedded function.

```
100 C = 12
110 S = 13
120 REM  The name of this program is FUNCTION
130 REM  This program generates a table of Y(i) values for a
140 REM  given function.
150 GOSUB 410
160 PRINT "This program will generate a table of values from a"
170 PRINT "given function.  "
180 PRINT
190 GOSUB 450
200 GOSUB 410
210 PRINT "ENTER the number of X data points and press Cr"
220 PRINT "N =";
230 INPUT N
240 GOSUB 410
250 PRINT "ENTER DATA:"
260 PRINT
270 FOR I = 1 TO N
280 PRINT "X(";I;") =";
```

Program 5 continued

```
290 INPUT X(I)
300 NEXT I
310 FOR I = 1 TO N
320 Y(I) = 2*(X(I)¢2)
330 NEXT I
340 GOSUB 410
350 PRINT
360 PRINT "X-Data";TAB(S);"Y-Data"
370 FOR I = 1 TO N
380 PRINT X(I),Y(I)
390 NEXT I
400 GOTO 480
410 FOR I = 1 TO 30
420 PRINT
430 NEXT I
440 RETURN
450 PRINT "Press ANY key to continue:"
460 A$=INKEY$: IF A$="" THEN 460
470 RETURN
480 END
```

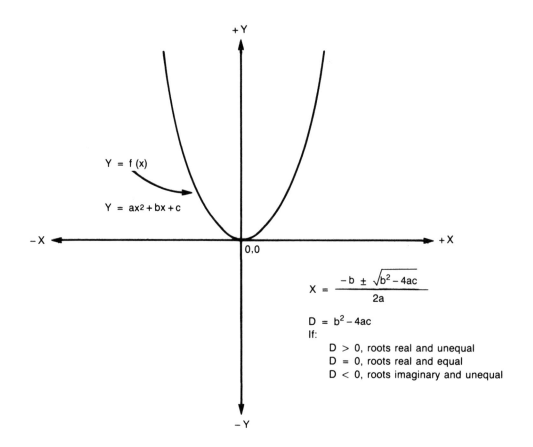

$+Y$

$Y = f(x)$

$Y = ax^2 + bx + c$

$-X$ $+X$

$0,0$

$-Y$

$$X = \frac{-b \pm \sqrt{b^2 - 4ac}}{2a}$$

$D = b^2 - 4ac$

If:

     $D > 0$, roots real and unequal

     $D = 0$, roots real and equal

     $D < 0$, roots imaginary and unequal

PROGRAM NO. 6: NAIVE DIFFERENTIATION I

This program is an example of naive differentiation. You will supply the program with the values of [x(i), and y(i)], and it will calculate the derivative using [y(i) − y(i-1)]/[x(i) − x(i-1)]. The

rigor with which the program makes the calculation depends upon several factors including the size of the x and y differentials. Normally, you will find that (except for a simple rectilinear "curve" of the form y = mx + b) the smaller the x and y differentials, the more accurate the derivative.

Program 6. Naive differentiation I.

```
5   REM    This program is named DIFFY1
100 REM    This program calculates the derivative of a small region of a curve
110 PRINT "This program uses the naive method for calculating the derivative"
120 PRINT "of a small region of a curve.  Four data points are required, two"
130 PRINT "adjacent points each on both X and Y axis (or T and Y axis if the"
140 PRINT "function is time varying"
150 GOSUB 430
160 GOSUB 430
170 PRINT "Enter Y1 and press Cr"
180 GOSUB 430
190 INPUT Y1
200 GOSUB 430
210 PRINT "Enter Y2 and press Cr"
220 GOSUB 430
230 INPUT Y2
240 GOSUB 430
250 PRINT "Enter X1 (or T1) and press Cr"
260 GOSUB 430
270 INPUT X1
280 GOSUB 430
290 PRINT "Enter X2 (or T2) and press Cr"
300 GOSUB 430
310 INPUT X2
320 D = (Y2 - Y1)/(X2 - X1)
330 GOSUB 430
340 GOSUB 430
350 PRINT "Derivative is: ";D
360 GOSUB 430
370 PRINT "Do another? Yes = 1, No = 2"
380 INPUT A
390 IF A = 1, THEN GOTO 100
400 IF A = 2, THEN GOTO 470
410 GOTO 360
420 GOTO 510
430 FOR X = 1 TO 5
440 PRINT
450 NEXT X
460 RETURN
470 FOR X = 1 TO 20
480 PRINT "Bye-Bye"
490 NEXT X
500 GOSUB 430
510 END
```

PROGRAM NO. 7: NAIVE DIFFERENTIATION II

This is another naive differentiation program and is subject to the same cautions as the previous program (no. 6). This program differs from the previous version in that it will allow you to enter any number of paired [x(i),y(i)] data points, and will print out the derivative for each using the naive method presented previously.

You are free to set the number of [x(i),y(i)] data points as you desire. You are cautioned, however, that when you select the number of point a dimensioned array is created that will chew up lots of memory space. Obviously, then, you will not want to claim an extremely large number of points unless you intend to use them.

This program is most useful for massaging tabulated data for which the equation is not known or easily deduced. Thus, program No. 7 is suited mostly for experimental or electronic-data acquisitions applications where there are more than a few data points. Presumedly, you would use a pocket calculator rather than a large microcomputer if there were but few data points, but then again, one achieves a certain awesome power if a computer is used.

There is a story about concerning one of the all-time greats in computer science. She entered the U.S. Naval service at an advanced point in her career and so had never encountered the bureaucratic mindset before. When she was asked to supply budget figures, the "bean-counter" was shocked, dismayed, and taken aback when she presented some handwritten figures on plain yellow legal paper: The bean-counter tossed her out on her tush. She then went to the computer room, commandeered a terminal and input the figures into a computer. The computer then typed them out on the line printer in exactly the same format that they were in on paper. Superprogrammer then took the striped computer paper back to the bean counter, and presented the "computer run:" the bean-counter triumphantly asked: "Now why didn't you do that before?"

Program 7. Naive differentiation II.

```
100 REM The name of this program is DIFFY2
110 REM This program computes the derivative of input data at several
120 REM points on a curve defined as Y0, Y1, Y2 ... etc.
130 GOSUB 540
140 PRINT "Determine the number of paired Y and X (or T) data points"
150 PRINT "that will exist in the curve"
160 PRINT
170 PRINT "ENTER THE NUMBER OF DATA POINTS (X and Y"
180 PRINT "MUST be equal!), and press CARRAIGE RETURN"
190 PRINT
200 INPUT "NUMBER OF DATA POINTS? ",N
210 GOSUB 500
220 DIM Y(N)
230 DIM X(N)
240 DIM DY(N)
250 DIM DX(N)
260 DIM D(N)
270 FOR I = 0 TO N-1
280 PRINT "ENTER Y(";I;") and press _R"
290 INPUT Y(I)
300 GOSUB 500
310 PRINT "ENTER X(";I;") and Press CR"
320 INPUT X(I)
330 GOSUB 500
340 NEXT I
350 FOR L = 1 TO N
360 DY(L) = Y(L) - Y(L-1)
```

```
370 DX(L) = X(L) - X(L-1)
380 NEXT L
390 FOR M = 1 TO N
400 D(M) = DY(M)/DX(M)
410 NEXT M
420 FOR P = 1 TO N
430 PRINT "DY/DX(";P;") = ";D(P)
440 NEXT P
450 GOSUB 500
460 PRINT "Do Another? Yes = 1, No = 2"
470 INPUT K
480 ON K GOTO 270,580
490 GOTO 460
500 FOR Z = 1 TO 5
510 PRINT
520 NEXT Z
530 RETURN
540 FOR Z = 1 TO 30
550 PRINT
560 NEXT Z
570 RETURN
580 GOSUB 500
590 PRINT "PROGRAM ENDED"
600 GOSUB 500
610 END
```

PROGRAM NO. 8: NAIVE INTEGRATION

This program calculates the integral of a curve of which you input the data points manually. The computer will prompt you to enter a number of [x(i), y(i)] data pair points. (Note: this is not the number of intervals, but rather the number of data points — the number of intervals is n-1). This program uses the Method of Rectangles shown in Fig. A-2 below. The area under the curve is broken up into a number of rectangular areas, which are then summed. The validity of this approach is partially determined by the number of data point pairs, and the size of each interval.

Program 8. Naive integration I.

```
100 REM    This program is named INTGTN1
110 REM    INTGTN1 will create an approximation of the area under a curve
120 REM    using the rectangles method.  It is a naive technique that is none
130 REM    the less useful.
140 IYDX = 0
160 LPRINT "* * * * * * * * * * *"
170 LPRINT "*                   *"
180 LPRINT "*  NAIVE INTEGRATION *"
190 LPRINT "*                   *"
200 LPRINT "*   (Method Number 1) *"
210 LPRINT "*                   *"
220 LPRINT "* * * * * * * * * * *"
230 GOSUB 760
240 LPRINT "PRINT OPTION"
250 LPRINT " 1.  Video Monitor"
```

Program 8 continued

```
260 LPRINT "  2.  Hardcopy Printer"
270 LPRINT "ENTER selection and press Cr"
280 GOSUB 760
290 INPUT P
292 LPRINT P
300 IF P ¶ 2, THEN GOTO 240
310 LPRINT "PRINT JUST INTEGRAL OR DATA TOO?
320 LPRINT
330 LPRINT "  1. Just print the integral value"
340 LPRINT "  2. Print both integral and data values"
350 LPRINT
360 LPRINT "ENTER selection and press Cr"
370 GOSUB 760
380 INPUT V
382 LPRINT V
390 GOSUB 760
400 LPRINT "The number of X and Y data points must be equal."
410 PRINT  "ENTER the number of data point pairs and press Cr"
420 GOSUB 760
430 INPUT N
432 LPRINT N
440 GOSUB 800
450 FOR I = 0 TO N-1
460 LPRINT "ENTER Y(";I;") and press Cr"
470 LPRINT
480 INPUT Y(I)
482 LPRINT Y(I)
490 LPRINT "ENTER X(";I;") and press Cr"
500 LPRINT
510 INPUT X(I)
515 LPRINT X(I)
520 NEXT I
530 GOSUB 800
540 FOR I = 1 TO N-1
550 DX(I) = X(I) - X(I-1)
560 NEXT I
570 FOR I = 1 TO N-1
580 A(I) = DX(I)*Y(I)
590 NEXT I
600 FOR I = 1 TO N-1
610 IYDX = IYDX + A(I)
620 NEXT I
630 IF P = 2, THEN GOTO 840
640 IF V = 1,THEN GOTO 690
650 LPRINT "X Values";"      ";"Y Values"
660 FOR B = 0 TO N-1
670 LPRINT TAB(3);X(B);"            ";Y(B)
680 NEXT B
690 GOSUB 760
700 LPRINT "Integral is: ";IYDX
710 LPRINT
720 LPRINT "Press any NUMBER key and then Cr to continue"
730 INPUT Z
732 LPRINT Z
740 GOSUB 800
```

```
750 GOTO 960
760 FOR S = 1 TO 1
770 LPRINT
780 NEXT S
790 RETURN
800 FOR S = 1 TO 2
810 LPRINT
820 NEXT S
830 RETURN
840 LPRINT "X Values";"      ";"Y Values"
850 IF V = 1, THEN GOTO 900
860 FOR B = 0 TO N-1
870 LPRINT TAB(3);X(B);"          ";Y(B)
880 NEXT B
890 LPRINT
900 LPRINT "Integral is: ";IYDX
910 LPRINT
920 LPRINT "Press any NUMBER key then Cr to continue"
930 INPUT Z
932 LPRINT Z
940 GOSUB 800
950 GOTO 960
960 LPRINT "FINISHED? (select one)"
970 LPRINT "1. Yes, let's get out of here"
980 LPRINT "2. No, let's do another one!"
990 GOSUB 760
1000 INPUT F
1001 LPRINT F
1010 IF F = 1, THEN GOTO 1040
1020 IF F = 2, THEN GOTO 100
1030 GOTO 960
1040 FOR D = 1 TO 3
1050 LPRINT "******************   BYE-BYE   ******************"
1060 NEXT D
1070 FOR D = 1 TO 3
1080 GOSUB 760
1090 NEXT D
1100 END
```

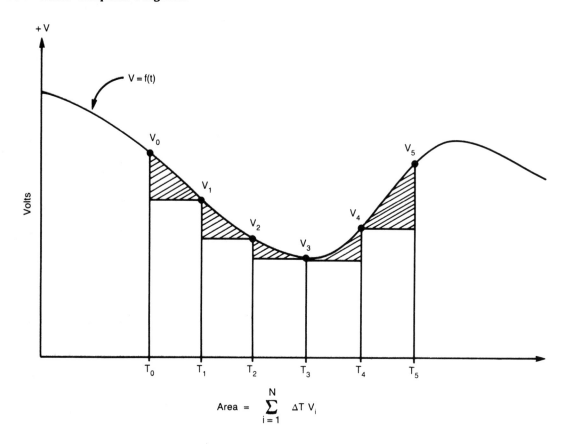

$$\text{Area} = \sum_{i=1}^{N} \Delta T\ V_i$$

PROGRAM NO. 9: NAIVE INTEGRATION II

This program is another version of naive integration. It is similar to program no. 8, except that it takes advantage of the fact that in most cases the intervals are equal throughout the x(i) range. You enter only the y(i) points, and the Delta-X that represents the common interval.

Program 9. Naive integration II.

```
100 REM    This program is named INTGTN2
110 REM    INTGTN1 will create an approximation of the area under a curve
120 REM    using the rectangles method.  It is a naive technique that is none
130 REM    the less useful.
140 IYDX = 0
150 GOSUB 800
160 PRINT "* * * * * * * * * * *"
170 PRINT "*                   *"
180 PRINT "*   NAIVE INTEGRATION   *"
190 PRINT "*                   *"
200 PRINT "*   (Method Number 1)   *"
210 PRINT "*                   *"
220 PRINT "* * * * * * * * * * *"
230 GOSUB 760
240 PRINT "PRINT OPTION"
250 PRINT "  1.   Video Monitor"
260 PRINT "  2.   Hardcopy Printer"
```

```
270 PRINT "ENTER selection and press Cr"
280 GOSUB 760
290 INPUT P
300 IF P ≠ 2, THEN GOTO 240
310 PRINT "PRINT JUST INTEGRAL OR DATA TOO?"
320 PRINT
330 PRINT "  1. Just print the integral value"
340 PRINT "  2. Print both integral and data values"
350 PRINT
360 PRINT "ENTER selection and press Cr"
370 GOSUB 760
380 INPUT V
390 GOSUB 760
400 PRINT "The number of X and Y data points must be equal."
410 PRINT "ENTER the number of data point pairs and press Cr"
420 GOSUB 760
430 INPUT N
440 GOSUB 800
450 FOR I = 0 TO N-1
460 PRINT "ENTER Y(";I;") and press Cr"
470 GOSUB 760
480 INPUT Y(I)
490 PRINT "ENTER X(";I;") and press Cr"
500 GOSUB 760
510 INPUT X(I)
520 NEXT I
530 GOSUB 800
540 FOR I = 1 TO N-1
550 DX(I) = X(I) - X(I-1)
560 NEXT I
570 FOR 1 = 1 TO N-1
580 A(I) = DX(I)*Y(I-1)
590 NEXT I
600 FOR I = 1 TO N-1
610 IYDX = IYDX + A(I)
620 NEXT I
630 IF P = 2, THEN GOTO 840
640 IF V = 1,THEN GOTO 690
650 PRINT "X Values";"      ";"Y Values"
660 FOR B = 1 TO N
670 PRINT TAB(3);X(B);"             ";Y(B)
680 NEXT B
690 GOSUB 760
700 PRINT "Integral is: ";IYDX
710 PRINT
720 PRINT "Press any NUMBER key and then Cr to continue"
730 INPUT Z
740 GOSUB 800
750 GOTO 960
760 FOR S = 1 TO 5
770 PRINT
780 NEXT S
790 RETURN
800 FOR S = 1 TO 30
810 PRINT
820 NEXT S
```

Program 9 continued

```
830 RETURN
840 LPRINT "X Values";"      ";"Y Values"
850 IF V = 1, THEN GOTO 900
860 FOR B = 1 TO N
870 LPRINT TAB(3);X(B);"          ";Y(B)
880 NEXT B
890 PRINT
900 LPRINT "Integral is: ";IYDX
910 PRINT
920 PRINT "Press any NUMBER key then Cr to continue"
930 INPUT Z
940 GOSUB 800
950 GOTO 960
960 PRINT "FINISHED? (select one)"
970 PRINT "1. Yes, let's get out of here"
980 PRINT "2. No, let's do another one!"
990 GOSUB 760
1000 INPUT F
1010 IF F = 1, THEN GOTO 1040
1020 IF F = 2, THEN GOTO 100
1030 GOTO 960
1040 FOR D = 1 TO 30
1050 PRINT "******************** BYE-BYE  ********************"
1060 NEXT D
1070 FOR D = 1 TO 3
1080 GOSUB 760
1090 NEXT D
1100 END
```

PROGRAM NO. 10: NAIVE INTEGRATION III

This method of naive integration is more complex than the previous methods and will yield somewhat better results. Like the previous methods, however, it is intended for persons who have tabulated data that must be manually entered.

Data are entered in [x(i), y(i)] pairs. You will be asked to enter the number of data pairs, so keep in mind that you are not being asked for the number of intervals (n-1). The program uses the method shown in Fig. A-3.

Program 10. Naive integration III.

```
100 REM The name of this program is INTGTN4
110 REM This program is an example of naive integration, but is superior
120 REM to others given previously because it will more nearly approximate
130 REM the true integral for small numbers of data points
140 GOSUB 760
150 PRINT "INTEGRATION -- METHOD No. 4"
160 GOSUB 720
170 GOSUB 720
180 PRINT "The number of X and Y data points must be equal"
190 PRINT
200 PRINT
```

```
210 PRINT "Input the total number of X and Y data points"
220 PRINT "and then press ENTER (Cr):"
230 INPUT N
240 GOSUB 760
250 PRINT "You have selected";N;"data points"
260 PRINT "Is this the correct number?"
270 PRINT "Yes = 1"
280 PRINT "No = 2"
290 INPUT L
300 IF L = 1, THEN GOTO 330
310 IF L = 2, THEN GOTO 160
320 GOTO 250
330 GOSUB 760
340 FOR I = 0 TO N-1
350 PRINT "INPUT Y(";I;")"
360 INPUT Y(I)
370 PRINT "INPUT X(";I;")"
380 INPUT X(I)
390 GOSUB 720
400 NEXT I
410 GOSUB 760
420 PRINT "You have finished enterring data"
430 PRINT "True = 1"
440 PRINT "False = 2"
450 INPUT Z
460 IF Z = 1, THEN GOTO 490
470 IF Z = 2, THEN GOTO 800
480 GOTO 410
490 IYDX = 0
500 FOR I = 1 TO N-1
510 DX(I) = X(I) - X(I-1)
520 A(I) = (DX(I)*Y(I-1)) + (DX(I)*((Y(I) - Y(I-1))/2))
530 NEXT I
540 FOR I = 1 TO N-1
550 IYDX = IYDX + A(I)
560 NEXT I
570 GOSUB 760
580 PRINT "OPTIONS MENU"
590 PRINT
600 PRINT "PRINT RESULTS ON:
610 PRINT
620 PRINT "1.   VIDEO MONITOR ONLY"
630 PRINT "2.   PRINTER ONLY"
640 PRINT "3.   BOTH VIDEO AND PRINTER"
650 INPUT F
660 GOSUB 760
670 IF F = 1, THEN GOTO 890
680 IF F = 2, THEN GOTO 1070
690 IF F = 3, THEN GOTO 1220
700 GOSUB 720
710 GOTO 580
720 FOR Z = 1 TO 5
730 PRINT
740 NEXT Z
750 RETURN
760 FOR Z = 1 TO 30
```

Program 10 continued

```
770  PRINT
780  NEXT Z
790  RETURN
800  GOSUB 760
810  PRINT "ERROR: The actual number of data does not agree"
820  PRINT "with the number enterred earlier"
830  PRINT
840  PRINT "Try Again"
850  PRINT "Press 1 and then Cr to continue:"
860  INPUT D
870  IF D = 1, THEN GOTO 100
880  GOTO 830
890  IF F = 3, THEN GOTO 910
900  IF F = 1, THEN GOTO 940
910  PRINT "OUTPUT ON BOTH PRINTER AND VIDEO MONITOR"
920  GOSUB 1260
930  GOTO 960
940  PRINT "OUTPUT ON VIDEO MONITOR ONLY"
950  GOSUB 1260
960  GOSUB 720
970  IF P = 2, THEN GOTO 1000
980  PRINT "INTEGRAL IS:";IYDX
990  GOTO 1060
1000 FOR I = 0 TO N-1
1010 PRINT "Y(";I;") = ";Y(I);"    X(";I;") = ";X(I)
1020 NEXT I
1030 PRINT
1040 PRINT
1050 PRINT "INTEGRAL IS:";IYDX
1060 IF F = 1, THEN GOTO 1320
1070 IF F = 3, THEN GOTO 1120
1080 LPRINT "OUTPUT ON PRINTER ONLY"
1090 GOSUB 720
1100 IF F = 3, THEN GOTO 1120
1110 GOSUB 1260
1120 IF P = 2, THEN GOTO 1150
1130 LPRINT "INTEGRAL IS:";IYDX
1140 GOTO 1210
1150 FOR I = 0 TO N-1
1160 LPRINT"Y(";I;") = ";Y(I);"    X(";I;") = ";X(I)
1170 NEXT I
1180 LPRINT
1190 LPRINT
1200 LPRINT "INTEGRAL IS:";IYDX
1210 GOTO 1320
1220 GOSUB 760
1230 PRINT "OUTPUT ON BOTH  PRINTER AND VIDEO MONITOR"
1240 GOSUB 720
1250 GOTO 920
1260 GOSUB 720
1270 PRINT "DATA SELECTION MENU:"
1280 PRINT "1.  OUTPUT INTEGRAL ONLY"
1290 PRINT "2.  OUTPUT INTEGRAL AND X/Y DATA"
1300 INPUT P
1310 RETURN
```

```
1320 GOSUB 720
1330 PRINT "1.  FINISHED"
1340 PRINT "2.  DO ANOTHER"
1350 INPUT R
1360 IF R = 2, THEN GOTO 100
1370 IF R = 1, THEN GOTO 1390
1380 GOTO 1320
1390 GOSUB 760
1400 FOR N = 1 TO 50
1410 PRINT"********** BYE-BYE ***********"
1420 NEXT N
1430 GOSUB 760
1440 GOSUB 760
1450 PRINT "FINISHED"
1460 END
```

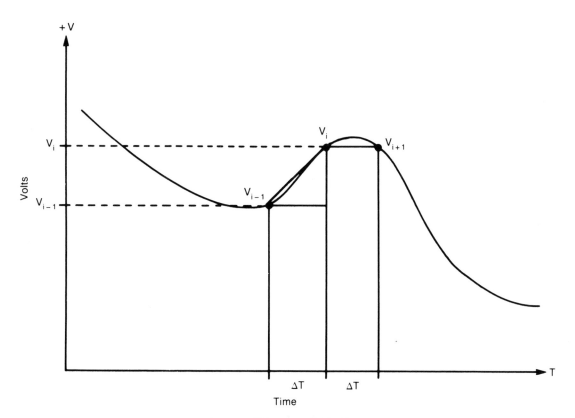

Area $= \Delta TV_{i-1} + \frac{1}{2}\Delta T \, (V_i - V_{i-1})$

PROGRAM NO. 11: INTEGRATION BY RIEMANN SUM

The method of Riemann sums integration is similar to naive integration but uses the expression

$$f(x) \ dx = (1/n) \ y(i)$$

The factor "n" is the number of intervals, from which you may deduce that the integral is more accurate if the number n is large. For the interval y(a) to y(b), the error will be not greater than:

$$Error = (1/n)[y(b) - y(a)]$$

Program 11. Integration by reimann sums.

```
100 REM  The name of this program is INTGTN5
110 REM  This program computes the integral using Reimann Sums
120 GOSUB 750
130 PRINT "INTEGRATION -- METHOD No. 4"
140 PRINT
150 PRINT "Yields result that is approximation of integral"
160 GOSUB 710
170 GOSUB 710
180 PRINT "The number of X and Y data points must be equal"
190 PRINT
200 PRINT
210 PRINT "Input the total number of X and Y data points"
220 PRINT "and then press ENTER (Cr):"
230 INPUT N
240 GOSUB 750
250 PRINT "You have selected";N;"data points"
260 PRINT "Is this the correct number?"
270 PRINT "Yes = 1"
280 PRINT "No = 2"
290 INPUT L
300 IF L = 1, THEN GOTO 330
310 IF L = 2, THEN GOTO 160
320 GOTO 250
330 GOSUB 750
340 FOR I = 0 TO N-1
350 PRINT "INPUT Y(";I;")"
360 INPUT Y(I)
370 PRINT "INPUT X(";I;")"
380 INPUT X(I)
390 GOSUB 710
400 NEXT I
410 GOSUB 750
420 PRINT "You have finished enterring data"
430 PRINT "True = 1"
440 PRINT "False = 2"
450 INPUT Z
460 IF Z = 1, THEN GOTO 490
470 IF Z = 2, THEN GOTO 790
480 GOTO 410
490 IYDX = 0
500 FOR I = 1 TO N-1
510 SUMY = SUMY + Y(I)
520 NEXT I
530 Z = X(N-1) - X(0)
```

```
540 Z = Z/N
550 IYDX = SUMY*Z
560 GOSUB 750
570 PRINT "OPTIONS MENU"
580 PRINT
590 PRINT "PRINT RESULTS ON:
600 PRINT
610 PRINT "1.   VIDEO MONITOR ONLY"
620 PRINT "2.   PRINTER ONLY"
630 PRINT "3.   BOTH VIDEO AND PRINTER"
640 INPUT F
650 GOSUB 750
660 IF F = 1, THEN GOTO 880
670 IF F = 2, THEN GOTO 1060
680 IF F = 3, THEN GOTO 1210
690 GOSUB 710
700 GOTO 570
710 FOR Z = 1 TO 5
720 PRINT
730 NEXT Z
740 RETURN
750 FOR Z = 1 TO 30
760 PRINT
770 NEXT Z
780 RETURN
790 GOSUB 750
800 PRINT "ERROR: The actual number of data does not agree"
810 PRINT "with the number enterred earlier"
820 PRINT
830 PRINT "Try Again"
840 PRINT "Press 1 and then Cr to continue:"
850 INPUT D
860 IF D = 1, THEN GOTO 100
870 GOTO 820
880 IF F = 3, THEN GOTO 900
890 IF F = 1, THEN GOTO 930
900 PRINT "OUTPUT ON BOTH PRINTER AND VIDEO MONITOR"
910 GOSUB 1250
920 GOTO 950
930 PRINT "OUTPUT ON VIDEO MONITOR ONLY"
940 GOSUB 1250
950 GOSUB 710
960 IF P = 2, THEN GOTO 990
970 PRINT "INTEGRAL IS:";IYDX
980 GOTO 1050
990 FOR I = 0 TO N-1
1000 PRINT "Y(";I;") = ";Y(I);"    X(";I;") = ";X(I)
1010 NEXT I
1020 PRINT
1030 PRINT
1040 PRINT "INTEGRAL IS:";IYDX
1050 IF F = 1, THEN GOTO 1310
1060 IF F = 3, THEN GOTO 1110
1070 LPRINT "OUTPUT ON PRINTER ONLY"
1080 GOSUB 710
1090 IF F = 3, THEN GOTO 1110
```

Program 11 continued

```
1100 GOSUB 1250
1110 IF P = 2, THEN GOTO 1140
1120 LPRINT "INTEGRAL IS:";IYDX
1130 GOTO 1200
1140 FOR I = 0 TO N-1
1150 LPRINT"Y(";I;") = ";Y(I);"    X(";I;") = ";X(I)
1160 NEXT I
1170 LPRINT
1180 LPRINT
1190 LPRINT "INTEGRAL IS:";IYDX
1200 GOTO 1310
1210 GOSUB 750
1220 PRINT "OUTPUT ON BOTH PRINTER AND VIDEO MONITOR"
1230 GOSUB 710
1240 GOTO 910
1250 GOSUB 710
1260 PRINT "DATA SELECTION MENU:"
1270 PRINT "1.   OUTPUT INTEGRAL ONLY"
1280 PRINT "2.   OUTPUT INTEGRAL AND X/Y DATA"
1290 INPUT P
1300 RETURN
1310 GOSUB 710
1320 PRINT "1.   FINISHED"
1330 PRINT "2.   DO ANOTHER"
1340 INPUT R
1350 IF R = 2, THEN GOTO 100
1360 IF R = 1, THEN GOTO 1380
1370 GOTO 1310
1380 GOSUB 750
1390 FOR N = 1 TO 50
1400 PRINT"********** BYE-BYE ***********"
1410 NEXT N
1420 GOSUB 750
1430 GOSUB 750
1440 PRINT "FINISHED"
1450 END
```

PROGRAM NO. 12: INTEGRATION OF AN EMBEDDED FUNCTION

This program allows integration of a function that you embed within the program (line 430). The program asks you the x(i) data and calculates the y(i) data. In line 430 I have used $Y = 2x$ so that you can run it and get an idea how the program works.

Program 12. Integration of an embedded function.

```
10 DIM Y(1000)
20 DIM X(1000)
30 IYDX = 0
40 Z = 12
50 REM   The name of this program is INTGTN6
60 REM   This program will compute the integral of a function
70 REM   embedded within the program.  The user must enter the
80 REM   correct function
90 S = 25
100 D = 27
110 GOSUB 700
120 PRINT TAB(S);"* * * * * * * * * * * * * * *
130 PRINT TAB(S);"*   INTEGRATION OF A FUNCTION   *"
140 PRINT TAB(S);"*      EMBEDDED WITHIN THE      *"
150 PRINT TAB(S);"*            PROGRAM            *"
160 PRINT TAB(S);"* * * * * * * * * * * * * * * *"
170 PRINT
180 PRINT TAB(D);"Copyright 1983 by J.J. Carr"
190 PRINT
200 PRINT
210 PRINT
220 GOSUB 740
230 GOSUB 700
240 PRINT "ENTER the number of X-data points and press Cr:";
250 INPUT N
260 GOSUB 700
270 FOR I = 0 TO N-1
280 PRINT "X(";I;") =";
290 INPUT X(I)
300 NEXT I
310 FOR I = 0 TO N-1
320 REM   User must enter the BASIC expression for the function
330 REM   to be integrated on next line
340 Y(I) = X(I)*2
350 NEXT I
360 PRINT
370 FOR I = 1 TO N-1
380 DX(I) = X(I) - X(I-1)
390 AA(I) = DX(I)*Y(I-1)
400 AB(I) = DX(I)*((Y(I)-Y(I-1))/2)
410 A(I) = AA(I) + AB(I)
420 NEXT I
430 FOR I = 1 TO N-1
440 IYDX = IYDX + A(I)
450 NEXT I
460 GOSUB 700
470 PRINT "X-Data";TAB(Z);"Y-data"
480 FOR I = 0 TO N-1
490 PRINT X(I),Y(I)
500 NEXT I
510 PRINT
520 PRINT "Integral is:";IYDX
530 PRINT
540 PRINT
550 GOSUB 740
```

Program 12 continued

```
560 GOSUB 700
570 PRINT "FINISHED?"
580 PRINT TAB(3);"1.  Yes"
590 PRINT TAB(3);"2.  No"
600 PRINT
610 PRINT
620 PRINT "SELECT ONE (1) from above:";
630 INPUT W
640 IF W ¶ 2, THEN GOTO 570
650 IF W = 2, THEN GOTO 30
660 GOSUB 700
670 PRINT "PROGRAM ENDED"
680 GOTO 770
690 GOTO 770
700 FOR I = 1 TO 30
710 PRINT
720 NEXT I
730 RETURN
740 PRINT "Press ANY key to continue:"
750 A$=INKEY$: IF A$="" THEN 750
760 RETURN
770 END
```

PROGRAM NO. 13: GEOMETRIC INTEGRATION OF AN EXPONENTIALLY DECAYING CURVE

The figure below shows a function that is well defined from T_0 to T_1 and enters a known exponential decay from T_1 onwards. In some instrumentation applications (for example, a biomedical cardiac output computer) the exponential decay portion is obscured by a recirculation artifact (which occurs in cardiac output measurements). This program will integrate such a function using a geometric method that creates a rectangle using points described in Fig. A-4.

Program 13. Geometric integration of a noisy exponential curve.

```
100   REM   The name of this program is INTGTN.GEO
110   REM   This program uses the method of GEOMETRIC INTEGRATION to
120   REM   calculate the area under an exponential decaying curve.
130   REM   It is used when the trailing edge of the curve is distorted
140   GOSUB 500
150   PRINT "* * * * * * * * * * * * * *"
160   PRINT "*                         *"
170   PRINT "*   GEOMETRIC INTEGRATION  *"
180   PRINT "*                         *"
190   PRINT "*     Method No.  1       *"
200   PRINT "* * * * * * * * * * * * * *"
210   GOSUB 460
220   PRINT "ENTER the beginning (i.e. peak) value of Y"
230   GOSUB 460
240   INPUT YP
```

```
250     GOSUB 460
260     PRINT "ENTER value of X when Y is peak"
270     GOSUB 460
280     INPUT X1
290     GOSUB 460
300     PRINT "ENTER value of X when Y is ";EXP(-1);"times peak Y value"
310     GOSUB 460
320     INPUT X2
330     IYDX = (YP)*(X2 - X1)
340 GOSUB 500
350     PRINT "Integral is approximately: ";IYDX
360     GOSUB 460
370     PRINT "Hit any NUMBER key and then Cr to continue"
380     INPUT Z
390 GOSUB 460
400     PRINT "FINISHED?  ENTER 1 and hit Cr"
410     PRINT "DO ANOTHER? ENTER 2 and hit Cr"
420     INPUT Z
430     IF Z = 1, THEN GOTO 540
440     IF Z = 2, THEN GOTO 100
450     GOTO 400
460     FOR S = 1 TO 5
470     PRINT
480     NEXT S
490     RETURN
500     FOR S = 1 TO 30
510     PRINT
520     NEXT S
530     RETURN
540 FOR S = 1 TO 30
550 PRINT "***************    GOOD-BYEiiiieeeeee!! ******************"
560 NEXT S
570 GOSUB 460
580 PRINT "GONE"
590 END
```

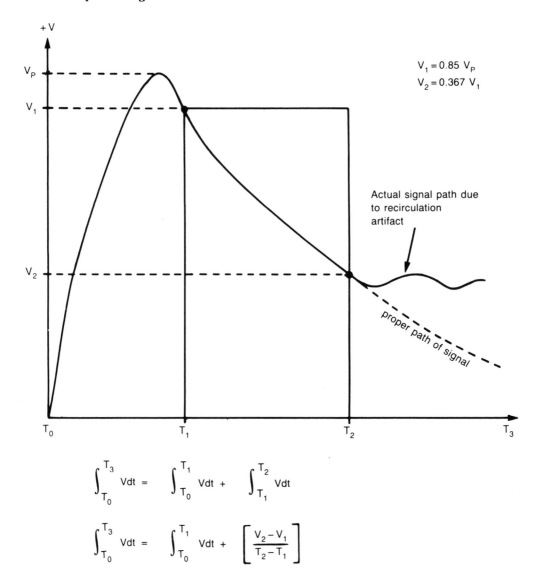

$$\int_{T_0}^{T_3} Vdt = \int_{T_0}^{T_1} Vdt + \int_{T_1}^{T_2} Vdt$$

$$\int_{T_0}^{T_3} Vdt = \int_{T_0}^{T_1} Vdt + \left[\frac{V_2 - V_1}{T_2 - T_1}\right]$$

PROGRAM NO. 14: CALCULATION OF THE VARIANCE AND STANDARD DEVIATION

This program calculates variance (s^2) and standard deviation (s) for a set of n data points.

This program permits the user two options. First, it allows you to print out or display either the variance/standard deviation or the variance/standard deviation and data.

Program 14. Variance and standard deviation.

```
100   REM   The name of this program is VARSD
110   REM   This program calculates the statistical variance for a group of
120   REM   X-data input from the keyboard.
130 DIM X(1000)
140 S = 30
150 A = 0
```

```
160 B = 0
170 C = 0
180 GOSUB 1230
190 PRINT TAB(S);"* * * * * * * * * * *"
200 PRINT TAB(S);"*                     *"
210 PRINT TAB(S);"*       VARIANCE      *"
220 PRINT TAB(S);"*          +          *"
222 PRINT TAB(S);"*  STANDARD DEVIATION *"
224 PRINT TAB(S);"*                     *"
230 PRINT TAB(S);"* * * * * * * * * * *"
240 GOSUB 1190
250 PRINT "This program calculates the statistical VARIANCE (S-squared) and"
255 PRINT "STANDARD DEVIATION (S) of X-data entered from the keyboard"
270 PRINT
280 GOSUB 1270
290 GOSUB 1230
300 PRINT TAB(W);"ENTER the number of X-data that will be used and press Cr"
310 PRINT
320 PRINT
330 PRINT "N =";
340 INPUT N
350 GOSUB 1230
360 PRINT "Now enter the X-data.  The screen will prompt you with ±X(i)',"
370 PRINT "where ±i' is an integer. UP TO 1000 X-DATA points are allowed"
380 PRINT
390 GOSUB 1270
400 GOSUB 1230
410 PRINT "ENTERING X-data: Press Cr after each entry"
420 PRINT
430 FOR I = 1 TO N
440 PRINT "X(";I;") =";
450 INPUT X(I)
460 NEXT I
470 FOR I = 1 TO N
480 A = A +(X(I))¢2
490 B = B + X(I)
500 NEXT I
510 C = (B¢2)/N
520 VAR = (A-C)/N
522 DEV = SQR(VAR)
530 GOSUB 1230
540 PRINT "Please select one DISPLAY options from menu below:"
550 PRINT "1.  Video Monitor"
560 PRINT "2.  Hardcopy Printer"
570 GOSUB 1190
580 PRINT "SELECTION IS:";
590 INPUT L
600 IF L ¶ 2, THEN GOTO 540
610 IF L = 2, THEN GOTO 760
620 GOSUB 1230
630 PRINT "Next, select the DATA OUTPUT option from menu below:"
640 PRINT TAB(5);"1.Display only variance & standard deviation"
650 PRINT TAB(5);"2.Display X-data,variance and standard deviation together"
660 PRINT
670 PRINT "ENTER selection and press Cr:";
680 INPUT D
```

Program 14 continued

```
690 IF D ¶ 2, THEN GOTO 630
700 IF D = 1, THEN GOTO 710 ELSE 1050
710 GOSUB 1230
720 PRINT "Variance S-squared is:";VAR
722 PRINT "Standard Deviation (S) is:";DEV
730 GOSUB 1190
740 GOSUB 1270
750 GOSUB 1300
760 GOSUB 1230
770 PRINT "PRINTER selected as the output display"
780 PRINT
790 PRINT "Next, select the DATA OUTPUT option from menu below:"
800 PRINT TAB(5);"1.  Display only variance and standard deviation"
810 PRINT TAB(5);"2.  Display X-data, variance and standard deviation together"
820 PRINT
830 PRINT "ENTER selection and press Cr:"
840 INPUT D
850 IF D¶2, THEN GOTO 790
860 IF D = 1, THEN GOTO 870 ELSE 920
870 GOSUB 1190
880 PRINT "OUTPUT DATA BEING DISPLAYED ON LINE PRINTER"
890 LPRINT "Variance S-squared is:";VAR
892 LPRINT "Standard Deviation (S) is:";DEV
900 GOSUB 1270
910 GOSUB 1300
920 GOSUB 1190
930 PRINT "OUTPUT DATA BEING DISPLAYED ON LINE PRINTER"
940 LPRINT "X(i)";TAB(11);"X(i)¢2"
950 LPRINT "____";TAB(11);"_____"
960 LPRINT
970 FOR I = 1 TO N
980 LPRINT X(I);TAB(12);X(I)¢2
990 NEXT I
1000 LPRINT
1010 LPRINT "Variance S-squared is:";VAR
1011 LPRINT "Standard Deviation (S) is:";DEV
1020 LPRINT
1030 GOSUB 1270
1040 GOSUB 1300
1050 GOSUB 1230
1060 PRINT "X(i)";TAB(11);"X(i)¢2"
1070 PRINT "____";TAB(11);"_____"
1080 PRINT
1090 FOR I = 1 TO N
1100 PRINT X(I);TAB(12);X(I)¢2
1110 NEXT I
1120 PRINT
1130 PRINT "Variance S-squared is:";VAR
1132 PRINT "Standard Deviation (S) is:";DEV
1140 PRINT
1150 PRINT
1160 GOSUB 1270
1170 GOSUB 1300
1180 GOTO 1430
1190 FOR I = 1 TO 5
```

```
1200 PRINT
1210 NEXT I
1220 RETURN
1230 FOR I = 1 TO 30
1240 PRINT
1250 NEXT I
1260 RETURN
1270 PRINT "Press any key to continue:"
1280 A$=INKEY$: IF A$="" THEN 1280
1290 RETURN
1300 GOSUB 1230
1310 PRINT "FINISHED?"
1320 PRINT "Yes = 1"
1330 PRINT " No = 2"
1340 INPUT F
1350 IF F ¶ 2, THEN GOTO 1310
1360 IF F = 2, THEN GOTO 140
1370 GOSUB 1230
1380 FOR I = 1 TO 30
1390 PRINT "*************¶¶¶¶ PROGRAM ENDED §§§§***************"
1400 NEXT I
1410 GOSUB 1190
1420 PRINT "GONE!"
1430 END
```

PROGRAMS 15 AND 16: CONVENTIONAL AND ORTHOGONAL LINEAR REGRESSION

The two programs listed in this section are alternate methods for calculating the slope and y intercept in linear regression analysis. The two methods are based on two different premises. In the conventional least-squares method of linear regression, minimize the distance from each data point to the line is minimized. The premise implicit in this method is that all error is along one axis. The orthogonal least squares method assumes that potential error exists in both x and y axis. In this method the same minimization is used, but the line between the curve y = a + bx and the data point is orthogonal (i.e., at right angles).

So why do you need two methods of linear regression? Engineers in particular are likely to have learned only one method (conventional). For many applications, the conventional method is fine because you can safely assume that all error is in one axis or the other. If you measure a voltage with respect to time, for example, you can measure the time (if required) to accuracies of 1 part in 10^9, so almost all of our error will be in the voltage measurement (of course, you can also measure the voltage very accurately, but that is a moot point in this case). But there are other situations where the error might be in either axis, and the y error and x error are independent of each other. Suppose you are creating a calibration curve for a pressure transducer for use in a medical research laboratory. Assume that the output display will be a strip chart (paper) recorder, which has an inherent 5-percent error. You might use a mercury manometer as the standard, and the paper recorder or other form of voltmeter as the data being calibrated. In this case the mercury manometer may have error and the voltmeter/chart will have error of its own.

These two methods of linear regression were shown to me by Mr. Art Ciarcowski of the U.S. Food and Drug Administration, Bureau of Medical

Devices, who also provided the example to follow.

The examples follow the listings for the respective programs. Both are calculated using the data table below. The acutal equation evaluated was:

$$y = 2 + 4x$$

Example Data.

X Data	Y Data
6.6	20.1
9.1	45.0
17.4	67.4
17.9	73.4
12.9	95.9
13.3	98.9
13.6	103.1
18.1	90.6
29.1	92.7
25.6	114.7
36.5	119.0
35.0	109.8
30.0	121.8
30.7	117.3
39.3	145.8
47.5	171.6
40.9	174.8
48.5	206.5

Program 15. Conventional least squares linear regression.

```
100   REM   The name of this program is LINREGR
110   REM   This program will compute the slope (b) and Y-intercept point of
120   REM   the line.  The Least Squares Method is used.
130   M = 30
140   A = 0
150   SUMX = 0
160   SUMY = 0
170   SUMX2 = 0
180   SUMY2 = 0
190   DIM X(1000)
200   DIM Y(1000)
210   GOSUB 1180
220   PRINT TAB(M);"* * * * * * * * * * *"
230   PRINT TAB(M);"*   LINEAR REGRESSION   *"
240   PRINT TAB(M);"*          by          *"
250   PRINT TAB(M);"*     LEAST SQUARES     *"
260   PRINT TAB(M);"*                       *"
270   PRINT TAB(M);"* * * * * * * * * * *"
280   GOSUB 1140
290   GOSUB 1220
300   GOSUB 1180
310   PRINT "Form is Y = (bX) -/+(a) -/+(error)"
320   PRINT
330   PRINT
340   PRINT "This program computes the slope (b) and Y-intercept point (a)"
350   PRINT "using the conventional Least Squares method of linear regression."
360   PRINT "The implicit assumption of this program is that all of the error"
370   PRINT "is associated with the X-data.  If your particular situation permits

380   PRINT "error in both X-data and Y-data, then it might be prefered to use"
390   PRINT "the ORTHOGONAL LEAST SQUARES method (see text of book)"
400   GOSUB 1140
410   GOSUB 1220
420   GOSUB 1180
430   PRINT "Enter the number of paired (X,Y) data points, N:"
440   PRINT
```

```
450 PRINT "N =";
460 INPUT N
470 GOSUB 1180
480 FOR I = 1 TO N
490 PRINT "X(";I;") =";
500 INPUT X(I)
510 PRINT "Y(";I;") =";
520 INPUT Y(I)
530 PRINT
540 NEXT I
550 COLOR 0,7
560 PRINT "THINKING  ---- PLEASE BE QUIET ---- NOISE DISTURBS MY ELECTRONS
    -----------------"
570 FOR I = 1 TO N
580 A = A + (X(I)*Y(I))
590 SUMX = SUMX + X(I)
600 SUMY = SUMY + Y(I)
610 SUMX2 = SUMX2 + (X(I)¢2)
620 SUMY2 = SUMY2 + (Y(I)¢2)
630 NEXT I
640 B = (SUMX*SUMY)/N
650 C = (SUMX¢2)/N
660 SLOPE = (A - B)/(SUMX2 - C)
670 YINT = (SUMY/N) - (SLOPE*(SUMX/N))
680 SLOPE = SLOPE*100
690 D = SLOPE -INT(SLOPE)
700 SLOPE = INT(SLOPE)
710 IF D = .5, THEN SLOPE = SLOPE + 1
720 IF D ¶ .5, THEN SLOPE = SLOPE + 1
730 IF D § .5, THEN SLOPE = SLOPE
740 SLOPE = SLOPE/100
750 YINT = YINT*100
760 D = YINT - INT(YINT)
770 YINT = INT(YINT)
780 IF D = .5, THEN YINT = YINT + 1
790 IF D ¶ .5, THEN YINT = YINT + 1
800 IF D § .5, THEN YINT = YINT
810 YINT = YINT/100
820 COLOR 7,0
830 PRINT
840 PRINT
850 PRINT "X-Data";TAB(11);"Y-data"
860 PRINT "_____";TAB(11);"_____"
870 FOR I = 1 TO N
880 PRINT X(I);TAB(12);Y(I)
890 NEXT I
900 PRINT
910 PRINT "Slope (b): ";SLOPE
920 PRINT "Y-intercept (a):";YINT
930 PRINT
940 PRINT
950 GOSUB 1220
960 GOSUB 1180
970 PRINT "FINISHED?"
980 PRINT "1.  Yes"
990 PRINT "2.  No"
```

Program 15 continued

```
1000 PRINT
1010 PRINT
1020 PRINT "ENTER SELECTION AND Press CARRAIGE RETURN"
1030 INPUT Z
1040 IF Z ¶ 2, THEN GOTO 970
1050 IF Z = 2, THEN GOTO 210
1060 GOSUB 1180
1070 COLOR 0,7
1080 FOR Q = 1 TO 30
1090 PRINT "**************** PROGRAM ENDED ****************"
1100 NEXT Q
1110 COLOR 7,0
1120 PRINT "Gone!"
1130 GOTO 1250
1140 FOR I = 1 TO 5
1150 PRINT
1160 NEXT I
1170 RETURN
1180 FOR I = 1 TO 30
1190 PRINT
1200 NEXT I
1210 RETURN
1220 PRINT "Press ANY KEY to Continue:"
1230 A$=INKEY$: IF A$="" THEN 1230
1240 RETURN
1250 END
```

Program 16. Conventional least squares orthogonal regression.

```
100  REM  The name of this program is ORTHOLNR
110  REM  This program will compute the slope (b) and Y-intercept point of
120  REM  the line.  The ORTHOGONAL LEAST SQUARES method is used.
130  M = 30
140  A = 0
150  SUMX = 0
160  SUMY = 0
170  SUMX2 = 0
180  SUMY2 = 0
190  DIM X(1000)
200  DIM Y(1000)
210  GOSUB 1150
220  PRINT TAB(M);"* * * * * * * * * * * * * * *"
230  PRINT TAB(M);"*                           *"
240  PRINT TAB(M);"*      LINEAR REGRESSION     *"
250  PRINT TAB(M);"*             by             *"
260  PRINT TAB(M);"*  ORTHOGONAL LEAST SQUARES  *"
270  PRINT TAB(M);"*                           *"
280  PRINT TAB(M);"* * * * * * * * * * * * * * *"
290  GOSUB 1110
300  GOSUB 1190
310  GOSUB 1150
320  PRINT "Form is (Y + Y-Error) = (bX) -/+(a) -/+(X-Error)
330  PRINT
340  PRINT
```

```
350 PRINT "This program assumes that error can exist in either X or Y data"
360 PRINT "For that kind of data, this Orthogonal Least Squares method is"
370 PRINT "regarded as superior to the Conventional Least Squares method."
380 GOSUB 1110
390 GOSUB 1190
400 GOSUB 1150
410 PRINT "Enter the number of paired (X,Y) data points, N:"
420 PRINT
430 PRINT "N =";
440 INPUT N
450 GOSUB 1150
460 FOR I = 1 TO N
470 PRINT "X(";I;") =";
480 INPUT X(I)
490 PRINT "Y(";I;") =";
500 INPUT Y(I)
510 PRINT
520 NEXT I
530 COLOR 0,7
540 PRINT "THINKING  ---- PLEASE BE QUIET ----NOISE DISTURBS MY ELECTRONS
----------------"
550 FOR I = 1 TO N
560 A = A + (X(I)*Y(I))
570 SUMX = SUMX + X(I)
580 SUMY = SUMY + Y(I)
590 SUMX2 = SUMX2 + (X(I)¢2)
600 SUMY2 = SUMY2 + (Y(I)¢2)
610 NEXT I
620 B = (SUMX*SUMY)/N
630 C = (SUMX¢2)/N
640 GOSUB 1230
650 SLOPE = SLOPE*100
660 D = SLOPE -INT(SLOPE)
670 SLOPE = INT(SLOPE)
680 IF D = .5, THEN SLOPE = SLOPE + 1
690 IF D ¶ .5, THEN SLOPE = SLOPE + 1
700 IF D § .5, THEN SLOPE = SLOPE
710 SLOPE = SLOPE/100
720 YINT = YINT*100
730 D = YINT - INT(YINT)
740 YINT = INT(YINT)
750 IF D = .5, THEN YINT = YINT + 1
760 IF D ¶ .5, THEN YINT = YINT + 1
770 IF D § .5, THEN YINT = YINT
780 YINT = YINT/100
790 COLOR 7,0
800 PRINT
810 PRINT
820 PRINT "X-Data";TAB(11);"Y-data"
830 PRINT "_____";TAB(11);"_____"
840 FOR I = 1 TO N
850 PRINT X(I);TAB(12);Y(I)
860 NEXT I
870 PRINT
880 PRINT "Slope (b): ";SLOPE
890 PRINT "Y-intercept (a):";YINT
```

Program 16 continued

```
900  PRINT
910  PRINT
920  GOSUB 1190
930  GOSUB 1150
940  PRINT "FINISHED?"
950  PRINT "1.  Yes"
960  PRINT "2.  No"
970  PRINT
980  PRINT
990  PRINT "ENTER SELECTION AND Press CARRAIGE RETURN"
1000 INPUT Z
1010 IF Z ¶ 2, THEN GOTO 940
1020 IF Z = 2, THEN GOTO 210
1030 GOSUB 1150
1040 COLOR 0,7
1050 FOR Q = 1 TO 30
1060 PRINT "**************** PROGRAM ENDED *****************"
1070 NEXT Q
1080 COLOR 7,0
1090 PRINT "Gone!"
1100 GOTO 1220
1110 FOR I = 1 TO 5
1120 PRINT
1130 NEXT I
1140 RETURN
1150 FOR I = 1 TO 30
1160 PRINT
1170 NEXT I
1180 RETURN
1190 PRINT "Press ANY KEY to Continue:"
1200 A$=INKEY$: IF A$="" THEN 1200
1210 RETURN
1220 END
1230 VARX = SUMX2 - (SUMX¢2)/N
1240 VARY = SUMY2 - (SUMY¢2)/N
1250 SDXY = A - B
1260 M = (4*(SDXY¢2))
1270 J = (VARY - VARX)¢2
1280 L = SQR(J + M)
1290 SLOPE = ((VARY - VARX) + L)/(2*SDXY)
1300 YINT = (SUMY/N) - (SLOPE*(SUMX/N))
1310 RETURN
```

Index